Springer Earth System Sciences

More information about this series at http://www.springer.com/series/10178

Margarita María Alconada-Magliano
Editor

Intensified Land and Water Use

A Holistic Perspective of Local to Regional Integration

Springer

Editor
Margarita María Alconada-Magliano
Facultad Ciencias Agrarias y Forestales
National University of La Plata
La Plata, Buenos Aires, Argentina

ISSN 2197-9596 ISSN 2197-960X (electronic)
Springer Earth System Sciences
ISBN 978-3-030-65442-9 ISBN 978-3-030-65443-6 (eBook)
https://doi.org/10.1007/978-3-030-65443-6

This Springer imprint is published by the registered company Springer Nature Switzerland AG
The registered company address is: Gewerbestrasse 11, 6330 Cham, Switzerland

I dedicated this book to those who have helped me to go through life with joy and renewed hope, and have made me who I am now, especially my parents and my wonderful daughters Sofìa, Elisa and Luciana, who have always encouraged me to seek a reflective and open-minded view of the world, which strengthens my determination to develop my profession. In this path of life full of gratitude and memories, my brothers and sisters, my nephews and nieces, my cousins, my friends and generous colleagues who have helped and inspired me, have always been by my side.

Preface

Introduction

Proper management and conservation of soil and water is essential to sustain life on earth. The consequences of inappropriate use of these resources have direct effects not only on their degradation and on food production, but also on other environmental problems closely linked, such as: climate change, natural disasters, hydrological cycle, quality of the air, energy production, loss of biodiversity, industrial and urban waste management.

Threats linked to land use that have an effect on the environment have been identified in various publications according to their type, occurrence, and magnitude, as well as their importance and relevance based on social, economic, and environmental factors characteristic of the different regions of the world and, associated to this the possibility of providing food and avoiding other conflicts (FAO 2011, 2016; OCDE-FAO 2019). In these and other publications related to the subject, there is general consensus on the basic principles of sustainable soil management to ensure food and quality water, through production management strategies that increase its efficiency and minimize environmental risk. However, holistic and interdisciplinary approaches are also required, which link those environmental aspects that are better known, with those that are not so well known, or even not taken into account but are evident as an integrating natural element for the rest of the elements of local and regional landscapes, such as the functioning of groundwater flow systems (Toth, 1962; 2016).

This book caters the need to be "*open to new proposals,*" to "*free ourselves from deeply rooted/ certainties,*" in order to adapt "*to the future on the grounds of evidence,*" as Prochnow (2020) points out. The irrefutable evidence indicates that the degradation and contamination of soil and water is constant and increasing, and that this affects the quality of life, therefore new holistic study approaches need to be incorporated in order to plan the sustainable use of these and other resources.

About this book

This book includes a selection of works that aim to show, from the viewpoint of different disciplines, the need for a *New Paradigm in the Study of the Environment*, in order to reverse a series of events and processes of soil and water degradation and contamination, which begin with an inadequate identification of the practices, actions, and inactions that generate them. This is one of the first books that integrate different disciplines of the natural sciences that are recognized for each thematic specialty and in different regions of the world, the need to incorporate the study of groundwater, as an inseparable element of the landscape. This holistic understanding of environments is what enables sustainable interventions, understood in its full meaning that includes environmental, productive, social, and economic aspects.

The themes are developed from a holistic perspective by leading specialists in sciences related to soil, water, vegetation, agronomy in general, with economic analysis and social implications. The authors, based on their detailed knowledge of a particular topic, reflect on shortcomings, inconsistencies, or limitations in the analyzes and interpretations of these topics, and in their relationships with other disciplines, to conclude on the need for a comprehensive vision of the functioning of natural systems intervened for the human.

However, not only the possible consequences in different environments are described, but also procedures or analysis criteria are provided to procure interventions that avoid or reverse the succession of events and processes that are often considered isolated or have been insufficiently linked, such as: land degradation and pollution, water pollution and "scarcity," climate change, loss of biodiversity, landslides, among other processes where, in general, anthropic action defines their intensity and direction.

The chapters are presented so that the reader begins by dimensioning the magnitude of the land degradation problems, their origins, and how this can be linked or be the origin of other natural catastrophes (Chapter 1). Likewise, in this last chapter, as well as in the others (Chapters 2 to 11), evaluation, prevention, and/or diagnosis procedures related to soil and water are proposed, applied to different case studies; also including analysis critical and reflective on technical, economic, social, and even epistemological aspects that prevent or hinder adequate and sustainable management of different environments.

What emerges as a general from the topics covered in this book is that studies related to soil, water, and vegetation (natural or cultivated), are frequently partially studied and/or not all the relationships that should be taken into account for their proper characterization in decision-making are considered. Consequently, the way in which the management, intervention, recovery, or enhancement of an environment is decided, as well as, the use of surface or groundwater, is insufficient to guarantee the preservation of natural resources and define the best possible use in a sustainable way. Similarly, natural disasters or extreme events, such as droughts, floods, landslides, accumulation of sediments, changes in the quantity and quality of water, including climate change and loss of ecosystem biodiversity, are often not fully understood or considered in all their relationships. This has its origins in various causes and is addressed in different chapters of this book, from different perspectives.

Thus, Chapter 1 highlights the inadequate cause–effect identification of the events indicated in the environment, which starts from not knowing or decoupling in the analyzes or interpretations, the elements that make up the landscape. The latter is particularly evident when analyzing the study of soils for the purpose of productive management in Chapter 8, or in the study of surface and groundwater, analyzed from different perspectives in various chapters. These perspectives include its study as an ecosystem in itself, with a biodiversity that can conquer and master the superficial ecosystems that it links, and that condition the quality of the water for different uses (Chapters 4, 5, 6, 9), and also, due to political, social conflicts and economic losses, which are generated from considering it inaccurately or incompletely and even not considering it (Chapters 7, 10, 11). In general, these chapters include as necessary, in order to establish development and management policies for soil and water, local and regional relations between regions of the same country and/or between countries.

Then, in this book, not only the state of progress in the disciplines included, but also holistic studies with a hydrological approach are proposed, applying different criteria in case studies from different parts of the world. In Chapters 1 and 2, procedures and models are presented that consider hydrological processes to predict the degradation of soil and water in general, and of salinization and sodification in particular, due to the incidence of the water table; and aspects of the vegetation associated with these environments are described (Chapter 3). In Chapters 4, 5, 6, 7, 8, 9, and 11, the proposed hydrological approach includes a dynamic vision of local and regional functioning, as it is allowed by Toth's Groundwater Flow Systems Theory, first published by this author in 1962 (József Tóth, 1962 until 2016). In this criterion, the surface basin, which is mainly linked to the topographic pattern, is distinguished from the *unitary basin*, which is the basic unit of flow systems and results from regional hydraulic continuity. The u*nitary basin* is not defined by topographic contrast but by hydrogeological (subsoil) contrast that generates flows that do not necessarily belong to the surface basin of interest. In Toth's theory, the basic diagnostic unit of groundwater is the Underground Flow System, which is defined as *"a natural and coherent unit, in space and time, consisting of groundwater of particular physicochemical quality, which circulates through materials geological with a geomorphological expression, with particular vegetation and soil."*

The dynamic functioning scheme proposed by the Tothian theory makes it possible to understand more precisely what happens in other fields of knowledge, as groundwater flows are the natural integrating element of other elements of the landscape. The water interdependence that exists between neighboring basins and the regional water continuity, also defined by other authors (Engelen and Jones 1986, Carrillo-Rivera 2000), is highlighted in this book when analyzing from aspects related to the use and management of the water from local and transboundary aquifers (Chapters 6, 7, 11), even in decision-making on land management and preservation of ecosystems (Chapters 5, 8, 9). This analysis vision is particularly important and evident in environments developed from sedimentary materials, as it happens with most of the soils where productive activities are carried out, and specifically important, when in addition, they are areas subject to the alternation of droughts and floods, with the presence of a phreatic surface close to the surface and the coexistence of water flows of different origins, which generate an intricate pattern of distribution of ecosystems (Chapters 5, 8, 9).

In summary, it can be affirmed through the valuable contributions that the authors made in this book, that the perspective of the functioning of the environment proposed by the Tothian theory allows: (a) *Regarding ecosystems in general*, define how they are linked and dependent on groundwater, and especially, their role as an ecosystem and in the definition of water quality. (b) *Regarding soil management,* define especially in areas subject to the alternation of drought and floods, how to take advantage of this functioning to help in extreme water condition; define the origin of the salinity and sodicity of soils due to natural or anthropic causes; foresee consequences on the environment when deforesting, draining, implementing irrigation; and select agricultural, livestock and/or forestry activities that adapt to and/or reverse a particular water condition. (c) With respect to water, it allows planning for water management in various uses and thus establishing development and preservation policies for the quality and quantity available, also having geopolitical importance in being able to clarify aspects related to transboundary aquifers.

Lastly, this book constitutes a contribution to Toth's own Theory, since it contributes through its application in particular areas (as case studies) to an exemplification of the relationships that arise between geomorphology, vegetation, type soil, and groundwater in different parts of the world.

La Plata, Argentina Margarita María Alconada-Magliano

Acknowledgments I am thankful to Dr. Jorge Rabassa for his editing work as well as to all the authors who generously and dedicatedly shared their experience in the presentation of their respective chapters. Finally, I want to thank Springer Publishing for giving us the opportunity to present our work with freedom.

An Overview of the Content of this Book

This book contains 11 chapters, whose relevant aspects that each of these includes are discussed below.

Chapter 1 Hydrological Approach for Evaluating Soil and Water Degradation Processes in a Changing Environment

In this chapter, it is described how the increased problems of soil and water degradation, are leading to growing risks and problems of food and water supply for an increasing World population, and are contributing to the loss of biodiversity and global climate changes. It is shown how those problems are mainly related to hydrological changes associated with inappropriate land use and management. The analysis of the processes involved in natural disasters like droughts, flooding, landslides, sedimentations, and related consequences, leads to the conclusion that they are presently more related to those hydrological changes derived from social and economic pressures than to the previewed global climate changes. There is proposed the use of a modeling approach (SOMORE), based on hydrological processes, to evaluate, preview, and prevent soil and water degradation processes and their catastrophic consequences. The author illustrates the application of such hydrological modeling approach, under different socioeconomic and biophysical conditions in different parts of the World.

Chapter 2 Prediction of Soil Salinization and Sodification Processes as Affected by Groundwater Under Different Climate and Management Conditions

Chapter 2 shows how the natural and man-induced development of salt-affected soils (saline and sodic), both under dry-land and irrigated conditions, are a consequence of hydrological and hydrogeochemical processes, where changes in groundwater depth and composition is usually a critical factor, affecting salt and water balances in the soil profile. The different processes leading to the development either of saline or sodic soils are herein analyzed, as well as their effects on soil properties and crop growth. The author proposes a modeling approach (SALSODIMAR), which may be used both for the diagnosis and prediction of soil salinization and sodification, as a basis for irrigation and drainage management to control their development. It is shown how this model has been successfully used for such purposes in different parts of Latin America, under dry-land and irrigated conditions, with variable soils, climate, irrigation, and drainage management.

Chapter 3 Soil Salinization and Sodification as Conditioners of Vegetation and Crops: Physiological Aspects of Plant Response to These Conditions

In this chapter, the authors addressed the negative effects of soil salinity and sodicity on plant growth and development. It highlights that naturally adapted species have physiological and morphological traits that allow them to thrive in these environments. The physiological mechanisms related to tolerance to salinity and alkalinity in plants are summarized. Likewise, the relationships that the literature describes between vegetation (natural or cultivated, herbaceous or arboreal), the environment (soil, groundwater, climate), and adaptive physiological mechanisms are analyzed. Forage species that may be constituted by their positive behavior in environments affected by salts are described as a potential source of new species to domesticate and as soil improvers, which requires an interdisciplinary and holistic view of the functioning of these environments.

Chapter 4 Groundwater and Its Role in Maintaining the Ecological Functions of Ecosystems—A Review

In Chapter 4, the integrative function of groundwater with the ecosystems is analyzed from a holistic perspective, and it is also considered as an ecosystem in itself. Scientific evidence of the role of groundwater due to its very high biodiversity in preserving water quality and ecosystem functioning is presented. The authors highlight the special importance that this comprehensive understanding has in different scenarios of global change, at different spatial and temporal scales. These analyzes were carried out addressing the current state of knowledge of groundwater, the importance of a comprehensive vision, showing the relevance of applying Toth's theory of groundwater flow systems to the implications of its management. Finally, recommendations are made to establish action guidelines for the sustainable management of groundwater from an environmental, social, and economic perspective.

Chapter 5 Groundwater Flows in Santa Fe Submeridional Lowlands. A Conceptual Model

In this chapter, a conceptual model of the functioning of groundwater flows in the "Bajos Submeridionales" region of the province of Santa Fe, Argentina, is proposed. This region with wetland areas and other associated ecosystems, has a great natural biodiversity that requires a holistic understanding to guarantee rational and sustainable management. The authors propose to analyze the different elements of the landscape according to the functioning scheme proposed in Toth's theory of groundwater flow systems. In this analysis, they find coexistence in the region of water flows of different origin, local, and regional. Likewise, they demonstrate that with this study approach it is possible to explain the complexity of the hydrogeological operation of this environment, characterized by the existence of very diverse geological materials, in a predominantly flat geomorphology, and subject to the alternation of droughts and floods. This analysis criterion is also the one that allows to define the sustainable productive management of this region, where livestock takes on particular significance.

Chapter 6 Hydrogeochemical Characterization of Groundwater and Its Interaction with Other Components of the Environment in Mexico

Chapter 6 has an exhaustive bibliographic review that is carried out on the origin and causes of the presence of trace elements, mainly As, F, Mn, and Fe, in groundwater, and their consequences on human health. The need for an environmental perspective in the understanding of health problems is raised, linking them to hydrogeological, geographical, pedological, atmospheric, and hydrogeochemical topics. Possible natural and anthropic causes are analyzed and systemic and dynamic study procedures are proposed to understand the functioning of elements of the natural and anthropic landscape, applying the theory of Toth's groundwater flow systems.

Experiences are presented, showing how the increasing deterioration of groundwater quality could be controlled, particularly analyzed in Mexico.

Chapter 7 Groundwater Flow Systems and Their Importance in the Assessment of Transboundary Groundwater: The Mexico–U.S.A. Case

Chapter 7 provides a detailed and empirical analysis of the proposals that emerge from several international legal frameworks to address the different technical, legal, and administrative issues that arise between countries that share aquifers. A critical analysis is made of the discrepancies that appear in the world in general and the Mexico–U.S. border region in particular, in aspects ranging from conceptual to methodological related to "*trans-boundary aqui*fer" and "*trans-boundary groundwater*" concepts. Based on the existing research, they support the need to deepen studies for management integrated, having a holistic understanding and dynamic functioning, which includes political geography and different environmental sciences, especially hydrogeology. In this matter, they reaffirm the relevance of the application of Toth's theory groundwater flow systems to produce scientific evidence that enables the incipient legal frameworks on transboundary groundwater.

Chapter 8 Landscape Functioning as a Basis for Establishing Sustainable Intervention: Soils and Groundwater Flows

In this chapter, a critical analysis is made on the scope and limitations of traditional soil studies. The consequences that partial studies of the environment have on its degradation are raised. The need to incorporate a holistic understanding of the functioning of the local soil in the local and regional landscape is shown through a case study: Northwestern Buenos Aires province, Argentina, characterized by the alternation of droughts and floods, and the presence of a shallow water table. It is demonstrated how by applying Toth's theory of groundwater flow systems, it is possible to understand the origin of soil properties and how the elements of the landscape, natural and anthropic, are related to each other. A new paradigm in the study of soil that includes the study of the functioning of groundwater, defining local and regional relationships, emerges as evident and necessary. This comprehensive and dynamic vision allows defining sustainable management by type of environment, and taking advantage of its functioning, for example, when selecting production schemes, location of forest plantations, among other practices of landscape intervention.

Chapter 9 Regional Groundwater Flow Systems: Their Role in Conserving the Marismas Nacionales Biosphere Reserve in Nayarit, Mexico

The relevance of the application of Toth's theory of groundwater flow systems is presented in this chapter, through a case study, the Marismas Nacionales de México, a RAMSAR site, so as to understand the environmental functioning of this region, characterized by an important hydrological complexity and associated ecosystems (swamps, mangroves, and other wetlands). The authors present a procedure for the study and interpretation of the available information, which integrates the observations generated from different disciplines in a simple and scientifically valid way. This holistic and dynamic understanding is proposed as the basis of studies for sustainable interventions and management.

Chapter 10 A Holistic Approach to the Estimation of Economic Losses Due to Water Stress in Agriculture in Buenos Aires Province, Argentina

In this chapter the absence of an adequate understanding and quantification of all the phenomena involved in the occurrence of extreme water phenomena, particularly droughts, is analyzed. It highlights the need to define water deficits with an integral vision, from a temporal, local and regional, and interdisciplinary perspective. It is encouraged to include in the private and state studies and planning, not only the climate but also the rest of the elements of the environment such as groundwater, soil, production systems, and implemented management. The magnitude of the economic loss caused by water stress events (generally unknown or underestimated) is exemplified in this chapter through a case study, using procedures widely used in the world, in the four main crops in the province of Buenos Aires, Argentina. The aim is to raise awareness, mainly among decision-makers and researchers, about the need to establish more precisely environmental–productive relationships, and their effects on the economy and political-administrative developments that could effectively contribute to mitigate the consequences of climate variability, moving in to manage drought and water scarcity and their related impacts toward a proactive management approach.

Chapter 11 Water Security and Groundwater: The Absence of Scientific Criteria in Groundwater Management Through Three Case Studies in Mexico

In Chapter 11, a critical analysis is made of the two main positions on the worldwide view regarding the problem of water (scarcity vs. inequity in distribution), and the absence in both positions of a proposal for scientific research that allows solving the true problem: "water insecurity." The political, administrative, and scientific situation that exists in Mexico is presented through three case studies, which show inconsistencies in the application of criteria to manage water at the national level, and in the aquifers that they share with other countries (transboundary aquifers). The authors find that there is a prioritization of political aspects over scientific ones, and a discretionary position in decisions for the use of water in Mexico. Likewise, they denote weaknesses in the characterization of aquifers and groundwater from a dynamic and functional perspective. They propose to reverse the situation raised, starting from considering the scientific criteria established by Toth's Theory on the functioning of groundwater flows, as other countries have implemented it. With this perspective of study, interdisciplinary, dynamic, and functioning of all environmental components it would enable effective protection of groundwater flows, and guarantee the "state's water security" and equitable distribution between users.

Tribute Professor Dr. Mario Alberto Hernández

September 18, 1942–June 22, 2018

In this book, a *tribute to our beloved mentor, recently deceased* Professor Dr Mario Hernández, is performed, for his paramount contributions in the fields of Hydrology and Hydrogeology of Latin America.

Mario Hernández was born in the Patagonian city of Comodoro Rivadavia, Argentina. He achieved both his Licenciado and Doctor degrees in Geology at the Universidad Nacional de La Plata. Later on, he attained Postgraduate degrees in General and Applied Hydrology in the Universidad Complutense of Madrid, Spain, another in Mathematical models in Hydrology and Geophysical Exploration, also in Madrid, and similarly, one on Remote Sensing in Hydrology at FAO, Rome, Italy.

His eminent career developed in the academic and scientific world, as well as in technical and consulting activities. In all these environments, he reached always the highest hierarchy and recognition. For several decades, Mario Hernández was Professor of Hydrogeology in the School of Natural Sciences of the University of La Plata, where he was distinguished as Emeritus Professor of this University after his retirement. He created and then became the Director of the Postgraduate degree in Ecohydrology in the same University. He offered Postgraduate courses at La Plata and many other universities in Argentina and throughout the entire continent.

He developed his scientific career as a researcher of CONICET, the National Research Council of Argentina. He was also Vice President of the Scientific Research Commission of the Buenos Aires province. He actively participated in the formation of legions of students, professionals, and technicians in many universities and scientific organizations of Argentina and Latin America. He was a dedicated member of several national and international associations and leader or advisor in plentiful research projects and doctoral dissertations.

His academic production is outstanding, with more than 180 research papers and books, having received prizes from Mexico, the Argentine National Academy of Sciences, the Argentine Geological Association, and the Latin American Association of Groundwater for Development, an institution of which he was the President in two four-year periods.

He was nominated as Honorary Member of the International Association of Hydrogeologists (IAH) for his groundbreaker contributions, most of them related to groundwater studies in the Pampas, the Argentine loess plains, and the investigation of water resources for the oil and gas industry.

In this book I, together with his uncountable colleagues and friends, would like to evoke his charming and delightful personality, his renowned professional and scientific knowledge, his everlasting standing next to democracy and human rights in Argentina and Latin America, as well as his amiable and enchanting family lifetime, always accompanied by his beloved wife, Nilda González, also a hydrogeologist, his lifelong partner in daily life and work.

His generosity, openhandedness, and professional lavishness have embossed in all of us, who have shared moments of his bountiful life, many unforgettable experiences and treasured awareness. He educated us in the belief that science should be used, taught and disseminated but not ever leaving aside all humanitarian principles that Mario Hernández had partaken with those who had the enduring honor and pleasure of following the pathway that he had opened.

Finally, it should be mentioned that, as recently as 2017, Springer-Nature had the valuable opportunity of publishing his book "Hydrogeology of a Large Oil-and-Gas Basin in Central Patagonia: the San Jorge Gulf Basin, Argentina" (Springer Brief Monographies in Latin American Studies). In that book, Mario Hernández enjoyed the infrequent satisfaction of co-authoring this book with his wife, Nilda González, and his son, Lisandro Hernández, also a hydrogeologist. A loving family of prominent hydrogeologists.

Margarita María Alconada-Magliano

Contents

Contributors

Yussef Ricardo Abud-Russell Colegio de Geografía, Facultad de Filosofía y Letras, Universidad Nacional Autónoma de México, Circuito Interior. Ciudad Universitaria S/N, Ciudad de México, México

Margarita María Alconada-Magliano Edafología, Facultad de Ciencias Agrarias y Forestales, Universidad Nacional de La Plata, La Plata, Argentina

José Joel Carrillo-Rivera Instituto de Geología, Universidad Nacional Autónoma de México, Circuito de la Investigación Científica S/N, Ciudad Universitaria, Ciudad de México, Mexico;
Instituto de Geografía, Universidad Nacional Autónoma de México. Investigación Científica. Ciudad Universitaria S/N., Ciudad de México, Mexico

Silvana Luisa Castro Instituto Nacional Del Agua, INA, Santa Fe, Argentina

Eduardo Luis Díaz Facultad de Ciencias Agropecuarias-Universidad Nacional de Entre Ríos, Entre Ríos, Argentina

Gonzalo Hatch-Kuri Programa de Maestría En Gestión Integrada de Cuencas, Facultad de Ciencias Naturales, Universidad Autónoma de Querétaro. Av. Junípero Serra, Santiago de Querétaro, Querétaro, México;
Maestría En Gestión Integrada de Cuencas, Facultad de Ciencias Naturales, Universidad Autónoma de Querétaro, S/N Santiago de Querétaro, Qro, Mexico

Rafael Huizar-Álvarez Instituto de Geología, Universidad Nacional Autónoma de México, Circuito de la Investigación Científica S/N, Ciudad Universitaria, Ciudad de México, México

Alessia Kachadourian Instituto de Ingeniería, Universidad Nacional Autónoma de México, Mexico City, Mexico

Debora Lithgow Red de Ambiente y Sustentabilidad, Instituto de Ecología, AC, Xalapa, Mexico

Néstor Alberto Mariano Instituto de Ambiente de Montaña y Regiones Áridas, Universidad Nacional de Chilecito, Chilecito, La Rioja, Argentina

Edgar Mendoza Instituto de Ingeniería, Universidad Nacional Autónoma de México, Mexico City, Mexico

Juanalberto Meza-Villegas Programa de Maestría En Ciencias Sociales, Facultad Latinoamericana de Ciencias Sociales. Colonia Héroes de Padierna, Tlalpan, Ciudad de México, México

Ildefonso Pla-Sentís Universitat de Lleida, Plaça de Víctor Siurana, Lleida, Spain

Andrés Alberto Rodríguez Consejo Nacional de Investigaciones Científicas y Técnicas (CONICET), Buenos Aires, Argentina;
Laboratorio de Fisiología de Estrés Abiótico y Biótico en Plantas, Unidad de Biotecnología 1, INTECH, Chascomús, Argentina

Raúl Jorge Rosa Centro Interdisciplinario de Investigaciones Aplicadas al Agua y al Ambiente. UNLP, Administración Agraria, Facultad de Ciencias Agrarias y Forestales, Universidad Nacional de La Plata, La Plata, Argentina

Julio César Sánchez-Angulo Programa de Maestría En Gestión Integrada de Cuencas, Facultad de Ciencias Naturales, Universidad Autónoma de Querétaro. Av. Junípero Serra, Santiago de Querétaro, Querétaro, México

Rodolfo Silva Instituto de Ingeniería, Universidad Nacional Autónoma de México, Mexico City, Mexico

Dora Cecilia Sosa Instituto Nacional Del Agua, INA, Santa Fe, Argentina

Edith Taleisnik Consejo Nacional de Investigaciones Científicas y Técnicas (CONICET), Buenos Aires, Argentina;
Instituto de Fisiología y Recursos Genéticos Vegetales, CIAP, INTA, Córdoba, Argentina;
Facultad de Ciencias Agropecuarias, Universidad Católica de Córdoba, Córdoba, Argentina

Elisabet Verònica Wehncke Centro de Investigación en Biodiversidad y Conservación, Universidad Autónoma del Estado de Morelos, Cuernavaca, Morelos, Mexico

Chapter 1
Hydrological Approach for Evaluating Soil and Water Degradation Processes in a Changing Environment

Ildefonso Pla-Sentís

Abstract Soil is fundamental to the needs of human life, and plays a central role in determining the quality of our environment. The importance to preserve soils in some crucial aspects for human life, like food production, earth hydrological cycle, biodiversity, and air composition, will be constantly increasing. The expansion and intensification of agricultural activities and increased number and size of populated areas, results in a changing environment, frequently associated with widespread soil and water degradation, due to inappropriate land use and management. Those degradation processes and the associated hydrological changes may result in increasing risks and problems of food and water supply for mankind, and in more frequent "natural" disasters like droughts, flooding, landslides, sedimentation, and contribute to the loss of biodiversity and global climatic changes. All these problems may be evaluated and previewed through modeling the hydrological and hydro-geochemical processes in order to achieve the required sustainability of the environment. This is possible through the integrated use and management of soil and water resources adapted to new social and economic pressures, and to the previewed climate changes. Some examples of such approach under different social-economical and biophysical conditions, in different parts of the world are presented herein.

Keywords Soil degradation · Water degradation · Natural disasters · Hydrology · Climate change

1.1 Introduction

The main factor attempting against the sustainability of agricultural production is soil and land degradation. The off-site effects of land degradation on increased risks of catastrophic flooding, sedimentations, landslides, etc., and on global climate changes are also of growing importance. Although land degradation is affected by soil and climate characteristics, it is mainly due to inappropriate use and management of the

I. Pla-Sentís (✉)
Universitat de Lleida, Plaça de Víctor Siurana, 1, 25003 Lleida, Spain
e-mail: ildefonso.pla@udl.cat

M. M. Alconada-Magliano (ed.), *Intensified Land and Water Use*, Springer Earth System Sciences, https://doi.org/10.1007/978-3-030-65443-6_1

natural resources soil and water, generally imposed by social and economic pressures. The processes of soil degradation caused by soil-climate-management interactions generally result in unfavorable and sometimes drastic changes in the soil hydrological processes.

The problems of soil and water degradation, and derived effects are increasing throughout the world. This is partially due to a lack of appropriate identification and evaluation of the degradation processes and of the relations cause–effects of soil degradation for each specific situation, and the generalized use of empirical approaches to select soil and water conservation practices. The main soil and water degradation processes include soil water erosion (surface and mass movements), soil sealing and crusting, soil compaction, soil and water salinization and sodification, and soil and water pollution. In addition to the negative effects on plant growth and on productivity and crop production risks, soil and land degradation processes may contribute, directly or indirectly to the degradation of hydrographic catchments. This negatively affects the production of hydroelectric power, and the quantity and quality of water supply for the population and for irrigation or other uses in the lower lands of the watershed. Catastrophic flooding, sedimentations, and landslides are also rooted in accelerated land degradation.

The processes of soil and water degradation are closely linked through unfavorable alterations in the hydrological processes determining the soil water balance and the soil water regime. They are also conditioned by the climatic conditions and by the use and management of the soil and water resources (Paustian et al. 2016). Although the close interaction between the conservation of the soil and water resources is increasingly being accepted, still in most of the cases they are evaluated separately, and consequently the prediction and prevention of the effects derived from their degradation are inadequate in many situations. This will become more important under the previewed effects of global climatic changes, which would mainly affect hydrological processes in the land surface, mostly related to the field water balance (Varallyay 1990).

Future climate changes, although still rather uncertain, will increase rainfall in some regions, while others might become drier, in a rather uneven spatial and time distribution. This may contribute to accelerate some land degradation processes leading to larger runoff and erosion, to increase risks of flooding, landslides, mass movements, and mudflows in tropical regions, and to induce higher risks of crop production in subtropical and temperate regions. But in any case, land use changes, including deforestations, and other human activities leading to soil degradation processes may affect more the soil hydrological processes and their effects on land degradation, than the previewed global climatic changes, or may increase the influence of these changes (Pielke et al. 2005; Bouwer 2011; Pla 2010). By the contrary, adequate land use and soil and crop management practices may make soils more resistant against the effects of climate changes and derived extreme events.

Global change is a term widely used to describe the effects of natural and human activities on Earth. The global change includes world scale changes in many aspects of the globe environmental systems, besides climate, including other physical, chemical, and biological processes. In many cases, the human systems are now the main drivers of such changes, induced by population growth and development, affecting resource use, energy consumption, land use and land cover, land degradation, and related consequences. At present nearly 50% of the land surface has been transformed by direct human action, and more than 50% of the accessible freshwater has been appropriated for human purposes. Additionally groundwater resources are being depleted rapidly in many places.

Frequently global changes connected with human activities are confused or only used in reference to global climate change, more specifically global warming, occurring primarily as a result of human induced enrichment of the atmosphere with greenhouse gases. But global change, either in its natural or human induced forms, extends well beyond climate change (Pla 2014b). Although we cannot forget the future potential effects of climate change (global warming), other kinds of changes are at present of more immediate and pressing concern from the point of view of human welfare (Fig. 1.1).

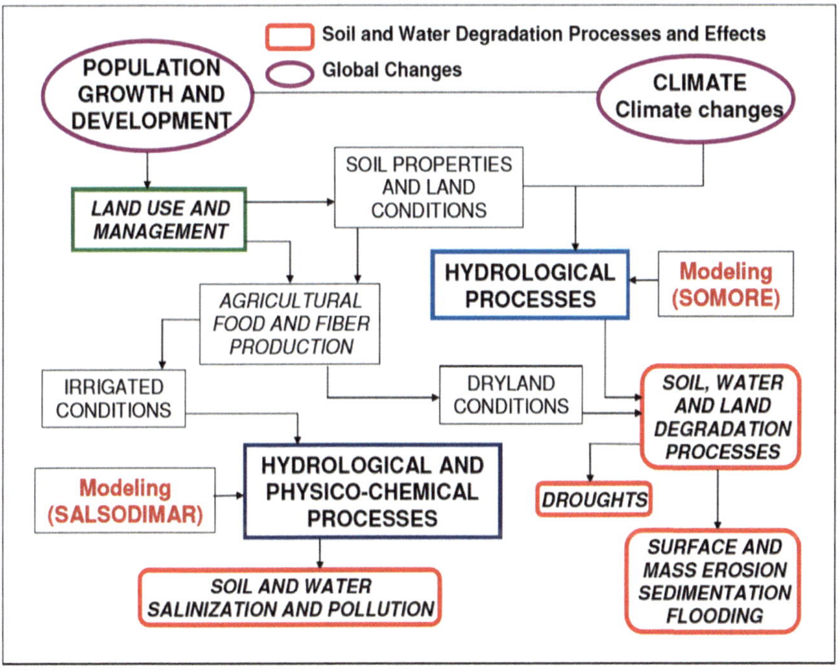

Fig. 1.1 Components and processes of a hydrological approach (Pla 2002, 2006, 2010) to evaluate and predict soil and water degradation processes and derived natural disasters (adapted from Pla 2018)

1.2 Hydrology and Soil Degradation Processes

The top layer of the soil is the one supporting most of the plant growth, and also the part with higher direct interactions with climate and vegetation, and more influenced by human activities. Climate is probably the main variable that has influence, directly or indirectly on the topsoil, and particularly the surface layer. Other surface processes are caused by the properties of the soil itself.

Unprotected soil surface is exposed to the direct impact of raindrops, causing disruption of soil aggregates and sealing and crusting effects. ***Sealing*** effects make reference to sharp decreases in water infiltration rates under wet conditions, while ***crusting*** effects refer to seals that have dried and hardened, offering resistance to seeding emergencies. Surface seals are thin soil layers with lower—sometimes by several orders of magnitude—saturated hydraulic conductivity than the underlying soil. The most important effect is the reduction in infiltration rates which may result in runoff and erosion, and inefficient use of rainwater in sloping lands, and plant injury due to waterlogging and reduced exchange of gases in flat lands. Although the hydraulic properties of the plow layer and deeper soil horizons are often used as a basis for deducing infiltration, in lands with scarce cover, infiltration and runoff are determined more by the changing soil surface conditions, than by internal soil physical properties. The amount of surface soil removed by runoff water depends to a large extent on the resistance of soil aggregates to be disrupted by the energy of raindrop impact.

Runoff induced by seal formation not only poses a problem with respect to soil erosion, but it is also water lost for storage of plant available water in the root zone, which may cause periods of water deficit for the plant, depending on the soil rooting depth and rainfall regime. In order to optimize the use of rainfall water and to control surface soil erosion, land and soil management practices have to be effective in reducing runoff and erosion, by imparting structural stability to the soil, improving water storage characteristics and reducing sealing. Protecting the soil surface with residues or cover crop against the impact of rainfall, and maintaining high levels of soil organic matter (SOM) in the surface soil, are the most effective methods of avoiding surface sealing. As a consequence of better infiltration there are possibilities of increased losses of water through the soil profile as internal drainage, and of larger transfer of pollutants from the soil to the groundwater.

Root growth is frequently restricted by excessive soil compaction. ***Compaction*** implies a decrease in volume, or increase in density, as a soil response to external forces. The result is a reduction in pore volume, and changes in pore size distribution, affecting air capacity and gaseous exchange, water retention, hydraulic conductivity, soil strength, and mechanical impedance to root growth. Compacted layers limiting root growth and water percolation may be natural or human induced. Excessive soil compaction by human activities is many times associated with the use of heavy machinery for tillage and other mechanical activities, and to more intensive land use. Shallow compacted layers generally become limiting barriers for root development and for deep percolation and drainage of excess infiltrated rainfall (Pla 1990). This may directly affect plant growth and crop production, and indirectly increase the risks of soil erosion, waterlogging, and water runoff losses. In some situations, compacted layers close to the soil surface may be loosened by tillage to enhance root growth and drainage, but their loosening effects are not lasting in most of the cases. For natural vegetation and rain-fed crops, primary consideration has to be given to the duration of the effective rainfall period, and consequently to the availability of soil water, which determines the possibility of successful establishment and sustainability of the cropping systems. The adverse effects of drought can be effectively minimized by selecting plants, crops and cropping systems with duration that match the secure moisture available period. Adequacy of moisture determines—when temperature is not limiting—the length of the growing period (LGP) (FAO 1981; Monteith and Virmani 1991).

The water balance models appear adequate to predict the reliability of the water supply for a plant during its growth, but most of them do not yet adequately consider surface sealing effects on water intake, waterlogging and runoff, and the effects of limiting shallow soil layers, natural or human induced, on root penetration, water retention at field capacity, and internal drainage.

Figures 1.2 and 1.3 show how the *length of the growing period* (LGP) (assuming no internal drainage limitations), in sloping lands would be influenced by the climatic

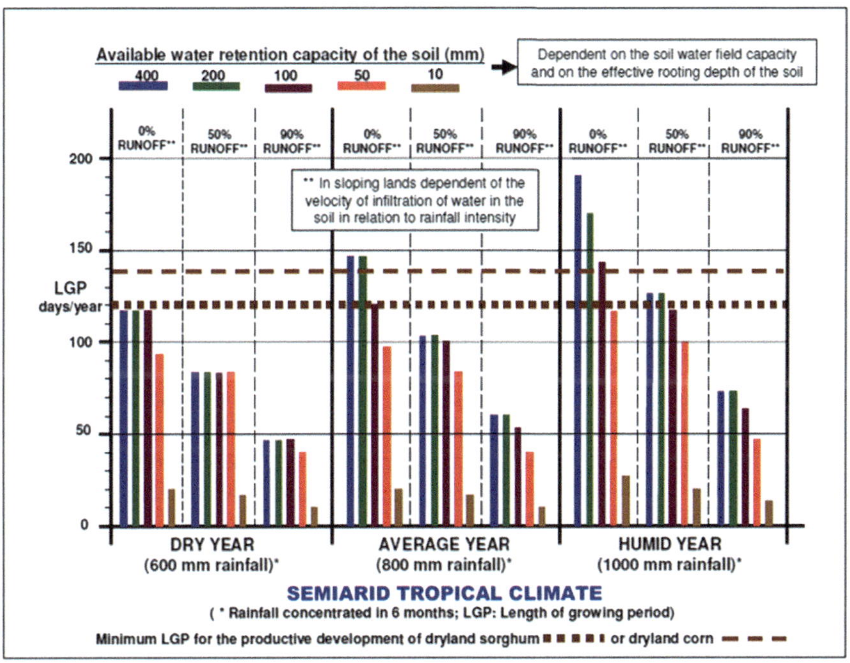

Fig. 1.2 Effects on the potential length of growing periods (LGP) of selected food crops (sorghum and corn) derived of semiarid tropical climate variations and changes, soil properties and degradation that affect hydrological processes, mainly those related to soil water retention and water runoff losses

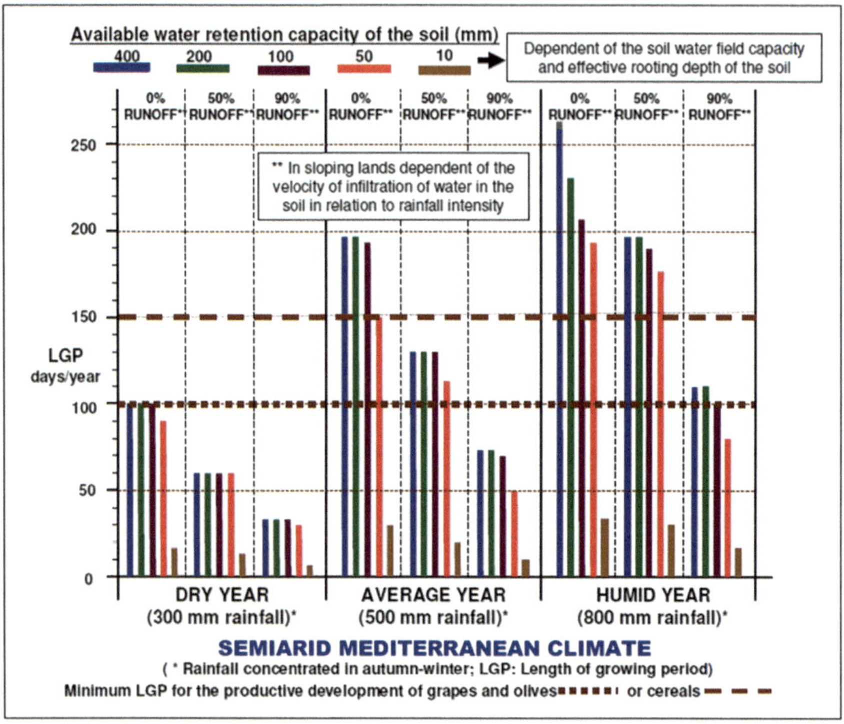

Fig. 1.3 Effects on the potential length of growing periods (LGP) of selected crops (grapes, olives and cereals) derived of semiarid mediterranean climate variations and changes, soil properties and degradation that affect hydrological processes, mainly those related to soil water retention and water runoff losses

conditions (rainfall and potential evapotranspiration), the distribution between infiltration and runoff (% of rainfall lost by runoff), and the soil storage capacity (available water capacity, determined by the effective rooting depth and water retention properties of the soil), for semiarid tropical and mediterranean climates. It may be seen that as the sealing effect increases (increasing % runoff), the influence of the rooting depth (*available water retention capacity*, AWC) on the LGP decreases. It is clear that the runoff and effective soil water capacity components of the water balance, both highly affected by soil degradation processes, cannot be neglected in the evaluation and prediction of the effects of those processes on water conservation and potential plant growth and crop production (Pla 1989; Pla et al. 2005).

Among the different land degradation processes, soil water erosion is the major threat to the conservation of soil and water resources. The processes of ***soil erosion***, caused by the interactions of soil, rainfall, slope, vegetation, and management, generally result on, or there are caused by unfavorable changes in the soil water balance and in the soil moisture regime, and in the possibilities of root development and activity.

Soil erosion processes have direct negative effects on plant growth and crop production, and off-site effects on increased risks of catastrophic floods, sedimentation, landslides, etc. Erosion is exacerbated by deforestation, by introduction of seasonal crops leaving the soil unprotected, by intensification or abandonment of agriculture, by overgrazing, and by improper maintenance of plantations and conservation structures. Besides surface erosion in gentle to moderate slopes, mass movements, and landslide erosion are common in more steep slopes (Pla 1992, 1993). In *surface erosion*, the soil particles detached by rainfall or running water, are transported by surface flowing water (surface runoff). *Mass movements* are the gravitational movements of soil material without the aid of running water (Crozier 1986). The hydrological processes leading to surface or landslide erosion are different (Pla 1992, 1997), and therefore, soil conservation practices very appropriate for controlling surface erosion processes may increase erosion danger by mass movements under specific combinations of climate, soil and slope.

Surface erosion is linked with intense precipitation events, high detachability of surface soil material and reduced infiltration. This reduction is induced by poor and weak surface soil structure and by poor cover of vegetation or plant residues in critical periods. Under these conditions, generally created by inadequate soil and crop management practices, the surface soil particles are detached by raindrop impact or by running water, and are transported down slope by runoff water, which flows more or less uniformly distributed on the soil surface, or concentrated in rills and gullies of different dimensions. Figure 1.4 shows examples in Argentina and Venezuela of the hydrological factors and processes involved in surface erosion processes.

Mass or landslide erosion generally affects soils with exceptional resistance to surface erosion due to excellent structural and hydraulic properties of the surface soil (Pla 1992). Sometimes mass erosion occurs on the steep slopes of gullies initially formed by surface erosion processes. Mass movements are generally initiated during and after concentrated and continuous precipitation events, and are associated with prolonged wet periods as a result of persistent antecedent rainfall, in soils with infiltration rates higher than internal drainage, which causes periodic saturation of the overlying soil (Pla 1997). These erosion processes are induced by the marked change in weight and consistency, decreasing cohesion among particles and micro-aggregates, of the surface soil overlying a layer retarding internal drainage. This retarding layer may be a natural pedogenetic pan, a lithic contact, or a compacted layer produced by inadequate tillage practices. The loss of cohesion and the fluid consistency after wetting close to saturation is more common in the surface layer of some soils like Ultisols and Andosols with very stable micro-aggregates. The water in the close to saturation surface soil is under a hydraulic gradient (depending on water supply and slope), and imparts lubrication to the underlying surface facilitating the sliding of the surcharged overlying soil material.

Landslides in hill slopes, which have been generally viewed as isolated catastrophic events resulting from infrequent rainfall and seismic events, are becoming major erosion and sediment transport processes worldwide, where deforestation (Sidle 1992) and drastic land use changes occurs, causing irreversible land degradation. Even in cases where landsliding is related only to extreme rainfall events

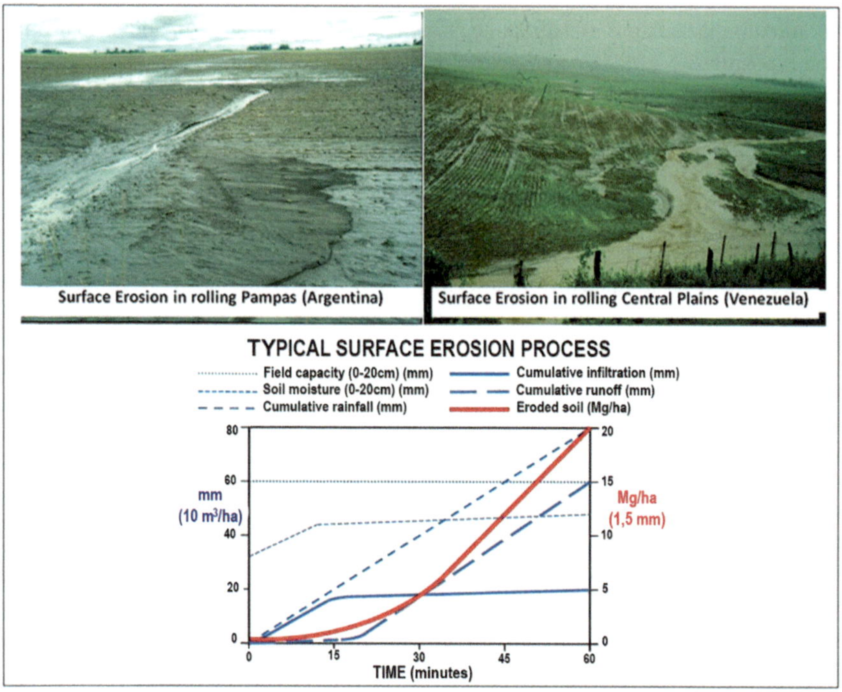

Fig. 1.4 Hydrological factors and processes involved in surface erosion processes in Argentina and Venezuela

with long return periods, the amount of soil removed and off-site effects generally outweigh the effects of the more studied continuous surface erosion processes. The influence of the effects of land use and management changes, including the injudicious application of soil and water conservation practices and structures, like terracing, in triggering landslides and mass erosion in general, is largely overlooked (Pla 2014a).

In deeper unconsolidated sedimentary or volcanic materials or in deeply weathered rocks, with decreasing permeability with depth, the accumulation of internal drainage water below the surface soil cover may lead with time to potential conditions for larger and deeper mass movements (Pla 2011). Change in weight and consistency of the surface soil, or deeper materials, cannot in themselves cause a landslide, but they do affect the susceptibility of a sloping land to triggering by some other factor, like earthquakes, removal of down slope (road cuts, etc.) or lateral support (gullies, cracks, etc.). In natural forested areas the possibilities of shallow landslides are generally much less than in clean cropped areas, and less than in pastures. Forests may have different stabilizing influences, but the main one is the mechanical reinforcement by tree roots, attaching potentially unstable surface soil to stable sub-strata, and providing a matted network which offers lateral attachment near the surface. Landslide erosion processes or mass movements in general, although occurring less frequently than surface erosion, may lead to much higher and more concentrated soil losses (Pla 1997), with more dangerous off-site effects. Figure 1.5 shows examples in Mexico and Colombia of the hydrological factors and processes involved in mass erosion processes.

Water, that is often the main limiting factor of plant growth, is also the main factor directly or indirectly responsible for soil and land degradation processes. These processes are strongly linked to unfavorable changes in the hydrological processes responsible for the soil water balance and for the soil moisture regime, which are affected by the climate conditions and variations, and by the changes in the use and management of soil and water resources (Pla 2002).

As it has been mentioned, the ***soil moisture regime***, determined by the changes in soil water content with time, is the main single factor conditioning moisture availability, plant growth and crop production. It is mainly conditioned by soil properties affecting the capacity and possibilities of infiltration, retention and drainage of rainwater, and the limitations to root growth under the particular rainfall characteristics (Pla 2002). These conditions may be modified by soil and plant management practices

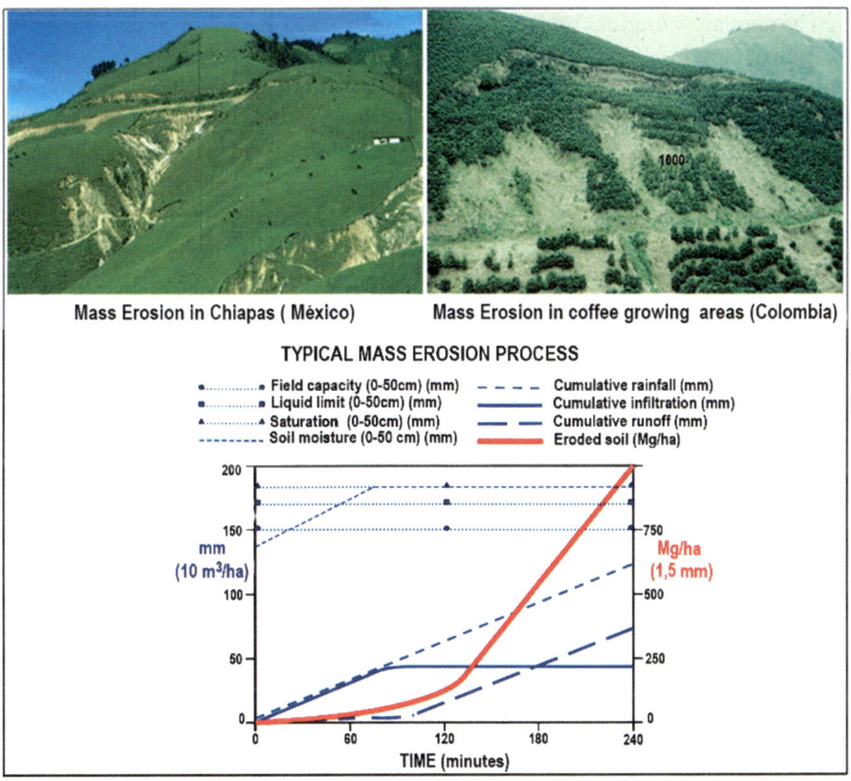

Fig. 1.5 Hydrological factors and processes involved in mass erosion processes in Mexico and Colombia

as tillage, irrigation, drainage, etc. Moisture availability is determined both by water gains from precipitation and water losses through runoff and evapotranspiration.

In most of the regions with semiarid to sub-humid temperate to tropical climates, the rainfall is highly variable among years and during the year, and usually occur in erratic storms of short duration and high intensities. Under non protected soil surface, associated with some intensive agricultural practices and overgrazing, extra precipitation in winter, occurring in intense episodes, may not be stored in the soil, but lost as runoff (Pla and Nacci 2001). These factors increase the risks of land degradation leading to *desertification processes*. The previewed effects of *global climate changes* would mainly affect *hydrological processes in the land surface*, mostly related to the soil water balance. In terms of *ecological and social impacts of climate change, changes in moisture availability are more important than changes in precipitation alone*. Low levels of moisture availability are associated with droughts and desertification. Reductions in mean annual rainfall leads to drier conditions, but increase in climate variability during the year, or increasing frequency of very dry years, could be equally or more important. Therefore, the term aridity for evaluating desertification, instead of only considering average rainfall conditions, would be more appropriate if it also considers variability through the whole hydrological cycle as well as climatic variations and fluctuations.

Human activities leading *to land degradation processes may affect more the soil hydrological processes than the previewed climate changes*, or may increase the influence of those changes (Pla 2010). *Forests* usually regulate stream flows, protect land from erosion, reduce flooding in adjacent areas, minimize the silting of rivers, canals and dams, and contribute to a *stable hydrology* essential for providing stable sources of water for human needs and irrigated agriculture. This water balance may be drastically upset by deforestation and forest fires, and especially by the consequent land degradation. Supply of available water may decrease irreversibly under unchanged soil properties and stable hydrological soil parameters due to reduced water income, increasing water consumption, or both. Under unchanged water income by rainfall, the hydrological parameters of soils may change irreversibly as a result of soil degradation (sealing, compaction, erosion, decreased water holding capacity, etc.), leading to the same effects of decreasing available water supply (see Figs. 1.2, 1.3). *Changes in land cover*, with decreasing water use, may lead to increasing groundwater levels, mainly in the lower parts of the landscape, with derived *effects of waterlogging and salinization processes* in some cases.

Irrigation causes drastic changes in the regime and balance of water and solutes in the soil profile, which may result in soil salinization, one of the processes of soil degradation leading to land desertification. The salinity problems are a consequence of *salt accumulation* in zones and depths where the soil moisture regime is characterized by *strong losses of water by evaporation and transpiration, and by reduced leaching of the remaining salts*. Under dryland conditions, this situation may be reached under shallow groundwater levels, where the water source comes by capillary rise. The salt accumulation may conduce to a partial or complete loss of soil capacity to provide the required amounts of water to plants, changing fertile lands to deserts (Pla 1996).

1.3 Evaluation and Prediction of Soil Degradation Processes

From the previous arguments, it follows that approaches based on water balance models are the more adequate to predict the soil moisture regime, related to the reliability of the water supply for a plant during its growth, and to external and internal drainage of excess water affecting runoff, erosion and groundwater level changes. This would be the main basis for determining the suitability of the land for various uses under given conditions of management. Research into the basic hydrological processes of land degradation, including climate and soil data is required. Research is also necessary in the hydrological changes as a result of various alternative land uses and agricultural systems and practices. The degree of aridization of soil may be quantitatively determined in terms of certain physical properties and water regime of soils (annual supply of available water in the root zone), using soil hydrological parameters (Pla 2006).

A hydrological approach to the evaluation and prediction of the conservation of soil and water against degradation processes would be essential for an adequate development, selection and application of sustainable and effective land use and management practices. The main objective must be to evaluate such hydrological processes, and to select and develop methodologies and techniques to correct or to control them under different conditions of soils, topography, and climate. This is required for suppressing or alleviating the negative effects of soil and water degradation on plant growth, on sustainable agricultural production, on the supply of water in adequate quantity and quality for the different potential uses, on the rise of groundwater levels and its consequences, and on catastrophic events such as flooding, sedimentation, landslides, etc.

Methodology for an adequate quantitative characterization and prediction of the affected soil hydrological properties is required to evaluate the present problems and to assess the vulnerability of soils to different degradation processes. Besides measurements under precise conditions in the laboratory, useful for understanding the hydrological processes, these have to be approximately quantified at field scale. If we intend to use the hydraulic functions to predict or to solve field problems, it is preferable to estimate them from field measurements and experiments. In structured soils, sampling and laboratory measurements are some times more difficult and time consuming than field measurements. Most of the results of experiments with repacked core samples cannot be directly transferred to quantify soil water behavior in the field. But field measuring techniques are often less accurate and more expensive than in the laboratory, because it is often difficult to rigorously establish boundary conditions, which fluctuate in space and time. This accuracy may be increased with a larger number of directly field-measured hydraulic properties. The objective has to obtain approximations acceptable within the limitations of the used methodologies, which can provide practical guidelines for field situations.

The methods and techniques applicable for predicting soil hydrological behavior under field conditions should allow simple and direct measurements, based on comprehensive physical relations, and should take into consideration the dynamic aspects of the soil hydrological properties, highly dependent on soil structure (Pla 1990; Nacci and Pla 1993). They include from simple, straightforward field techniques, usually providing rough estimates of soil hydraulic properties, to rather complicated techniques for more accurate time-consuming measurements, requiring sophisticated skills and costly equipment. The simple field techniques must be preferred (Pla 1990), because of operational considerations, and because they are more able to be adapted to the required sample volume and spatial variation—there are possible more replicate measurements—of soil hydraulic properties under field conditions. Although modern indirect techniques like remote sensing, computerized data processing, GIS, and simulation models may help in the required evaluations, they will always require updated and accurate direct measurements or estimations of soil hydraulic parameters. Specially needed are better and simpler methods to monitor important hydraulic properties of soils and their dynamics on a field scale, for both diagnostic and prediction purposes. These properties should be also quantified in terms of the dynamic action of root growth (Larson and Pierce 1994).

It is generally accepted that for evaluation and prediction of land degradation there are required long-term experiments on a catchment basis. Although this would be desirable, it is not possible if short-term solutions are required, as it is usually the case. There are mandatory new approaches based on the evaluation of soil hydrological properties together with historical rainfall records, under different scenarios of changing climate, soil properties, topography, and land and crop management (Pla 1998). Research on soil degradation has to concentrate more on hydrological and soil degradation processes for interpretation of land degradation problems, with the help of computer-based programs that can be applied to different environments (Pla 1998). This process-based approach makes the extrapolations more soundly based, and may allow them to select or to develop a more adequate package of technologies to reduce soil degradation, while being social and economically acceptable. In any case it is desirable to keep the information that must be obtained simple, so that basic objectives can be achieved. Pressure to change sustainable traditional systems of use and management of land resources must be considered additional parameters (social economic factors) to the biophysical factors in soil degradation processes.

1.4 Modeling Hydrological Processes to Predict Soil Degradation

Probabilities and risks of soil degradation and its influence on crop production and environmental damage (Pla 1994), may be partially satisfied with the use of modeling, where the large number of important variables involved in the degradation processes, and their interactions, may be integrated. Direct measurements of runoff and soil loss in the traditional erosion field plots, generally associated to get some of the inputs required by the so-called Universal Soil Loss Equation (USLE) (Wischmeier and

Smith 1978), is a slow and costly process. The latter is due to the high variability of climate and soils in time and space, which makes it not practical in places where the resources are scarce and there are required short-term solutions. Therefore, the prediction of water erosion is presently generally done using mostly empirical, and much less process-based methods and models (Nearing et al. 1994). Among these, the USLE and its derivatives and adaptations, has been the most widely used worldwide. These models require local information not available in many cases, which frequently is being substituted by information generated through sub-models and regression equations developed under conditions very different to the ones where they are applied (Richter and Streck 1998). These may lead to great errors in the prediction of soil erosion and selection of conservation practices, with catastrophic results and economical losses in investments and conservation structures.

The presently used empirical models must be replaced with process-based event models, which require a better understanding of changing hydrological properties as influenced by soil management, cropping sequences, vegetation, and climate (Foster and Lane 1987). These models must allow a detailed quantification of hydrological processes for both actual and potential conditions, answering major questions about problems of soil degradation and crop production, related to different alternatives of land management (Pla 1997, 1998). While they are developed, the main benefit of these models is the identification of gaps of knowledge and data, and the understanding of the degradation processes. Process-based prediction models, based on equations that represent fundamental hydrological and erosion processes, including rainfall, infiltration, internal drainage, and runoff, may solve the limitations of the empirical soil loss prediction models, like site specificity, limited transferability and others.

Simulation models based on hydrological processes may be very helpful to integrate and to convert the measured or estimated soil, climate, plant and management parameters into predicted soil water balances and soil moisture regimes for each particular combination of them, actual or previewed. These models may be very simple, or they can be extremely complex, requiring many resources (time, equipment, manpower) and input information which is seldom available, or difficult to determine, or non representative, making less complex models often more suitable for practical purposes. Simulation errors derived from estimation errors in soil properties and the sampling costs are generally lower when simple models are used for predicting water balance in space (Leenhardt et al. 1994). Additionally, simpler models require fewer input data, and therefore they allow larger samples and sampling densities for a given field measurement.

The models used in predicting crop performance, and soil degradation processes derived from the impact of land use and management practices in the soil hydrology, must include weather (mainly rainfall) variability in space and time, and soil properties and their spatial variability. The required data about soil properties are those influencing water entry and retention in the soil, limits of water retention capacity of the soil, loss of water by evaporation, and environment for root growth. Methods to approximate the needed soil properties from existing soil taxonomic data (pedotransfer functions, PTF) have been proposed when there are no possibilities for direct measurement of those soil attributes. But when the used correlations (PTF) have not been obtained or validated in situ, the possibilities of large errors in the output of the models are very high. The weather data required for modeling are the ones influencing potential water supply to the plants and to runoff and evapotranspiration. These include daily values of rainfall, and monthly values of solar radiation and of maximum and minimum temperatures. When long-term weather records or daily rainfall are not available, an alternative procedure would be to use stochastic time-series modeling, to generate a sequence of weather data similar to historical sequences (Ritchie et al. 1990).

The flow diagram of Fig. 1.6, which was the basis for the development of the simulation model SOMORE (Pla 1989, 1992, 1997, 2006), simulates the evolution of the soil water balance in the soil profile with a time step of one day, using easily obtainable soil and meteorological data as input. It may be used to predict the soil moisture regime, including waterlogging, rainfall losses by surface runoff, surface and internal drainage, and groundwater levels, under different conditions of soils, topography, climate, vegetation, crops, and management. The model accounts for infiltration of rainfall into the soil as limited by surface sealing effects and limiting layers (natural or induced by management) close to the soil surface, and for internal drainage or subsurface runoff as affected by rainfall infiltration, effective root depth and saturated hydraulic conductivity of the limiting soil layer. The predictions may be used to identify the more probable degradation processes, and for the selection of the

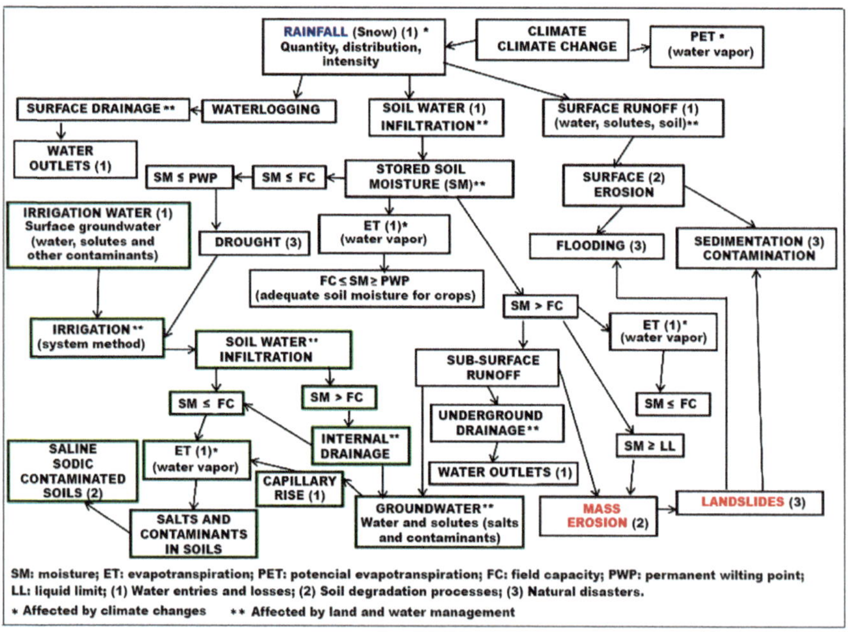

Fig. 1.6 Flow diagram of the model SOMORE (Pla 1997, 2002, 2006) integrating the hydrological processes related with soil degradation processes and effects

best alternatives, with more probabilities of success, of soil and water conservation practices for each combination of soils, climate, and topography.

The predicted soil moisture regime may be interpreted in relation to problems of drought or aeration in the overlying soil, at different times and growth stages of natural vegetation or crops, and also in relation to irrigation requirements, possibilities of tillage operations, and erosion hazard by different processes. To preview the possible influences of different combinations of soil and water management on the soil moisture regime, there is required a previous identification and evaluation of the main critical factors affecting problems of soil degradation and of water supply to crops. The variable annual rainfall data, with a particular return period, are used to simulate the behavior of a particular condition or management system in different years, and therefore, based on that previewed behavior, it is possible to select or design, with a probabilistic approach, the best system of soil and water management to control soil degradation. It is also possible to predict the soil degradation processes and effects, and the problems of water supply to crops, with different return periods, for each condition or proposed land use and management. The selection of certain return periods is important, because they largely determine the requirements of management practices and conservation structures in relation to costs and benefits, for different levels of risk and probabilities of failures. A particular season or year is described, or analyzed, in relation to the long-term variability, based on rainfall records from the past. The prediction of concentration of surface and subsurface runoff, and of the conditions of soil moisture would permit to preview which days or periods of the year would have the greatest flood, erosion, and sedimentation hazard, and what would be the most probable erosion process (Pla 1992, 1993, 1997, 1998). This is more useful for designing erosion control strategies than the use of empirical models which have proved not to be able to predict the time and probabilities of occurrence of concentrated runoff and erosion, and much less landslides or mass movements in general.

The prediction of concentration of surface and subsurface runoff, and of the conditions of soil moisture would permit to preview which days or periods of the year would have the greatest flood, erosion and sedimentation hazard, and what would be the most probable erosion process.

1.5 Examples of Application

As examples of the potential use of modeling hydrological processes for evaluating and predicting soil and water degradation processes, and for guiding soil and water conservation practices, three situations with different soils, climate, topography, cropping and management conditions are presented. The same approach could be used for any other climatic conditions and combination of soil and management parameters.

The ***first case*** (Figs. 1.7 and 1.8) refer to rolling lands (4–10% slopes) with sandy loam ***Alfisols***, in the Center Plains of *Venezuela*, under a ***tropical semi arid climate***,

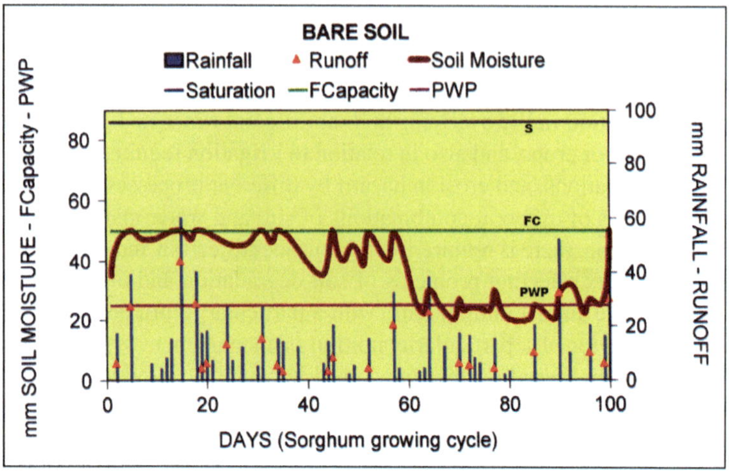

Fig. 1.7 Soil moisture regime during the growing period of sorghum in a bare tilled Alfisol, in land with 4–10% slope, under a semiarid tropical climate

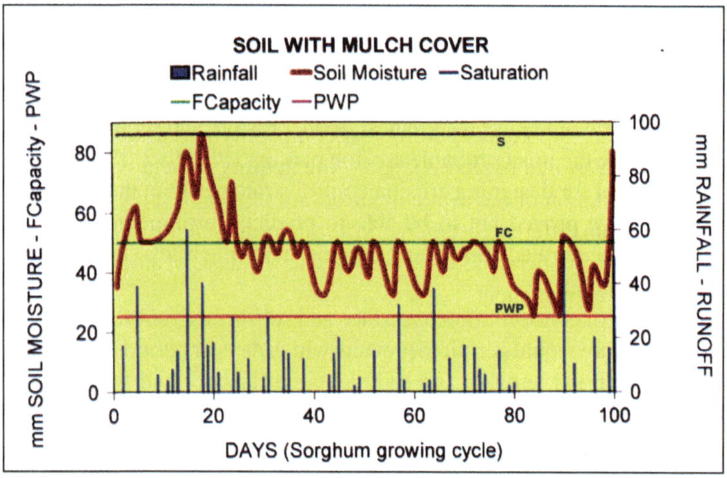

Fig. 1.8 Soil moisture regime during the growing period of sorghum in an Alfisol 4–10% slopes, covered with sorghum residues, under a semiarid tropical climate

characterized by a strong seasonal distribution and high variability of rainfall from one year to the other, and within the same year. The traditional use for pastures have changed to continuous cropping of *rainfed sorghum*, with a length of growing period of 90–100 days. The plant residues are usually used as forage for cattle during the dry season. The main constraints for a high and sustained productivity have been identified as soil moisture deficits and surface soil water erosion (Pla 1989, 1997). Sealing effects on bare soil appear to be the main cause of concentrated runoff during intense storms, causing water and soil losses. The root growth is limited by the presence of an argillic horizon at 20–40 cm depth (which gets closer to the soil surface after accelerated erosion), and shallow (10–15 cm) clean tillage using mostly disk harrows.

Figure 1.7 shows the daily moisture regime in a bare tilled soil during the growing period of sorghum, under average rainfall (RP: 2 years) and shallow 20 cm root depth. The runoff and potential soil erosion—accompanied by flooding and sedimentation in the lower parts of the landscape—are more critical in the first 1/3 of the growing period, while water deficits are concentrated in the last 1/3, coinciding with the critical reproductive and grain formation period. From that we may expect a good vegetative growth, but reduced grain production.

Figure 1.8 shows how the soil moisture regime would be affected by soil cover with sorghum residues, preventing runoff (and erosion) and water deficit, even with the relatively shallow (20 cm) rooting depth and average rainfall. In this case we may conclude that the marked surface soil sealing effect is clearly the main cause of concentrated runoff (30–50% of the total rainfall in the rainy season), of the erosion and of the moisture deficit in bare soil with moderate slopes. The same procedure could be used to explore the potential soil water balances and moisture regimes for other different combinations of climate, crops, effective soil depth, and management practices.

In the ***second case*** (Figs. 1.9 and 1.10), there was studied the interaction of changes in land use and management, and in climate, with land degradation processes associated with unfavorable effects on hydrological processes in dryland vineyards in NE *Spain* (Penedés). The water use of grapevines through the growing season is characterized by lessened requirements in the periods before bloom and after harvest until fall (autumn), and a maximum consumption in the middle part of the growing season. If the reserve water capacity of the soil in the rooting zone is not enough, reduced amounts of rainfall during the main growing season of grapevines (June–August) may lead to a long-term soil water deficit, which can affect growth, production and maturation, in spite of the natural survival capacity of grapevines under drought conditions. There were evaluated problems of soil water supply to the plants through the different growing periods in the year, and of runoff and surface erosion (Pla and Nacci 2001), derived from changes in hydrological behavior under the new leveling, planting and management practices. The study areas are in lands with silty loam ***Inceptisols***, highly calcareous (developed in calcilutites), in undulated hilly topography (4–20% slope), cultivated with rain-fed grapevines for high quality wine and cava production, under a ***mediterranean semi arid climate***, with an average annual rainfall of approximately 600 mm, very irregularly distributed (greatest rains in autumn-winter, a very dry summer). Likewise, there is great variability in the totals from one year to another (400–750 mm).

Rainfall is typified by many storms in autumn, and occasionally in spring of high concentration and intensity. Climate change may increase the irregularity of this rainfall, the frequency of dry years and the probability of extreme events, phenomena that have been observed in the last 25 years. In order to decrease costs of the scarcely available manual labor, to increase production and to speed all operations, the current

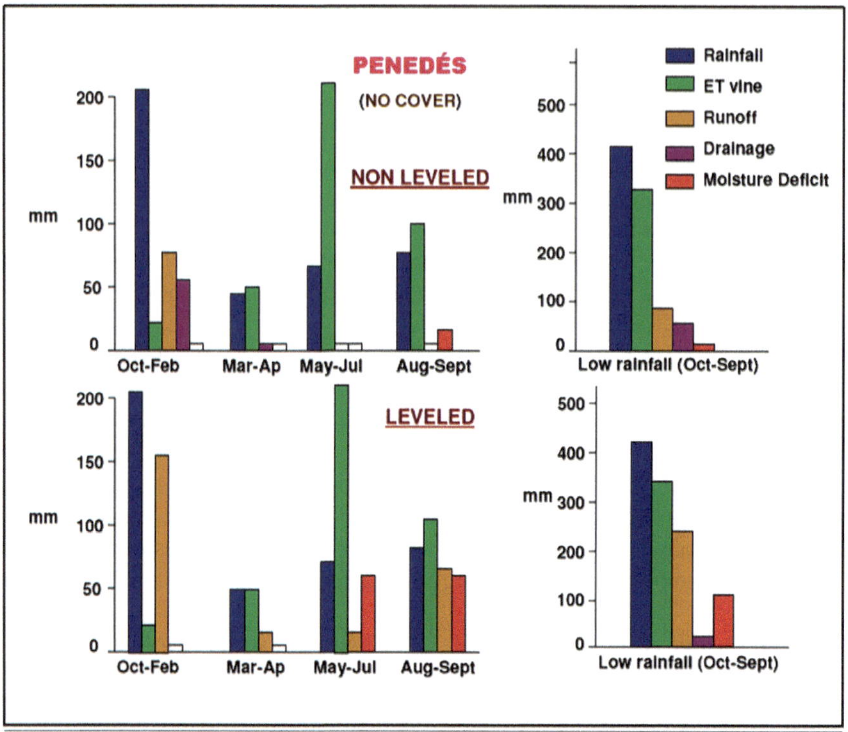

Fig. 1.9 Rainfall and soil water balance components in rain fed vineyards of the Penedés region in years with low rainfall (Return period: 5 years) in non leveled and leveled land, without use of green cover

trend is toward full mechanization of all practices, including harvesting, which has required to change the plantation pattern, and to perform heavy land leveling operations to smooth the slopes and to decrease the irregularities of the original topography. This has caused big changes in the surface drainage network and on the effective soil rooting depth and surface soil properties (Pla and Nacci 2003) resulting in drastic effects on the soil moisture regime. The major effects are on surface runoff, surface erosion, and in the retention of rainfall water in the soil for utilization by the grapevines (Figs. 1.9 and 1.10). It may be appreciated that the risks of erosion and drought would be much higher in the leveled lands. It is also shown that the only possibility to have a green cover between the vine rows, is during the resting period, and that if a cover was maintained (usually it dries out in the April–May drier period) for the rest of the year it would need to be killed with a selective herbicide, not toxic to the vines. It is evident that the use of a green cover crop in the resting period would increase the possibilities of drought in the critical budburst, bloom and veraison period (May–July) in drier years, and in soils with lower available water retention capacity. A positive effect of the green cover crop would be a reduction in

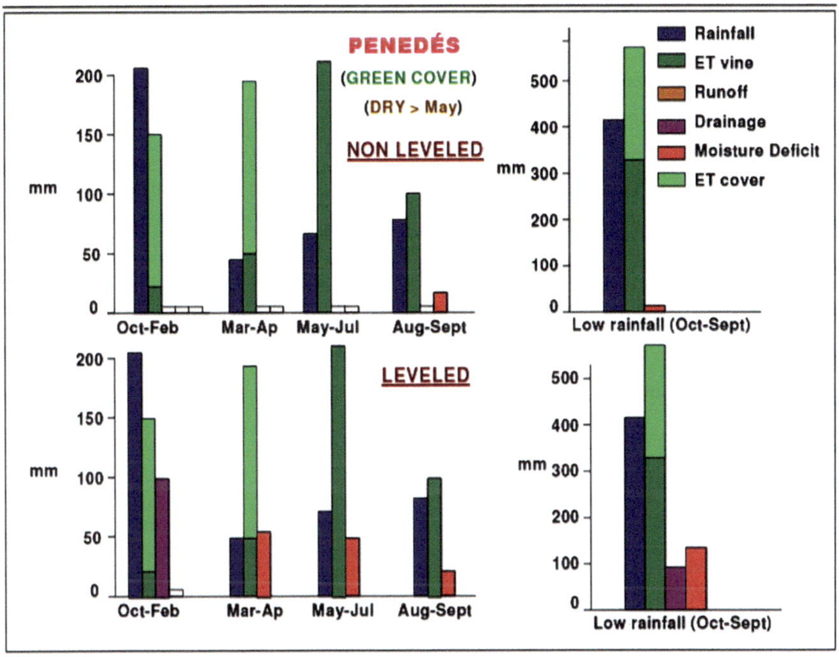

Fig. 1.10 Rainfall and soil water balance components in rain fed vineyards of the Penedés region in years with low rainfall (Return period: 5 years) in non leveled and leveled land, with use of green cover

the water runoff losses and in the accompanying soil water erosion, especially in the more humid years with concentrated rainfall.

The ***third example*** (Fig. 1.11) has to do with sloping lands with ***Ultisols***, developed on clay rocks, with very steep slopes (30–50%), in an area under a ***tropical humid climate***, in Western *Venezuela*. The land, originally with a dense humid tropical forest, has been deforested to be used for pastures, usually overgrazed. Under these new conditions, and due to the presence at 30–40 cm depth of an argillic horizon with a marked decrease in saturated hydraulic conductivity, there have been developed catastrophic *mass erosion processes*, mainly landslides, in years and periods with concentrated and continuous rainfall events (Pla 1992). The surface soil, with micro-aggregates very stable to wetting and raindrop impact, do not show any sealing effect, and maintain a minimum rain water infiltration rates much higher than the underlying soil below 30 cm depth. The combination of a year with high rainfall (RP: 10 years), and the restriction in internal drainage below 30 cm depth (Fig. 1.11), create conditions in two periods (30–40, and 135–170 days) of the rainy season, which may lead to accelerated erosion with landslides and mass movements. In this case it may be concluded that the problem of erosion or runoff has nothing to do with surface sealing effects, but it is mainly caused by the combination of high surface infiltration rates, restricted drainage at relatively shallow depth, loss of the anchorage effect of

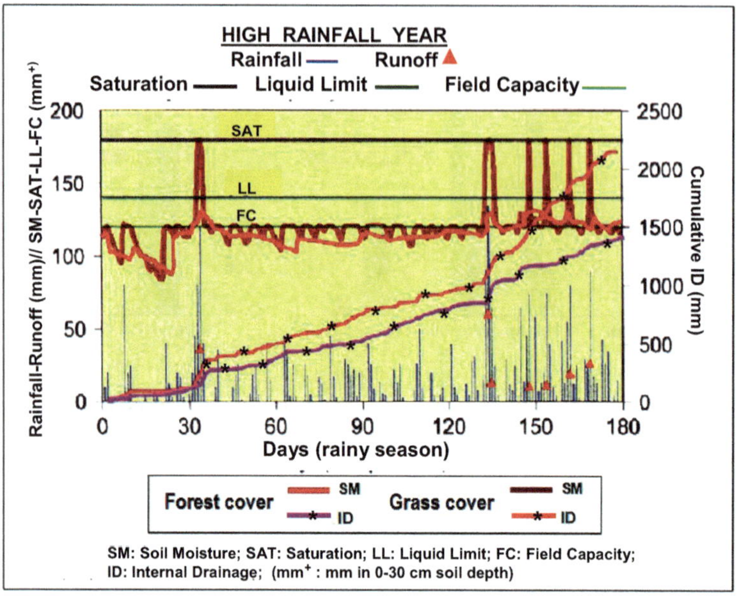

Fig. 1.11 Soil water moisture regime, runoff and internal drainage regimes in steep lands with Ultisols, a year exceptionally rainy (RP: 10 years), with natural forest and with grass covered deforested land, under a humid tropical climate

roots from the permanent natural vegetation or crops, and concentration of rainfall events.

1.6 Synthesis

Figure 1.12 shows in a very simplified way, the interactions of the different conditions and processes leading to potential land degradation processes and derived effects, both on-site and off-site. Likewise, are shows the steps to be followed for land use planning leading to effective soil and water conservation practices for sustainable land use and management, under different and variable biophysical and socioeconomic scenarios.

1.7 Final Considerations

Soil and water degradation are closely linked due to unfavorable alterations in the hydrological processes determined by the balance and regime of the soil water. Likewise, are conditioned by the climatic conditions and by the use and management of

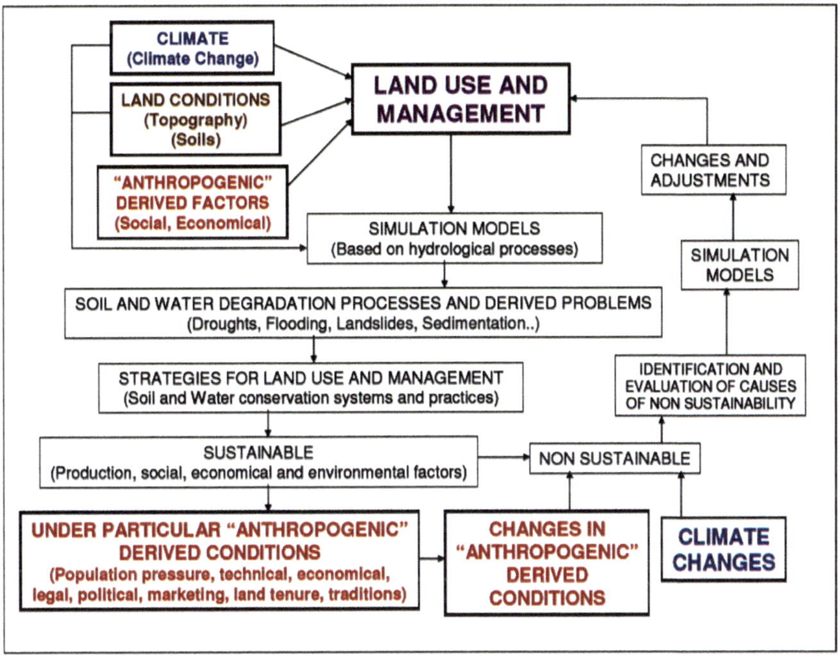

Fig. 1.12 Interactions of the different conditions and processes leading to potential land degradation processes and derived effects, and steps to be followed for land use planning leading to sustainable land use and management (adapted from Pla 2018)

the soil and water resources. This will become more important under the previewed effects of global climatic changes, but in any case, land use changes, including deforestations, and other human activities leading to soil degradation processes may affect more the soil hydrological processes and their effects on land degradation, than the previewed global climatic changes, or may increase the influence of these changes. Moreover, adequate land use and soil and crop management practices may make soils more resistant against the effects of climate changes and derived extreme events.

In general, it may be concluded that hydrological approaches, based on appropriate in situ evaluations of soil hydrological properties and processes, under different scenarios of changing climate, soil properties, and land use and management, complemented with the use of simple simulation water balance models based on those processes, may be very useful, and even indispensable, to predict and to identify the biophysical causes of land degradation at local, national and regional levels. This is a required previous step for an adequate planning of more sustainable land use and management for crop production and environmental protection, and for the selection and development of short and long-term strategies and technologies to reduce or to control land degradation processes, and the related social economic and security problems.

References

Bouwer LM (2011) Have disaster losses increased due to anthropogenic climate change? Bull Amer Meteor Soc 92(1):39–46
Crozier MJ (1986) Landslides: causes, consequences and environment. Croom Helm. London, Dover NH, p 252
FAO (1981) Informe del Proyecto de Zonas Agroecológicas. Volumen 3. Metodología y Resultados para América del Sur y Central. FAO. Rome, Italy, p 252
Foster GR, Lane LJ (1987) Beyond the USLE: advancements in soil erosion prediction. In: Boersma LL et al (eds) Future developments in soil science research. Soil Sci Soc Am, Madison, USA, pp 315–326
Larson WE, Pierce FJ (1994) The dynamics of soil quality as a measure of sustainable management. In: JW Doran et al. (eds) Defining soil quality for a sustainable environment. Soil Sci Soc Am Sp Publ. Madison, USA 35:37–52
Leenhardt D, Voltz M, Bornand M (1994) Propagation of the error of spatial prediction of soil properties in simulating crop evapotranspiration. Eur J Soil Sci 45:303–310
Monteith JL, Virmani SM (1991) Quantifying climatic risk in the semiarid tropics: ICRISAT experience. In: RC Muchow, JA Bellany (ed.) Climatic risk in crop production: models and management for the semiarid tropics and subtropics CAB International, Wellingford, UK, pp 183–204
Nacci S, Pla I (1993) Técnicas y equipos simplificados para evaluar propiedades físicas importantes para el manejo de suelos en los trópicos. In: R Villegas, D Ponce de León (eds) Memorias XI Congreso Latinoamericano and II Congreso Cubano de la Ciencia del Suelo, V 1, SLCS-INICA, La Habana, Cuba, pp 199–217
Nearing MA, Lane LJ, Lopes VL (1994) Modeling soil erosion. In: Lal R (ed) Soil erosion research methods, 2nd edn. SWCS-ISSS, Ankeny, USA, pp 127–158
Paustian K, Lehmann J, Ogle S, Reay D, Robertson GP, Smith P (2016) Climate-smart soils. Nature 532:49–57
Pielke RA, Agrawala S, Bouwer LM, Burton I, Changnon S, Glantz MH, McCarthy JJ (2005) Clarifying the attribution of recent disaster losses: a response to Epstein and McCarthy. Bull Am Meteor Soc 86(10):1481–1484. https://doi.org/10.1175/BAMS-86-10-1481
Pla I (1989) Soil water constraints for dryland corn and sorghum production in Venezuela. In: P Unger et al. (eds) Challenges in dryland agriculture. Amarillo/Buchland, USA, pp 140–144
Pla I (1990) Methodological problems to evaluate soil physical degradation. Trans. 14th Int Congress of Soil Sci Soc Kyoto, Japan I, pp 95–100
Pla I (1992) La erodabilidad de los Andisoles en Latino América. Suelos Ecuatoriales 22(1) Sociedad Colombiana de la Ciencia del Suelo, Bogotá, Colombia, pp 33–43
Pla I (1993) Erosión de suelos de ladera del trópico Andino y Centroamericano. In: F. Munevar et al. (eds) Manejo Integrado de Recursos Naturales en Ecosistemas Tropicales para una Agricultura Sostenible ICA Santa Fe de Bogotá, Colombia, pp 21–36
Pla I (1994) Soil degradation and climate-induced risks of crop production in the tropics. Trans 15th ISSS Congress ISSS. Acapulco, Mexico 1:163–188
Pla I (1996) Soil salinization and land desertification. In: Rubio et al. (eds) Soil degradation and desertification in mediterranean environments, Geoforma Ed, Logroño, Spain, pp 105–129
Pla I (1997) A soil water balance model for monitoring soil erosion processes and effects on steep lands in the tropics. In: I Pla (ed) Soil erosion processes on steep lands. Special Issue of Soil Technology, Elsevier, Amsterdam, The Netherlands 11(1), pp 17–30
Pla I (1998) Modelling hydrological processes for guiding soil and water conservation practices. In: Rodríguez A et al (eds) The soil as a strategic resource: degradation processes and conservation measures, Geoforma edn. Logroño, Spain, pp 395–412
Pla I (2002) Hydrological approach to soil and water conservation. In: Rubio et al. (eds) Man and soil at the Third Millenium, Geoforma Ed, Logroño (España) I, pp 65–87

Pla I (2006) Hydrological approach for assessing desertification processes in the mediterranean region. In: WG Kepner et al (eds) Desertification in the mediterranean region. A Security Issue. Springer, Heidelberg, Germany, pp 579–600
Pla I (2010) Sustainable water management under climate change. In: M Zlatic (ed) Global change—challenges for soil management. Advances in GeoEcology 41 Catena Verlag GMBH, Reiskirchen, Germany, pp 22–36
Pla I (2011) Evaluación y Modelización Hidrológica para el Diagnóstico y Prevención de "Desastres Naturales". Gestión y Ambiente 14. IDEA-UN-Medellín, Colombia 3:17–22
Pla I (2014a) Advances in soil conservation research: challenges for the future. SJSS. Spanish J Soil Sci 4(3):265–282. SECS (Spain)
Pla I (2014b) Conflicting issues in controlling land degradation and global climate changes. In: M Zlatic, S Kostadinov (eds) Challenges: sustainable land management-climate change. Advances in GeoEcology 43, Catena Verlag GMBH. Reiskirchen, Germany, pp 192–210
Pla I (2018) Anthropogenic and climate change factors in present and future soil and water conservation problems. In: Zlatic and Kostadinov (eds) Soil and water resources protection in the changing environment. Advances in GeoEcology 45, Catena Verlag GMBH, pp 258–268
Pla I, Nacci S (2001) Impacts of mechanization on surface erosion and mass movements in vineyards of the Anoia-Alt Penedés Area, Catalonia, Spain. In: Scott et al. (eds) Sustaining the global farm, Purdue University-USDA, ARS, West Lafayette, USA, pp 812–816
Pla I, Nacci S (2003) Tradicional compared to new systems for land management in vineyards of Catalonia (Spain). In: Roose et al, (eds) Techniques Traditionnelles de GCES en Milieu Mèditerranien. (E.). Bulletin Reseau Erosion 21: 213–223. Montpellier (France)
Pla I, Ramos MC, Nacci S, Fonseca F, Abreu X (2005) Soil-moisture regime in vineyards of Catalunya (Spain) as influenced by climate, soil and land management. In: J Benítez, F Pisante (eds) Integrated soil and water management for orchard development. Land and Water Bulletin 10, FAO, Rome, Italy, pp 41–49
Richter J, Streck T (1998) Modeling processes in the soil as a tool for understanding and management in soil and water conservation. In: LS Bhushan et al (ed) Soil and water conservation: challenges and opportunities, vol I, New Delhi (India), pp 234–241
Ritchie JT, Godwin DC, Singh U (1990) Soil and water inputs for the IBSNAT crop models. In: IBSNAT symp. Proc. Decision support system for agrotechnology transfer, University of Hawaii, Honolulu, USA, pp 31–46
Sidle RC (1992) A conceptual model of changes in root cohesion in response to vegetation management. J Environ Qual 20(1):43–52
Varallyay GY (1990) Influence of climatic change on soil moisture regime, texture, structure, and erosion. In: Scharpenseel HW et al (eds) Soil on a warmer earth Elsevier. The Netherlands, Amsterdam, pp 39–49
Wischmeier WH, Smith DD (1978) Predicting erosion losses: a guide to conservation planning. USDA Agricultural Handbook N° 537, Washington DC, USA

Chapter 2
Prediction of Soil Salinization and Sodification Processes as Affected by Groundwater Under Different Climate and Management Conditions

Ildefonso Pla-Sentís

Abstract Salt-affected soils, both saline and sodic, may develop both under dryland and irrigated conditions, with negative consequences in the environment, in the crop production and in the animal and human health. The processes of sodification have generally received less attention and are less understood than the development of saline soils. In both of them hydrological processes are involved in their development, but in the case of sodic soils we have to consider some additional chemical and physicochemical reactions. This is especially true where we have to include the effects of the groundwater level and composition. Both the addition of irrigation water and the changes in depth and composition of groundwater, may cause great changes in the water and solute balances in the soil profile. Modeling may be very useful for the diagnosis and prediction of such changes, and for the selection of the best practices and systems of irrigation and drainage for a more efficient use of irrigation water, reducing the losses and contamination of surface and groundwater, and controlling the soil salinization and sodification. There is presented an adaptation of the model SALSODIMAR, including new specific hydrological components of the water and solute balances, to make it useful to predict the processes of both the dryland salinization and sodification processes originated by groundwater, and the combined effects of irrigation and groundwater, with or without vegetation.

Keywords Salinization · Sodification · Hydrological processes · Groundwater · Modeling

2.1 Introduction

Salinization and sodification are the main processes of soil degradation developing both under dryland and irrigated conditions. The growing development of irrigated agriculture is necessary for the sustainable production of the food required by the increasing World's population. Such development is limited by the increasing scarcity

I. Pla-Sentís (✉)
Universitat de Lleida, Plaça de Víctor Siurana, 1, 25003 Lleida, Spain
e-mail: ldefonso.pla@udl.cat

M. M. Alconada-Magliano (ed.), *Intensified Land and Water Use*, Springer Earth System Sciences, https://doi.org/10.1007/978-3-030-65443-6_2

and low quality of the available water resources and by the competitive use of those resources for other purposes. There are also increasing problems of contamination of surface and groundwater to be used for other purposes, by the drainage effluents of irrigated lands. Taking into consideration the great investments required for the development of irrigated agriculture, the high contribution of this kind of agriculture to the World´s food production and the increasing scarcity and cost of available water resources, the degradation of irrigated lands through soil salinization and sodification becomes very important from the economic, social, and environmental points of view. Besides, large and increasing proportions of the World's irrigated land (25–50%, depending on the evaluations) are affected by excessive salinity and sodicity.

In general, in salt-affected soils the water deficits make the survival of natural vegetation and crops difficult or impossible. These moisture deficits are due to the difficulties for the plants to use the water stored in the soil ("*saline soils*"), or to the difficulties for root development and for water infiltration into the soil ("*sodic soils*"). Although those processes may occur, and have occurred under natural conditions, they become accelerated, leading to secondary salinization, when mainly in arid and semiarid environments, the soil water regime is drastically changed with the introduction of irrigation and/or drainage.

In many countries irrigation has become a very important component of food production, sometimes the most important. The area with irrigation in the World has increased from 50 million hectares (Mha) in the year 1900 to 100 Mha in 1950, and to 300 Mha today. The decrease of productivity has also contributed to this phenomenon. The decrease of productivity and the increasing risks of dryland agricultural production on lands have already been affected by other degradation processes, mainly water erosion. The yearly loss of productivity, and desertification of irrigated lands, amounts to 1.5 Mha around the World, but the salinity problems, in different degrees, presently affect almost 25–50% (depending on the evaluations) of all the irrigated land. Although the affected areas by salinity are much less than the ones affected by other degradation processes like erosion, the social, economical, and environmental effects are of the same magnitude, as a consequence of the high value and productivity of irrigated lands, and their coincidence with areas of large urban and industrial developments. In arid and semiarid climates, the scarcity and erraticity of rainfall, together with the high evapotranspiration rates, makes the water and salt balances favorable for the processes of soil salinization, especially under poor drainage conditions.

The ***soil salinity*** problems are a consequence of salt accumulation in zones and depths where the soil moisture regime is characterized by strong losses of water by evaporation and transpiration, and by reduced leaching of the remaining salts. The ***soil sodicity*** problems appear as a consequence of changes in the composition and concentration of those salts, with changes in the equilibrium exchangeable cations, leading to higher relative accumulation of exchangeable sodium percentage (ESP) (Rengasamy and Sumner 1998). Both processes of salinization and sodification are therefore influenced by soil water and solute balances (Pla 1997), depending upon climate, crops, soils, groundwater depth, irrigation and groundwater composition, and irrigation and drainage management (Pla 1983).

To predict salinization problems, the main prerequisite is to identify the source of salts and to characterize the main factors determining the regime of water and salts in the soil. This is not easy, because the hydrological and chemical processes involved in the process of salinization and, especially, of sodification, are usually very complicated (Pla 1998). Therefore, we have to simplify some of them to be able to develop models that can be put into practice.

2.2 Processes of Development of Salt-Affected Soils

In general, all soils with problems directly or indirectly derived from the amounts and kinds of salts in solution are referred as "***salt-affected soils***" (Pla 1983). The resulting problems may be very different, depending on the geochemical processes involved in the development of salinization. Salt-affected soils, both saline and sodic, may develop both under dryland and irrigated conditions (Pla 2015), affecting negatively the physical and chemical soil properties, the crop production and the animal and human health (Fig. 2.1).

This applies both to the salinization developed through natural processes (primary salinization), and to the salinization induced by human intervention (secondary salinization). In both cases, the main responsible factors are the concentration and the relative composition of salts in the surface and groundwater, and the changes they may suffer in soil solution. These changes are dependent on the soil moisture regime, which is a consequence of the drainage and climate conditions. Drainage is the result of the hydraulic properties of the soil profile, of the groundwater depth, and of the landscape position. Rainfall and evapotranspiration are the main climatic factors to be considered. The development of secondary salinization, in different soils and climates, is generally caused by drastic changes in the soil moisture regime due to the introduction of irrigation with drainage restrictions.

Both in surface and groundwater, and in soil solution, most of the salts are a combination of the cations Ca^{2+}, Mg^{2+}, and Na^{+}, and of the anions HCO_3^-, Cl^- and SO_4^{2-}. In some highly fertilized soils, the anion NO_3^- may also accumulate in soil solution. The main natural source of the predominant salts is the weathering of minerals in the earth crust, the rainfall in coastal areas, and the dissolution of fossil salts in some geological formations of marine origin. The human intervention brings additional salts to the soils through irrigation water, residual waters, and fertilization. The differences in amounts and kinds of salts accumulated in soil solutions result in "salt-affected soils" of varied chemical, physical and physicochemical properties, having different management requirements for their prevention, use and reclamation. Based on the main effects on soils and crops we may classify the "salt-affected soils" in "saline soils" and "sodic soils." Although in both of them, hydrological processes are involved in their development, in the case of sodic soils we have to consider some additional chemical and physicochemical reactions, making their modeling and prediction more difficult.

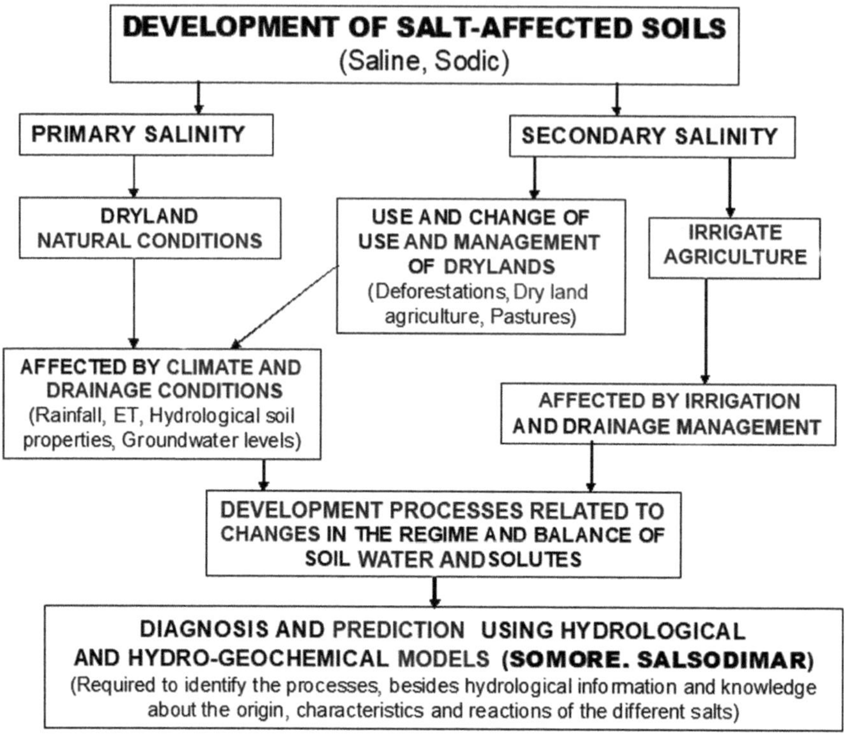

Fig. 2.1 Common factors and processes in the development of salt-affected soils under dryland and irrigated conditions

Saline soils are those where the salt content and osmotic pressure of the soil solution do not allow the absorption by the crop of a great portion of the soil water, and do not show any direct effect on the soil physical properties. The main consequence is the partial or complete reduction in plant growth due to physiological water deficits. For practical purposes, the salt concentration is expressed in terms of electrical conductivity (units of dS/m: deciSiemen/meter at 25 °C) in soil saturation extract (USDA 1954). One dS/m is approximately equivalent to a salt concentration in solution of 10 meq/l and to an osmotic pressure of 36 KPascals. It is well known that the soil moisture stress for plants is composed by the matric stress, which increases when the soil moisture decreases, and the osmotic stress, which increases when the salinity in soil solution increases. Both stresses are more or less additive. Therefore, one approach to reduce the effects of salinity would be to maintain the matric stress as low as possible, through frequent or continuous irrigation. Another approach would be to plant salt-tolerant crops, which are able to grow and produce economical yields even at high soil moisture stresses, through adjustments in their transpiration rates or in the osmotic pressure in their cells. As a consequence of the selective accumulation of some specific electrolytes in soil solution, in some occasions, specific nutritional or toxic effects are associated or precede the more general osmotic stress effects. This is the case with some sensitive crops, when chloride, sodium, and sometimes boron, reach critical levels in soil solution.

Sodic soils are dominated by Na (and by Mg in some cases) on their cation exchange sites. The sodicity status in the soil may be expressed by the Sodium Adsorption Ratio (SAR) (USDA 1954):

$SAR = Na^{+}s /((Ca^{2+}+Mg^{2+})s)^{1/2}$ $(mmol/liter)^{1/2;}$ (s: solution)
($Na^{+}s$; $(Ca^{2+}+Mg^{2+})s$: mmol/liter in soil solution)

or by the Exchangeable Sodium Percentage (ESP):

$ESP = 100$ ($Na^{+}e$ (meq/100 g)/CEC (meq/100 g)); (e: exchangeable)

Although both related one to the other, it is more appropriate to measure the RAS for practical reasons (ESP is very difficult to measure correctly in most of the sodic soils), and because in many sodic soils, under very different conditions, makes it even a more reliable index of sodicity than ESP. Depending on different factors, soils are considered sodic if the ESP is 5–40%. Although it is generally reported (following the value given in USDA 1954) that sodic soils are the ones with ESP (or SAR) higher than 15, it has been found that sodicity problems may develop at values ranging from 5–40%, depending on texture and mineralogy of the soils and on the accompanying electrolyte concentration. Traditionally the "*sodic soils*" have been called "*alkali soils*," although these include only the sodic soils with presence and accumulation of Na bicarbonates and carbonates and pH higher than 8.5–9.0. There are other soils with properties or sodic soils with lower pH and lower relative levels of Na than the so called "alkali soils" (Pla 1968).

Sodicity produces changes on the soil physical properties, both by dispersion and plugging of soil pores by the moving clay particles and soil pore blockage by swelling clays. When surface soil disperses, the clay and silt particles clog surface pores, resulting in soil sealing, reduced infiltration and surface waterlogging. This affects land use and plant growth by decreasing the permeability of water and air through the derived soil waterlogging, and impeding root penetration. The impacts of these mechanisms are affected by several soil factors, mainly texture, clay mineralogy, total salinity and pH (Rengasamy and Olsson 1991). Dispersion affects more soils with illite and kaolinite clays, at very low values of SAR if the salinity levels are also low, while swelling effects are more common in soils with smectites.

The relationship between soil salinity and its flocculating effects, and sodicity and its dispersive or swelling effects on soil physical properties, specially the ones related with soil infiltration rates and hydraulic conductivity, is required to predict how specific soils will behave under different predicted combinations of salinity (C_{SE}) and sodicity (SAR_{SE}) (Fig. 2.2).

Those relationships, already reported by Quirk and Scofield (1955) and by Reeve (1960), depend highly on clay type, soil chemical reactions and soil texture (Frenkel et al. 1978; Oster and Shainberg 2001). Among the main soil chemical reactions affecting those salinity/sodicity relationships are the ones involving bicarbonates

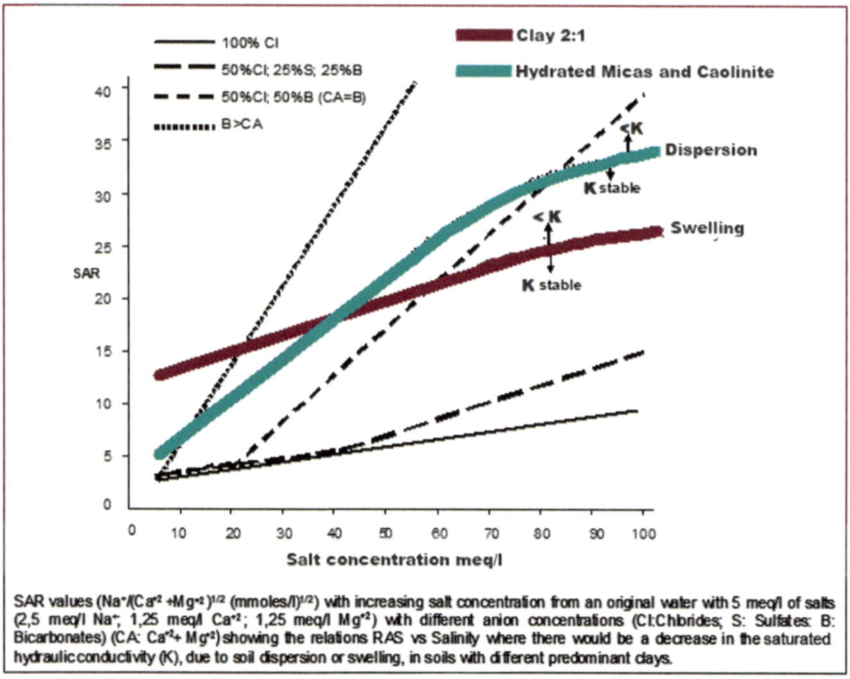

Fig. 2.2 Relationships between salinity and sodicity in the soil solution, and their effects on clay dispersion or swelling and in the soil hydraulic conductivity, for different anionic composition of the original water and different soil clays (Adapted from Pla 2015)

and carbonates of Ca, Mg, and Na, and Ca sulfates, leading to salt precipitation or dissolution, with changes in the soil salinity and sodicity levels (Pla 1967, 1998). Many errors in the evaluation and prediction of sodicity problems and effects are due to the non correct consideration of those chemical reactions under different relations among those cations and anions (Fig. 2.2). The main mistakes are done not considering the effects of sodium bicarbonate accumulation in soil solution, coming from irrigation water or groundwater, or produced by reactions under anaerobic conditions as the one reported in Fig. 2.3.

Conditions					
IRRIGATION WATER OR GROUNDWATER					
Concentration:	(High)	(Medium)	(Low)		
EC	>2dS/m	1-2 dS/m	<1dS/m		
Composition	Cl>S>B	S≥Cl>B	B≥S>Cl		B>S>Cl (B>CA)
	Na ≥CA	CA>Na	CA>Na		Na≥CA
DRAINAGE	(Variable)		(Very restricted)		(Restricted)
Soil Perm (I)	1-50 mm/hour		< 1mm/hour		< 5mm/hour
Groundwater depth#	< 1,5 m		< 0,5 m		< 1,0 m
CLIMATE	(Ar.-DSAr.)		(Ar.-DSAr.)	(DSAr.-SH.)	(Ar.-HSAr.)
IMA (P/ETP)	< 0,5		< 0,5	0,5-1,0	< 0,8
LGP (P>(ETP/2))	< 120 days		< 120 days	120-270 days	<180 days
Resulting problem					
SOIL SOLUTION (SE)	(Very Saline)		(Mod. Saline)	(Sligh. Saline)	(Var. Salinity)
Concentration (EC)	> 8 dS/m		> 4 dS/m	< 4 dS/m	> 2 dS/m
Composition	Cl>>S>>B	Cl≥S>>B	S>Cl>B	S≥B>Cl (*) →	B≥S>Cl
	Na>CA	Na≥CA	Na>CA	Na>>CA	Na>>CA
pH	< 8,5		< 8,5	> 7,5	> 8,5
PRECIP. SALTS	CAC + CaS		CAC + CaS	CAC	CAC
POTENTIAL PROBLEM	**SALINITY**			**SODICITY**	

(*) Change in composition under anaerobic conditions ($2Na^+ + SO_4^{=} + 2C + 2H_2O = S^{=} + 2NaHCO_3$)
Permanent groundwater depth or presence of soil layers restricting internal drainage. Depths for medium to fine texture soils. May be shallower for coarse texture soils.

EC: Electrical Conductivity; Cl: Chlorides; S: Sulphates; B: Bicarbonates; Na: Sodium; CA: Calcium+Magnesium; CAC: Ca+Mg Carbonates; CaS: Calcium Sulphate; SE: Saturation Extract; I: Infiltration rate; IMA: Aridity Index; LGP: Length of Growing Period; P: Rainfall; ETP: Potential Evapotranspiration; Ar: Arid Climate; DSAr: Dry Semi-Arid Climate; SH: Sub-Humid Climate; HSAr: Humid Semi-Arid Climate

Fig. 2.3 More common conditions leading to the development of different kinds of saline and sodic soils (adapted from Pla 2014)

2.3 Prediction of Soil Salinization and Sodification Processes

The problems of secondary salinization, leading to saline or sodic soils, are due to poor water management (irrigation and mainly drainage) in relation to a particular combination of climate, soil, crop, fertilization practices, groundwater level and salinity, quality of irrigation water, and irrigation system (Pla 1988) (Fig. 2.3). Today there are known methods and technical possibilities to reclaim salt-affected soils, but in general they are too costly. When socio economic problems justify the reclamation, still it is possible to have difficulties to do it, derived from the scarcity of water of good quality for leaching, or from potential problems of contamination of surface and groundwater used for irrigation or for domestic and industrial purposes (Halliwell et al. 2001; Leal et al. 2009). Therefore, it would be more convenient and economical to pre-establish, through appropriate predictive models, which would be the best alternatives for irrigation and drainage water management to prevent salinization or sodification problems, for each combination of climate, soil and available—quantity and quality—irrigation water. As indicated above, this would still be more important if there is a high competition for the use of scarce resources of good quality water, when the quality of the available water is poor, or when it is necessary to reduce the effluents of drainage water to a minimum, and to control groundwater levels and quality.

Both the addition of irrigation water and the changes in the depth and composition of groundwater may cause drastic changes in the water and solute balances in the soil profile. Modeling may be very useful for the diagnosis and prediction of such changes, and in the selection of the best practices and systems of irrigation and drainage for a more efficient use of irrigation water and for reducing the losses and contamination of surface and groundwater, and controlling the soil salinization and sodification. Current available computer models provide only a limited ability to predict correctly those impacts of salinity and sodicity under different management conditions (Suárez 2005). It is necessary to include in them the interaction of many physical and chemical processes, for predicting short and long term consequences of varied management practices (Pla 1997).

Most of the presently used concepts and approaches (static water quality indices, fixed limits for soil salinity, and sodicity critical levels, empirical predictive models based on statistical relations or in laboratory methods not reflecting field conditions), although can characterize a particular system, can only be applicable to the conditions under which they were developed, but cannot be reliable if extrapolated to different situations.

Models predictions would have to incorporate the effect of both the soil chemical (solution and exchange composition) changes, and soil and water management, as a function of irrigation and groundwater composition; and also the balance of water and solutes in the soil as influenced by climate, irrigation system, groundwater depth, and crop water uptake. These model requirements are particularly important for soil sodification, where to reach an equilibrium level of exchangeable sodium may take a long time (up to several decades) to occur, but at the same time the resulting effects are very difficult to reverse (Pla 2015).

Hydrological studies, including water and salt balances, and water table fluctuation analysis, will be needed to preview the consequences of land use and vegetation changes on soil salinization and specially sodification processes, and to guide the requirements of irrigation and drainage management to prevent them (Pla 2006). It has been found that vegetation changes can have strong effects on water dynamics, and on salt and sodium accumulation and distribution at different temporal and spatial scales under shallow groundwater.

The proposed model "SALSODIMAR" (Pla 1968, 1983, 1988, 1997) is based on an *independent balance of the salts and ions more common in irrigation waters, in groundwater and in soil solution*, and takes into consideration the processes and effects derived of the interaction among the compositions of the irrigation and groundwater, the evapotranspiration, the reactions of solution, precipitation and cation exchange, the soil hydrological properties and the effective leaching fraction (Figs. 2.3, 2.4). It was recently adapted to include *new specific hydrological components* of the water and solute balances, to make it useful to predict the processes of both the dryland salinization and sodification processes originated in the very *variable groundwater depth and composition*, and the combined effects of irrigation and groundwater, with or without vegetation or mulch cover (Table 2.1). The effects of hydrological factors affecting the water and solute balances, the soil characteristics and the irrigation management were included in the model (Pla and Dappo 1977; Pla 1983, 1986), and all was programmed in EXCEL for practical use. As such, it has been successfully used and tested at different levels, under tropical and subtropical semi arid to sub-humid climate conditions (Vargas 2001; De Paz et al. 2004; Guerrero et al. 2004; Pla 1986, 1988, 2006; Ramírez 2012; Ramírez et al. 2014; Sánchez 2013).

2.4 Applications of the Model SALSODIMAR. Case Studies

In Table 2.2 five examples of the use of the proposed modeling approach to predict the type and approximate levels of soil salinization and sodification have been presented. Those problems have developed under different combinations of the main hydrological and chemical factors and processes, associated with various irrigation and drainage management under tropical, subtropical and temperate climate conditions in Latin America. In all cases the salinity problems are related to the depth and composition of the groundwater.

Those ones related to the development of saline soils correspond to irrigated lands, under tropical-subtropical climate conditions, with sugarcane in Barahona (Dominican Republic) (Fig. 2.5), and with vegetable crops in Guantánamo (Cuba) (Fig. 2.6). In the two cases, the levels, control and distribution of water with furrow irrigation systems, are very deficient, with non-very effective drainage systems and conditions. The salinity of the groundwater is much higher in Guantánamo, due to the presence of marine sediments at relatively shallow depths. Irrigation waters are of relatively low salinity and good quality both in Barahona and Guantánamo.

Sodic soils are those ones developed in the other three cases: Cauca Valley (Colombia, Fig. 2.7), Western Plains (Venezuela, Fig. 2.8) and Low Pampa (Argentina, Fig. 2.9). In all cases there was a good adjustment between measurement (M) and calculated (CS) through the SALSODIMAR Model (Table 2.2).

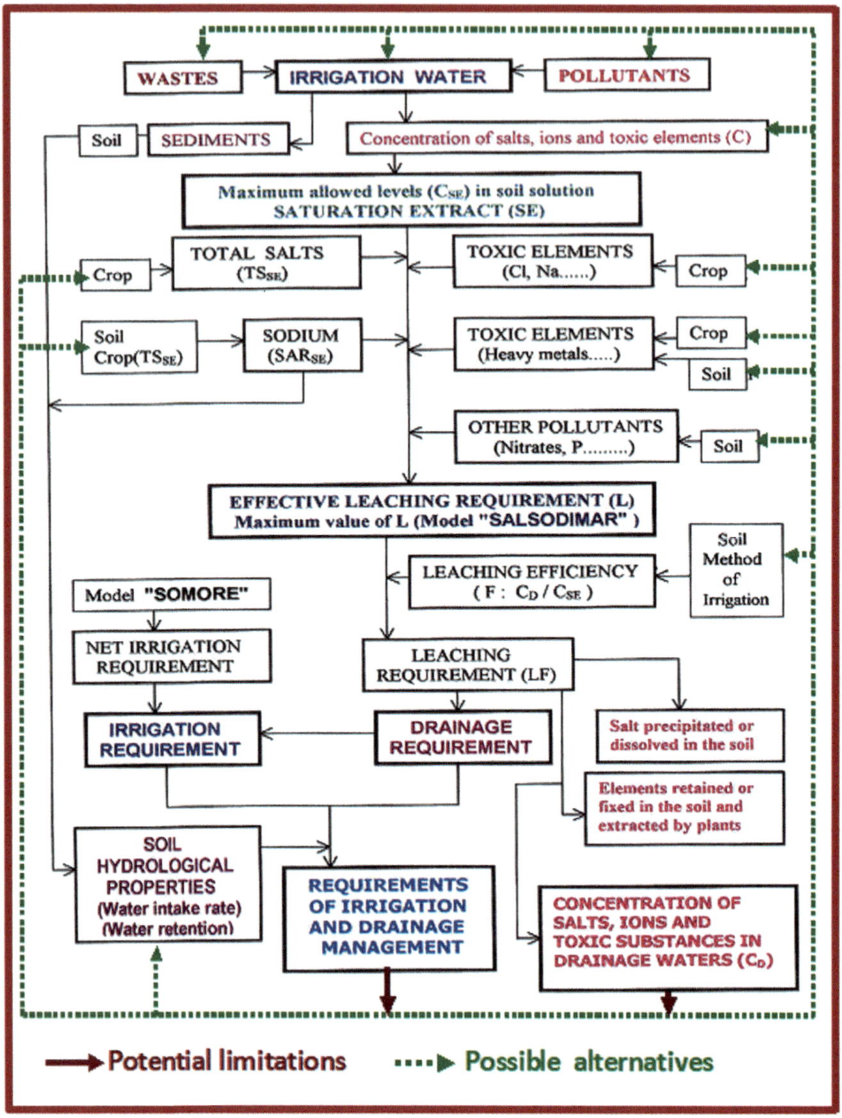

Fig. 2.4 Flow diagram of a modeling approach based on a balance of salts and sodium in irrigated lands (SAR: Sodium Adsorption Ratio) (Model SOMORE: Pla 2006) (Model SALSODIMAR: Pla 1997)

Table 2.1 Calculation of the water and salt balances in the surface soil as affected by groundwater, irrigation, and soil cover, to predict salinization and sodification processes under different conditions

IRRIGATION CONDITIONS (SOIL COVER WITH CROPS)

(NET UPWARD FLOW)

$H_G = H_{ET} - H_T - H_P - H_R$ (If: $H_{ET} - H_T - H_P - H_R \geq 0$);
$L_G = (H_P + H_R)/(H_E - H_T - H_P - H_R)$
At equilibrium: If: $CA_G \geq B_G$
If: $CAB_G/L_G \geq 10$ and $CaS_G/L_G \geq 30$
$CA_{SE} = ((CA_G - CAB_G - CaS_G)/L_G) + 40$; $Na_{SE} = Na_G/L_G$
$CAC_p = (CAB_G/L_G) - 10$; $CaS_p = (CaS_G/L_G) - 30$
If: $CAB_G/L_G \geq 10$ and $CaS_G/L_G < 30$
$CA_{SE} = ((CA_G - CAB_G)/L_G) + 10$; $Na_{SE} = Na_G/L_G$
$CAC_p = (CAB_G/L_G) - 10$
If: $CAB_G/L_G < 10$ and $CaS_G/L_G < 30$
$CA_{SE} = CA_G)/L_G$; $Na_{SE} = Na_G/L_G$
If: $CA_G < B_G$
$CA_{SE} = CA_G$; $Na_{SE} = Na_G/L_G$
$CAC_p = (CAB_G/L_G) - CA_G$

$C_{SE} = CA_{SE} + Na_{SE}$
$SAR_{SE} = Na_{SE}/(CA_{SE}/2)^{1/2}$

(NET DOWNWARD FLOW)

If: $H_{ET} - H_T - H_P - H_R < 0$
$H_D = H_P + H_R - H_{ET}$; $L_D = (H_P + H_R - H_{ET})/H_R$
At equilibrium: If: $CA_R \geq B_R$
If: $CAB_R/L_DE \geq 10$ and $CaS_R/L_DE \geq 30$
$CA_{SE} = ((CA_R - CAB_R - CaS_R)/L_DE) + 40$; $Na_{SE} = Na_R/L_DE$
$CAC_p = (CAB_R/L_DE) - 10$; $CaS_p = (CaS_R/L_DE) - 30$
If: $CAB_R/L_DE \geq 10$ and $CaS_R/L_DE < 30$
$CA_{SE} = ((CA_R - CAB_R)/L_DE) + 10$; $Na_{SE} = Na_R/L_DE$
$CAC_p = (CAB_R/L_DE) - 10$
If: $CAB_R/L_DE < 10$ and $CaS_R/L_DE < 30$
$CA_{SE} = CA_R)/L_DE$; $Na_{SE} = Na_R/L_DE$
If: $CA_R < B_R$
$CA_{SE} = CA_R$; $Na_{SE} = Na_R/L_DE$
$CAC_p = (CAB_R/L_DE) - CA_R$

$C_{SE} = CA_{SE} + Na_{SE}$
$SAR_{SE} = Na_{SE}/(CA_{SE}/2)^{1/2}$

DRYLAND CONDITIONS

(BARE SOIL)

$H_G = H_E - H_P$ (If: $H_E - H_P \geq 0$); $L_G = H_P/(H_E - H_P)$
At equilibrium: If: $CA_G \geq B_G$
If: $CAB_G/L_G \geq 10$ and $CaS_G/L_G \geq 30$
$CA_{SE} = ((CA_G - CAB_G - CaS_G)/L_G) + 40$; $Na_{SE} = Na_G/L_G$
$CAC_p = (CAB_G/L_G) - 10$; $CaS_p = (CaS_G/L_G) - 30$
If: $CAB_G/L_G \geq 10$ and $CaS_G/L_G < 30$
$CA_{SE} = ((CA_G - CAB_G)/L_G) + 10$; $Na_{SE} = Na_G/L_G$
$CAC_p = (CAB_G/L_G) - 10$
If: $CAB_G/L_G < 10$ and $CaS_G/L_G < 30$
$CA_{SE} = CA_G)/L_G$; $Na_{SE} = Na_G/L_G$
If: $CA_G < B_G$
$CA_{SE} = CA_G$; $Na_{SE} = Na_G/L_G$
$CAC_p = (CAB_G/L_G) - CA_G$

$C_{SE} = CA_{SE} + Na_{SE}$
$SAR_{SE} = Na_{SE}/(CA_{SE}/2)^{1/2}$

(SOIL COVER WITH VEGETATION)

$H_G = H_{ET} - H_T - H_P$ (If: $H_{ET} - H_T - H_P \geq 0$);
$L_G = H_P/(H_E - H_T - H_P)$
At equilibrium: If: $CA_G \geq B_G$
If: $CAB_G/L_G \geq 10$ and $CaS_G/L_G \geq 30$
$CA_{SE} = ((CA_G - CAB_G - CaS_G)/L_G) + 40$; $Na_{SE} = Na_G/L_G$
$CAC_p = (CAB_G/L_G) - 10$; $CaS_p = (CaS_G/L_G) - 30$
If: $CAB_G/L_G \geq 10$ and $CaS_G/L_G < 30$
$CA_{SE} = ((CA_G - CAB_G)/L_G) + 10$; $Na_{SE} = Na_G/L_G$
$CAC_p = (CAB_G/L_G) - 10$
If: $CAB_G/L_G < 10$ and $CaS_G/L_G < 30$
$CA_{SE} = CA_G)/L_G$; $Na_{SE} = Na_G/L_G$
If: $CA_G < B_G$
$CA_{SE} = CA_G$; $Na_{SE} = Na_G/L_G$
$CAC_p = (CAB_G/L_G) - CA_G$

$C_{SE} = CA_{SE} + Na_{SE}$
$SAR_{SE} = Na_{SE}/(CA_{SE}/2)^{1/2}$

H_G: Water coming from the water table, reaching the surface soil by capillary rise (mm); **H_E**: Water loss by potential evaporation in a bare soil surface (mm); **H_{ET}**: Water loss by potential evapo-transpiration in a soil surface covered by vegetation (mm); **H_T**: Water loss by transpiration of the vegetation or crop covering the soil surface (mm); **H_P**: Effective rainfall (infiltrated in situ) (mm); **H_R**: Water applied by irrigation (mm); **C_G**: Salt concentration in the ground-water (meq/liter); **C_R**: Salt concentration in the irrigation water (meq/liter); **C_D**: Salt concentration in the drainage water (meq/liter); **C_{SE}**: Salt concentration in the saturation extract of the surface soil (meq/liter); **SAR**: Sodium Adsorption Ratio $(Na/(CA)^{1/2})$ $(meq/liter)^{1/2}$; **SE**: Saturation Extract of the soil; **E**: Leaching efficiency; **G**: Ground-water; **CA**: $Ca^{++} + Mg^{++}$ (meq/liter); **Na**: Na^{+} (meq/liter); **B**: HCO_3^- (meq/liter); **S**: $SO_4^{=}$ (meq/liter); **CAB** = CA if B ≥ CA; **CAB** = B if B < CA; **CaS** = Ca-B-CACl if CaS ≥ 0 (CaCl = Ca – B – S if CaCl ≥ 0); **CAC_P**: Precipitated (Ca+Mg) carbonates (meq/liter); **CaS_P**: Precipitated Ca Sulphates (Gypsum) (meq/liter)

Table 2.2 Measured and calculated (model SALSODIMAR) resulting salinity and sodicity under the different conditions of climate, soils, crop, irrigation system, and irrigation water or groundwater composition

LOCATION	SOIL (TEXTURE)	RAINFALL (ET) mm / year	CROP (IRRI. SYST)	IRRIG. WATER (GR.WATER) EC	SAR	RESULTING SOIL pH	EC_{SE} M	EC_{SE} CS	SAR_{SE} M	SAR_{SE} CS
BARAHONA (REP. DOM.)	Entisol (SCL-SL)	580* (1450)	Sugar cane (Furrows)	0.8 (5.2) (GD) (NC): 120 cm (GD) (PC): 180 cm	1,9 (10)	(NC) : 7.7 (PC) : 7.8	14 3	12 4	13 6	15 (SAL) 7
GUANTANAMO (CUBA)	Entisol (CL)	800* (1800)	Vegetables (Furrows)	0.8 (26) (GD): 100-120 cm	2.5 (32)	7.2	12	10	20	24 (SAL)
CAUCA VALLEY (COLOMBIA)	Mollisol (SL-CL)	820* (1620)	Sugar cane (Furrows)	0.5 (B>CA) (2.5) (B>CA) (GD): 50-120 cm	2.4 (25)	8.8	2.5	2.2	21 (B>CA)	16 (SOD)
WEST. PLAINS (VENEZUELA)	Vertisol (CL-C)	940* (1590)	Rice (Flooding)	0.6 (B>CA) (2.0) (B>CA)	2.1 (13)	9.7	12	13	54 (B>>CA)	65 (SOD)
LOW PAMPA (ARGENTINA)	Mollisol (SCL)	982* (NC): (500)	Pasture (No irrigation)	(1.2) (B>CA) (GD): 0-160 cm	(6,1)	(NC): 9.3 (DP): 8.7	12 3.5	 3.0	48 (SOD) 22 (B>>CA)	 26 (SOD)

C: Clay; CL: Clay loam; SL: Silty loam; SCL: Silty clay loam; ET: Evapo-Transpiration; IRRIG.WATER: Irrigation Water; IRRI.SYST: Irrigation System; GR.WATER: Groundwater; EC: Electrical Conductivity dS/m; SAR: Sodium Adsorption Ratio; SE: Saturation Extract; GD: Groundwater Depth; NC: No Plant Cover; PC: Plant Cover; DP: Degraded Pasture; M: Measured (0-30cm); CS: Calculated; SAL: Saline Soil; B: Bicarbonates; CA: Ca+Mg; * Average values (variable and irregular distribution of rainfall)

SALINE SOIL

Barahona (Dominican Republic):
Semi-arid humid tropical climate. Silty Loam texture Entisols. Furrow irrigation of sugar cane.
Irrigation water low in salts and sodium, mainly bicarbonates and sulphates.
Poor irrigation and drainage management. Moderate infiltration rates.
Variable (100-200 cm) groundwater depths. Precipitated gypsum in the soil.

Fig. 2.5 Saline soil in Barahona (Dominican Republic) with furrow irrigated sugar cane

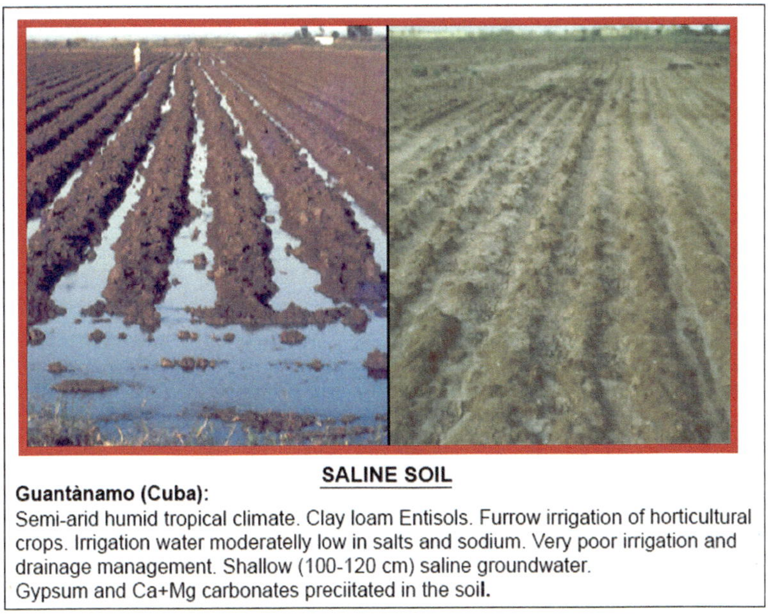

Fig. 2.6 Saline soil in Guantanamo (Cuba) with furrow irrigated vegetables

Fig. 2.7 Sodic soil in the Cauca Valley (Colombia) with furrow irrigated sugar cane

Fig. 2.8 Sodic soil in the Western Plains (Venezuela) with flood irrigated rice

In the Cauca Valley (Colombia) (Fig. 2.7), with dry sub-humid tropical climate, the lands are under furrow irrigated sugarcane, with a non-effective and poorly maintained drainage system. The irrigation water, low in salinity, has a small content of sodium bicarbonate (González 2001; Sánchez 2013). The use of this water in a fine textured soil (SL-CL) and with the implemented management, led to sodification and soil deterioration that corresponds to what the Model SALSODIMAR estimates (Table 2.2).

In the Western Plains (Venezuela) (Fig. 2.8), with dry sub-humid tropical climate, the lands, with clay Vertisols, are continuously cropped with rice under flooding, with irrigation water in the dry season and with rainfall water in the humid season. To decrease water percolation losses and to control weeds, the surface soil is puddled with rotary shaking under saturated conditions (Table 2.2). The irrigation water is low in salinity, with a small content of sodium bicarbonate. The analysis of the soil was performed in samples taken under dry conditions, in between rice crops (Pla 1985). The management of the crop, water quality, and irrigation system in this environment led to soil sodification. The measured values corresponded with those estimated by the Model.

The sodic soils in the Low Pampa (Argentina) (Fig. 2.9), with sub-humid temperate climate, have developed in non-irrigated lands under dryland pasture (natural pasture) (usually overgrazed) for cattle production (Casas and Pittaluga

Fig. 2.9 Sodic soil in the Low Pampa (Argentina) with dryland pasture

1990). Under a very variable rainfall amounts and distribution, the groundwater is generally very shallow, with temporal surface flooding some years. In the region, groundwater flows of different origin, of intermediate and local type (responding to far and local rainfalls, respectively) coexist, which modify the depth of the water table (Alconada-Magliano et al. 2011). However, the shallow groundwater levels may have been influenced by the extensive recent agricultural use of the neighbor higher Pampas lands, where natural pastures, with higher ET rates, have been substituted by crops, mainly soybeans, with lower ET rates. The groundwater is both saline and rich in sodium bicarbonate.

Problems of surface soil salinization and sodification are generally higher in lands where the plant cover is poorer due to overgrazing. In this case, there was also, a reasonably good correspondence between the measured and the predicted (calculated) values of salinity and sodicity (Table 2.2).

It is necessary to take into consideration that the measured values were done in not well controlled commercial fields, where although having been under about the same management at least during the last 10–20 years, probably they have not reached equilibrium levels of salinity and specially of sodicity.

2.5 Final Considerations

It is shown here how a modeling approach (SALSODIMAR), based on the balance of water and soluble components, taking into consideration the involved hydrological and physicochemical processes, of both the irrigation water and groundwater under different water and land management conditions, may be adapted for the diagnosis and prediction of salinity and sodicity problems, and for the selection of alternatives for their management and amelioration. It may be concluded that the prediction of soil salinity and sodicity problems, derived from increased use of low quality irrigation waters, including more or less treated waste-waters, in poorly drained soils, with shallow fluctuating groundwater levels, with variable salinity content and composition, require adequate simulation modeling. These models have to be based on modeling hydrological processes responsible for water and solute balances in the soil, as influenced by climate, crops and irrigation and drainage management. Soil chemical and physicochemical reactions affecting the relationships of salinity and sodicity levels, have also to be included in modeling.

The proposed adaptation of the model SALSODIMAR that includes all those requirements, resulted to give reasonably good predictions in the preliminary evaluations of salinity and sodicity in the different case studies included in this paper. Additional research under field conditions would be needed for further improvement of the model predictions.

References

Alconada-Magliano MM, Fagundo Castillo JR, Carrillo Rivera JJ, Hernández PG (2011) Origin of flooding water through hydrogeochemical identification, the Buenos Aires plain. Argentina Environmental Earth Sciences 64(1):57–71

Casas RR, Pittaluga A (1990) Anegamiento y salinización de los suelos en el noroeste de la provincia de Buenos Aires. In: Manejo de tierras Anegadizas. PROSA, FECIC. Gráfica Guadalupe. Buenos Aires (Argentina)

De Paz JM, Visconti F, Zapata R, Sánchez J (2004) Integration of two simple models in a geographical information system to evaluate salinization risk in irrigated land of the Valencian Community, Spain. Soil Use Manag 20:333–342

Frenkel H, Goertzen JO, Rhoades JD (1978) Effects of clay type and content, exchangeable sodium percentage, and electrolyte concentration on clay dispersion and soil hydraulic conductivity. Soil Science Society America Journal, Madison 42(1):32–39

González A (2001) Efecto de la aplicación de aguas bicarbonatadas sobre las propiedades físicas y químicas de algunos suelos del Valle del Cauca y Tolima. Unpublished Doctoral Thesis, Universidad Nacional de Colombia, Palmira, p 138

Guerrero JP, Pla I, Camacho RG (2004) Génesis de un suelo sódico alcalino en Ghaguaramas Venezuela. Agronomía Trop 54(4):433–460

Halliwell DJ, Barlow KM, Nash DM (2001) A review of the effects of wastewater sodium on soil physical properties and their implications for irrigation systems. Aust J Soil Res 39(6):1259–1267

Leal RMP, Herpin U, Fonseca A, Firme L, Montes C, Melfi A (2009) Sodicity and salinity in a Brazilian Oxisol cultivated with sugarcane irrigated with wastewater. Agric Water Manag 96(2):307–316

Oster JD, Shainberg I (2001) Soil responses to sodicity and salinity: challenges and opportunities. Aust J Soil Res 39(6):1219–1224
Pla I (1967) Evaluación Cuantitativa de los Efectos de los Bicarbonatos en el Agua de Riego sobre las Propiedades Químicas y Físicas de los Suelos y Factores que Influyen sobre la Magnitud de Tales Efectos Rev Fac Agron. Alcance N° 12. Maracay (Venezuela), p 88
Pla I (1968) Evaluation of the quality of irrigation waters with high bicarbonate content in relation to the drainage conditions. Trans. 9th Int. Congr Soil Sci Soc Adelaide (Australia) 1:357–370
Pla I (1983) Sistema integrado agua-cultivo-suelo-manejo para evaluar la calidad de agua de riego. In: Isotopes and radiation techniques in soil physics and irrigation studies, IAEA Vienna (Austria): 191–206
Pla I (1985) Origen, distribución y diagnóstico de suelos afectados por sales en Venezuela. Rev Fac Agron XIV (1–2):125–150. Maracay (Venezuela)
Pla I (1986) Diagnostic criteria for soil and water salinity in Venezuela. Agrokémia és Talajtan 35 (3–4):431–440. Budapest (Hungary)
Pla I (1988) Riego y desarrollo de suelos afectados por sales en condiciones tropicales. Soil Technology 1(1):13–35
Pla I (1997) Evaluación de los procesos de salinización de suelos bajo riego. Edafología SECS (Spain): 241–267
Pla I (1998) Modelling the processes of dissolution and precipitation of salts in soils as related to management of irrigation and drainage. Proc. 16th World Congress of Soil Science ISSS Montpellier (France) I:98–104
Pla I (2006) Hydrological approach for assessing desertification processes in the mediterranean region. In: Kepner et al. (eds) Desertification in the mediterranean region: a security issue, Springer: 579–600
Pla I (2014) Nuevas experiencias en la evaluación y diagnóstico de procesos de salinización y sodificación de suelos en América Latina. Suelos Ecuatoriales 44(2):125–137
Pla I (2015) Advances in the prognosis of soil sodicity under dryland and irrigated conditions. International Soil and Water Conservation Research WASWAC (China) 2(4):50–63
Pla I, Dappo F (1977) Field testing of a new system for qualifying irrigation waters. Proceedings of the International Conference on Managing Saline Waters for Irrigation: 376–387 Lubbock. Texas (USA)
Quirk JP, Schofield RK (1955) The effect of electrolyte concentration on soil permeability. J Soil Sci 6:163–178
Ramírez JA (2012) Evaluación general de la salinidad y modelización de los riesgos de salinización en suelos del Valle del Cauca: Uso práctico del modelo SALSODIMAR para evaluar el riesgo de salinización en tierras bajo riego de la parte plana del Valle del Cauca (Colombia). Unpublished Master's Thesis in Agricultural Sciences. Universidad Nacional de Palmira (Colombia)
Ramírez JA, Madero E, Pla I, Zapata R (2014) Salinidad en suelos dedicados a caña de azúcar en el Valle del Cauca, Parte 1: Salinidad actual (Unpublished report)
Reeve RC (1960) The transmission of water by soils as influenced by chemical and physical properties. Int Congr Agr Eng Trans 5:21–32
Rengasamy P, Olsson KA (1991) Sodicity and soil structure. Aust J Soil Res 22:413–431
Rengasamy P, Sumner ME (1998) Processes involved in sodic behavior. In: Sodic soils: distribution, properties, management, and environmental consequences. ME Sumner and Naidu R, Editors. New York: Oxford University Press, pp 35–50
Sánchez EG (2013) Manejo de un agua de riego bicarbonatada sódica para caña de azúcar en un suelo mal drenado, Valle del Cauca, Colombia. Unpublished Master's Thesis in Agricultural Sciences, Facultad de Ciencias Agropecuarias, Universidad Nacional de Colombia, http://bdigital.unal.edu.co/12722/1/7010015.2013.pdf
Suárez DL (2005) UNSATCHEM 3.1: computer model for prediction of water and multi-component solute transport. Proceedings of the International Salinity Forum, Managing Saline Soils and Water: Science, Technology, and Soil Issues. Riverside, CA, USA, pp 163–168

USDA (1954) Diagnosis and improvement of saline and alkali soils. In: LA Richards (ed) USDA Handbook 60, Washington DC, USA, p 166. https://www.ars.usda.gov/ARSUserFiles/20360500/hb60_pdf/hb60complete.pdf
Vargas GE (2001) Salinización inducida en los suelos agrícolas de Canarias: caracterización y prognosis. Unpublished Doctoral Thesis, Universidad de La Laguna, Tenerife (Spain), p 427

Chapter 3
Soil Salinization and Sodification as Conditioners of Vegetation and Crops: Physiological Aspects of Plant Response to These Conditions

Andrés Alberto Rodríguez and Edith Taleisnik

Abstract The type and characteristics of native and introduced vegetation prevailing in any environment or landscape depends on the features of the elements that integrate it (soil, groundwater, and geomorphology), as well as the way it is hydrologically associated to other environments. The negative effects of soil salinity and sodicity on plant growth and development condition the type of plants and their relative success in areas affected by these factors. Adapted species feature physiological and morphological traits that enable them to prosper in those environments. Mechanisms related to salinity and alkalinity tolerance in plants are summarized in this chapter. Those mechanisms, along with morphological adaptations (not addressed here), participate in determining the vegetation occupying areas affected by salinity and/or sodicity. Species-specific responses may result from distinct trait evolution induced by stress. Strong collaboration among scientists of different disciplines is required to undertake complementary holistic studies that can integrate information on plant functional traits with the environmental features of natural ecosystems.

Keywords Plant stress adaptation · Salinity · Sodicity · *Chloris gayana*

A. A. Rodríguez (✉) · E. Taleisnik
Consejo Nacional de Investigaciones Científicas y Técnicas (CONICET), Buenos Aires, Argentina
e-mail: andresrodriguez@conicet.gov.ar

A. A. Rodríguez
Laboratorio de Fisiología de Estrés Abiótico y Biótico en Plantas, Unidad de Biotecnología 1, INTECH, Intendente Marino km 8, Chascomús, Argentina

E. Taleisnik
Instituto de Fisiología y Recursos Genéticos Vegetales, CIAP, INTA, Camino a 60 Cuadras km 5.5, X5020ICA Córdoba, Argentina

Facultad de Ciencias Agropecuarias, Universidad Católica de Córdoba, Córdoba, Argentina

M. M. Alconada-Magliano (ed.), *Intensified Land and Water Use*, Springer Earth System Sciences, https://doi.org/10.1007/978-3-030-65443-6_3

3.1 Introduction

In natural ecosystems, plant establishment is conditioned by the elements that define an environment which determine the type and characteristics of native and introduced vegetation. Soil conditions, namely salinity and sodicity, are among the environmental elements that have a strong bearing on vegetation identity.

In Argentina, soils affected by salts are widely distributed and according to FAO-UNESCO, it is the third largest land area affected by halomorphism in the world, after Russia and Australia (Lavado and Taboada 2017). Additionally, natural land cover changes introduced by human activities have had drastic consequences on water dynamics and soils salinization, which in turn, influence natural vegetation and crop establishment (Nosetto et al. 2008). The Australian dryland salinity is a well-studied example of the consequences of alteration of vegetation on water fluxes at landscape scale and subsequent soil salinization (Marchesini et al. 2017). These authors cited information indicating that almost 40% of Australian dry forests were converted to crops and pastures during the latter part of the twentieth century, a change that led to the salinization of more than 2 million hectares of productive lands in the southwest part of the country. In the most important productive crop area in the country, the Argentine Pampas region (Fig. 3.1), the surface with halomorphic soils occupies 20% (160,000 km^2) of the total area (Imbellone et al. 2010).

The distribution of native vegetation in saline areas is controlled by salinity levels and the depth of the water tables (Villagra et al. 2017). It has been suggested that a strong local soil salinity gradient is one of the main environmental forces that may have driven the upsurge of C4 species in Argentine grasslands with positive to neutral water balance (Feldman et al. 2008). Salt-affected soils support the natural occurrence of halophytic native plant species that can complete their life cycle under those conditions (Flowers and Colmer 2015). Halophytic species express inherent traits

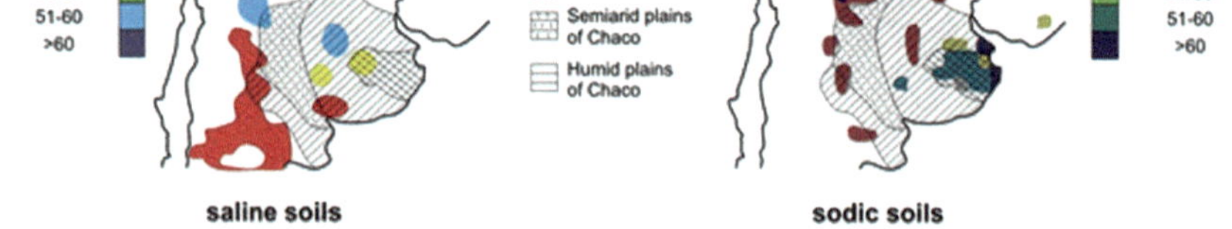

Fig. 3.1 Saline (a) and sodic (b) soils in Argentina (adapted from FAO/UNESCO Soil Map of the World). Natural ecoregions are marked by different line patterns (adapted from Ferrero and Villalba [2009], Sierra and Montecinos Urbina [1990], and Rascovan et al. [2013])

that enable them to thrive in these soils, which limit growth and development of non-adapted plants, and are potential sources not only of new species for domestication and soil reclamation, but also of tolerance mechanisms (Marinoni et al. 2019).

Whereas native vegetation reflects environmental constraints, including salinity, the negative effects of salinity, alkalinity and sodicity on plant growth and crop yields have been recognized as agricultural challenges since ancient times. Acknowledgement of variability in salt tolerance among cultivated species also comes from a long time ago. As an example, in the fertile plains of southern Iraq, increasing soil salinization due to over-irrigation has been held responsible for decreasing yields of wheat and barley and the gradual replacement of salt-sensitive wheat for the more salt-tolerant barley in the period spanning from 3500 to 1700 BC (Jacobsen and Adams 1958). A gradual migration of the population to the northern part of the alluvial plain may have been linked to increasing soil salinity, which thus played an important part in the decline of Sumerian civilization.

3.2 Development of a Theoretical Framework to Interpret the Physiological Responses of Plants to Salinity

Halophytic species have inherent adaptive traits that enable them to prosper under salinity, yet, as mentioned above, this condition is perceived by most crop plants as *stress*. In plants, stress is considered to be any environmental condition that negatively affects growth, development or any of their underlying processes (Levitt 1972). Stress reduces yield of crop plants and decreases survival chances in non-crop plants. The characteristics of the responses to stress depend on the nature of the stressing agent (type, intensity and combination among stress agents), the genotype and the plant developmental stage. Levitt's fundamental book (Levitt 1972) organized previous knowledge and proposed the theoretical bases to interpret the responses of plants to environmental stress, which is equivalent to what is also known as *abiotic stress*.

Bibliography about the physiological bases of plant responses to salinity is available since the early years of the twentieth century (e.g., the work on cotton by Harris (1929) and therein cited earlier literature). A review by Hayward et al. (1949) analyzed some of the research carried out from the end of the nineteenth century up to that time, discussing ways to evaluate plant responses to salinity and proposing tolerance mechanisms. The term *tolerance* will be used in this chapter, rather than *resistance* in relation to abiotic stress, because plants are sessile organisms that necessarily coexist with environmental stress conditions. The term *resistance* is used in responses to biotic stresses where resistant plants restrict the entry or significantly inhibit the multiplication of the pathogen within them, thus limiting coexistence with the stress situation. The book *Diagnosis and improvement of saline and alkali soils*, published in 1954 by the US Salinity Laboratory, ARS, USDA (Richards 1954), summarized an important part of the knowledge about the characteristics of soils affected by these conditions, and the tolerance or the susceptibility of some crops, as known at the time. That book includes technical details on how to characterize and analyze these types of soils, tips for their recovery, and it contains a chapter where the authors discuss plant responses to salinity and the specific responses to various types of salts in the substrate. Also, it proposes tolerance criteria, lists halophytic plant species and indicates the relative tolerance of various crops. Years later, other researchers from the same laboratory analyzed the performance of many crops under saline conditions (Maas and Hoffman 1977) and reported inter- and intraspecific variability in plant responses to salinity. Maas and Hoffman (1977) also reported that susceptibility to salinity varies along the ontogeny and that responses described at a specific stage do not necessarily predict responses at other stages. Those authors also indicated that the responses of crop performance to salinity depend on the length of the exposure period, on the fertility and aeration of the soil, as well as on environmental conditions. Salinity and sodicity levels that affect plant growth and crop productivity also depend on the type of salts present, soil characteristics (texture, composition, etc.), presence of groundwater (quality, and depth), among other environmental factors. The information on crop salt tolerance summarized by Maas and Hoffman (1977) is eloquent in this respect, highlighting the wide range of salt tolerance found in crops. Pla (1979, 2014) remarked that the choice of the most appropriate crops and/or practices needs to be evaluated on an individual basis, taking into account experience and observations. While most authors suggest that 4 dSm^{-1} electrical conductivity in the soil solution distinguish saline from non-saline soils, lower values have also been indicated (Bohn et al. 1993).

Since the second half of the twentieth century, studies about the physiological and molecular causes underlying plant responses to salinity have increased significantly (Flowers 2004). The recognition that halophytic plant species have adaptive mechanisms to salinity promoted active research on those plants, as summarized by Flowers et al. (1977). At about the same time, a review by Greenway and Munns (1980) organized concepts on salinity tolerance in non-halophytes, outlining that saline stress imposes water and ion limitations on plants that in turn can cause nutritional imbalances and toxicity. Later, it was shown that plants initially respond mainly to low water potentials associated with saline substrates, while ion-specific effects are observed later, as the concentration of potentially toxic ions that are in excess in saline substrates, such as Na^+ and Cl^-, builds up internally (Munns and Termaat 1986), causing, in turn, alterations in the accumulation of essential elements such as K^+ (Wu et al. 2015), Ca^{2+} and N (Läuchli et al. 2004). These alterations normally lead to modifications in carbon metabolism and electron transport processes that cause abnormal increases in the production of reactive oxygen species (Miller et al. 2010). Excess reactive oxygen species generation exert adverse effects and create a situation called oxidative stress that stem from their interaction with different macromolecules (Rodríguez and Taleisnik 2012). On account of these effects, three types of mechanisms can contribute to plant salinity tolerance (Rajendran et al. 2009, Roy et al. 2014), operating singly or, more frequently, in combination. In the first type, the tolerance to substrate low water potential can be achieved by accumulation of organic and inorganic solutes (Zhang et al. 1999) that decrease internal water potential and generate the necessary gradient to insure plant water uptake. The second type involves control of potentially toxic ions uptake and compartmentalization, and also the maintenance of intracellular concentrations of essential nutrients suitable for normal metabolism. The third type of salinity tolerance mechanisms confers tolerance to toxic ions in leaf tissues (James et al. 2008). This group of mechanisms involves processes that control oxidative stress by increasing the activity of reactive oxygen species detoxification systems.

Functional traits are morpho-physio-phenological traits which impact fitness indirectly via their effects on growth, reproduction and survival (Violle et al. 2007). The mentioned mechanisms, along with morphological adaptations, participate in determining the vegetation occupying saline areas, as species-specific responses may result from distinct trait evolution induced by salt stress (Liu et al. 2019).

3.3 Perennial Pasture Grasses: Relevant Native or Naturalized Components of the Saline Landscape

Inventories of salt-tolerant species found in Argentine rangelands are provided in recent publications (Pensiero et al. 2017; Villagra et al. 2017). These species are found in natural ecosystems established in saline soils, often in combination with drought or floods, high temperatures, and excessive irradiation. The advancement of the agricultural frontier necessarily alters natural ecosystems. Semiarid sedimentary plains occupied by dry forest ecosystems often display low groundwater recharge rates and accumulation of salts in the soil profile (Jayawickreme et al. 2011). Such salt accumulation may have been predominantly caused by mechanisms that exclude groundwater solutes uptake by tree roots (Jobbágy and Jackson 2004). The transformation of these natural systems to rain-fed agriculture has led to rising water tables and a slow, but steady, process of groundwater and soil salinization in vast areas of Australia (see references cited by Amdan et al. 2013). In the semiarid plains of Chaco, a sub-region within the Great Chaco region in Argentina (Fig. 3.1), unprecedented deforestation rates have taken place. The replacement of perennial native shrubs and trees by annual crops and pastures with lower transpiration capacities, leading to increases of water table levels, groundwater salinization and eventual soil salinization in discharge areas (Turner and Ward, 2002), as was seen in the Espinal sub-region (Fig. 3.2) within the Argentine Pampas region (Jayawickreme et al. 2011).

Perennial forage grasses are particularly relevant for the productive incorporation of soils affected by salts since they can contribute to reduce the negative on- and off-site impacts of salinity particularly in pasture systems that are predominantly composed by these plant species (Rogers et al. 2005). This publication included extensive lists of potential species for these areas. Published information on salinity tolerance in perennial forage grasses is considerably less than for annual cultivated grasses such as rice, wheat and maize. *Chloris gayana* is a salt-tolerant perennial fodder grass which has been extensively introduced in saline areas throughout the

Fig. 3.2 *C. gayana* in saline soils. **a** Salt-affected soils (saline and sodic) environment in groundwater flow discharge area (Flooding Pampa, Argentina); **b** Details of growing *C. gayana* stolons; **c** *C. gayana* along with naturalized plants of the halophyte *D. spicata*. Photos by the authors

world (Loch and Lees 2001), and in rangelands it coexists with *Distichlis spicata* (Fig. 3.2), a renowned salt-tolerant species (Amen et al. 1970; Díaz et al. 2013).

Salinity tolerance mechanisms in *C. gayana* have been recently summarized by Taleisnik and Pérez (2016). The presence of actively secreting salt glands on the leaf epidermis of this species is one of the traits that contribute to its salt tolerance (Céccoli et al. 2015; de Luca et al. 2001) and has been used for breeding purposes (Zorin and Loch 2007). Table 3.1 shows some adaptive traits in *C. gayana*.

Table 3.1 Some functional traits in *C. gayana* that may render potential positive consequences in diverse environments

Functional trait	Potential consequences for various environments			
	Discharge areas	Saline soils	Arid soils	References
C4 metabolism			Adaptability	Johnson and Hatch (1970)
High water use efficiency		Rapid cover	Rapid cover	Snyman (1994)
Main water absorption from shallow water table	Flood control			Chiacchiera et al. (2016)
Osmolyte accumulation in the growing section of leaf blades		Adaptability	Adaptability	de Luca et al. (2001)
High salt tolerance	Adaptability improvement of soil physical characteristics			Richards (1954) Otondo et al. (2015)
High capacity for salt uptake	Water table salinity control	Soil salinity control		de Luca et al. (2001)
Salt extrusion through salt glands		Salt tolerance		de Luca et al. (2001)

3.4 Sodic Soils Combine Agrophysical, Physiological, and Hydrological Limitations to Plant Growth and Development

Sodic soils are extensively present in Argentina (Taleisnik and Lavado 2017). As high pH, sodicity, and salinity often occur together, it is generally difficult to separate these factors in determining natural vegetation. Nonetheless, there are floristic inventories in areas dominated by high pH in the Flooding Pampa (or Salado River Basin, as it is also named) grassland (Perelman et al. 2001). As with natural halophytes, such species are potential sources of high pH-tolerance mechanisms.

Sodic and saline-sodic soils combine high exchangeable Na percentage with high pH (Lavado and Taboada 2017). Agricultural productivity of alkaline soils is restricted by their agrophysical, physiological and hydrological properties (Luna et al. 2016). *Agrophysical* properties are related to the high swelling capability of soil colloids, which deform during shrinking, leading to low porosity and oxygen deficiency. *Physiological* factors are associated with high concentrations of water-soluble salts and nutrition imbalance, with micronutrient deficiencies (Fe and Zn) being a typical consequence of alkaline soil pH (Marschner 1995). *Hydrological* factors refer to the low water permeability of these soils; alkali soils are less permeable to water than non-alkali soils and a high content of unavailable moisture also characterizes them. It has been reported that the effects of alkalinity on the growth of *Panicum coloratum* were less damaging in a hydroponic medium than in soil (Luna et al. 2016) indicating that alkalinity and hypoxia reduced growth non-synergistically. These results underscore that studies of plant response to alkaline substrates carried out in aerated nutrient solutions can only partially address the complexity of this stress and experimental approaches to analyze the physiological responses of plants to alkalinity should contemplate both structural and nutritional limiting aspects. As mentioned above, alkaline soils are typical of the depression of the Salado River basin (Fig. 3.1) located within the Argentine Pampas region, one of the most important livestock breeding areas in Argentina. *Lotus* species have traditionally been cultivated as forage in this area. It was reported that different *Lotus japonicus* ecotypes present divergent responses to saline, sodic and saline-sodic stresses (Bordenave et al. 2019). In particular, Fe-deficiency has been identified in some *L. japonicus* ecotypes grown in alkalinity (Babuin et al. 2014). In a similar study with a larger number of *Lotus* genotypes it was reported that alkalinity tolerance depended on Fe uptake mechanisms and structural alterations in the roots (Campestre et al. 2016). In sorghum, growth in alkaline substrate both in hydroponics and in soil indicated that responses to both alkaline and Fe substrates followed parallel trends (Luna et al. 2018). Lower induction of expression for genes for phytosiderophore synthesis and transport in alkaline-susceptible genotypes may be related to a reduced Fe availability, leading in turn to alteration in photochemical and biochemical reactions involving Fe. Thus, those results provide support to the concept that susceptibility to Fe-deficiency and alkalinity conditions are associated and highlight some of the physiological traits that underlie this association.

3.5 Final Considerations

This chapter summarized some mechanisms that contribute to the successful establishment of plants in saline and sodic soils. These mechanisms result from inherent genetic information, putative epigenetic controls, and expressed through complex molecular networks. As pointed out by Ismail and Horie (2017) application of appropriate molecular and genomic tools can contribute solutions to increase plant stress tolerance. Yet, it is evident that effective transfer of scientific knowledge requires the application of reliable phenotyping methods (Hackl et al. 2014) and adequate identification of physiological characteristics linked to plant performance under *stress* (Ghanem et al. 2015) and, above all, strong collaboration among scientists of different disciplines, breeders, and producers (Passioura 2010). Accurate information about the target environments (Pensiero et al. 2017) and field trials should be an inherent part of any project aimed at introducing new stress-tolerant germplasm.

Complementary holistic studies are essential to integrate information on plant functional traits with the environmental features of natural ecosystems (Díaz et al. 2016). Information provided by plant physiologists for this purpose requires agreement on phenotyping protocols and field evaluations to substantiate putative contributions of specific mechanisms to plant performance.

References

Amdan M, Aragón R, Jobbágy E, Volante JN, Paruelo JM (2013) Onset of deep drainage and salt mobilization following forest clearing and cultivation in the Chaco plains (Argentina). Water Resour Res 49(10):6601–6612

Amen RD, Carter GE, Kelly RJ (1970) The nature of seed dormancy and germination in the salt marsh grass *Distichlis spicata*. New Phytol 69(4):1005–1013

Babuin M, Campestre M, Rocco R, Bordenave C, Escaray F, Antonelli C, Calzadilla P, Gárriz A, Serna E, Carrasco P, Ruiz O, Menéndez A (2014) Response to long-term NaHCO3-derived alkalinity in model *Lotus japonicus* ecotypes Gifu B-129 and Miyakojima MG-20: transcriptomic profiling and physiological characterization. PLoS One 9(5):e97106

Bohn HL, Mc Neal BL, O´Connor GA (1993) Química del suelo. Limusa, México, 370p

Bordenave CD, Rocco R, Maiale SJ, Campestre MP, Ruiz OA, Rodríguez AA, Menéndez AB (2019) Chlorophyll a fluorescence analysis reveals divergent photosystem II responses to saline, alkaline and saline–alkaline stresses in the two *Lotus japonicus* model ecotypes MG20 and Gifu-129. Acta Physiol Plant 41(9):167

Campestre MP, Antonelli C, Calzadilla PI, Maiale SJ, Rodríguez AA, Ruiz OA (2016) The alkaline tolerance in *Lotus japonicus* is associated with mechanisms of iron acquisition and modification of the architectural pattern of the root. J Plant Physiol 206:40–48

Céccoli G, Ramos J, Pilatti V, Dellaferrera I, Tivano JC, Taleisnik E, Vegetti AC (2015) Salt glands in the Poaceae family and their relationship to salinity tolerance. Bot Rev 81(2):162–178

Chiacchiera S, Bertram N, Taleisnik E, Jobbágy E (2016) Effect of water table depth and salinity on growth dynamics of Rhodes grass (*Chloris gayana*). Crop Pasture Sci 67(8):881–887

de Luca M, García Seffino L, Grunberg K, Salgado M, Córdoba A, Luna C, Ortega L, Rodríguez A, Castagnaro A, Taleisnik E (2001) Physiological causes for decreased productivity under high salinity in Boma, a tetraploid *Chloris gayana* cultivar. Aust J Agric Res 52(9):903–910

Díaz FJ, Benes SE, Grattan SR (2013) Field performance of halophytic species under irrigation with saline drainage water in the San Joaquin Valley of California. Agric Water Manag 118:59–69

Díaz S, Kattge J, Cornelissen JH, Wright IJ, Lavorel S, Dray S, Reu B, Kleyer M, Wirth C, Prentice IC, Garnier E, Bönisch G, Westoby M, Poorter H, Reich PB, Moles AT, Dickie J, Gillison AN, Zanne AE, Chave J, Wright SJ, Sheremet'ev SN, Jactel H, Baraloto C, Cerabolini B, Pierce S, Shipley B, Kirkup D, Casanoves F, Joswig JS, Günther A, Falczuk V, Rüger N, Mahecha MD, Gorné LD (2016) The global spectrum of plant form and function. Nature 529:167

FAO/UNESCO (2020) Soil map of the world. www.fao.org

Feldman SR, Bisaro V, Biani NB, Prado DE (2008) Soil salinity determines the relative abundance of C3/C4 species in Argentinean grasslands. Glob Ecol Biogeogr 17:708–714

Ferrero ME, Villalba R (2009) Potential of *Schinopsis lorentzii* for dendrochronological studies in subtropical dry Chaco forests of South America. Trees 23(6):1275

Flowers TJ (2004) Improving crop salt tolerance. J Exp Bot 55(396):307–319

Flowers TJ, Colmer TD (2015) Plant salt tolerance: adaptations in halophytes. Ann Bot 115(3):327–331

Flowers TJ, Troke PF, Yeo AR (1977) The mechanism of salt tolerance in halophytes. Annu Rev Plant Physiol 28(1):89–125

Ghanem ME, Marrou HLN, Sinclair TR (2015) Physiological phenotyping of plants for crop improvement. Trends Plant Sci 20(3):139–144

Greenway H, Munns R (1980) Mechanisms of salt tolerance in non halophytes. Annu Rev Plant Physiol 31(1):149–190

Hackl H, Hu Y, Schmidhalter U (2014) Evaluating growth platforms and stress scenarios to assess the salt tolerance of wheat plants. Funct Plant Biol 41(8):860–873

Harris A (1929) The correlation between the soil salinity and flowering date in cotton. J Agric Res 28(2):109–112

Hayward HE, Wadleigh CH, Norman AG (1949) Plant growth on saline and alkali soils. In: Advances in Agronomy, vol 1. Academic Press, pp 1–38

Imbellone PA, Giménez J, Panigatti JL (2010) Suelos de la región pampeana. Procesos de formación. Editorial INTA, Buenos Aires, 280p

Ismail AM, Horie T (2017) Genomics, physiology, and molecular breeding approaches for improving salt tolerance. Annu Rev Plant Biol 68:405–434

Jacobsen T, Adams RM (1958) Salt and silt in ancient Mesopotamian agriculture. Science 128(3334):1251–1258

James RA, von Caemmerer S, Condon AG, Zwart AB, Munns R (2008) Genetic variation in tolerance to the osmotic stress component of salinity stress in durum wheat. Funct Plant Biol 35(2):111–123

Jayawickreme DH, Santoni CS, Kim JH, Jobbágy EG, Jackson RB (2011) Changes in hydrology and salinity accompanying a century of agricultural conversion in Argentina. Ecol Appl 21(7):2367–2379

Jobbágy EG, Jackson RB (2004) Groundwater use and salinization with grassland afforestation. Glob Change Biol 10(8):1299–1312

Johnson HS, Hatch M (1970) Properties and regulation of leaf nicotinamide–adenine dinucleotide phosphate–malate dehydrogenase and 'malic' enzyme in plants with the C4-dicarboxylic acid pathway of photosynthesis. Biochem J 119(2):273–280

Läuchli A, Lüttge U, Cramer G (2004) Sodium-calcium interactions under salinity stress. In: Läuchli A, Lüttge U (eds) Salinity: environment-plants—Molecules. Springer, Netherlands, pp 205–227

Lavado RS, Taboada MA (2017) Génesis y propiedades de los suelos halomórficos. In: Taleisnik E, Lavado RS (eds) Ambientes salinos y alcalinos de la Argentina. Recursos y aprovechamiento productivo. Orientación Gráfica Editora and Universidad Católica de Córdoba, Buenos Aires, pp 9–28

Levitt J (1972) Responses of plants to environmental stresses. Academic Press, New York, p 698

Liu M, Liao H, Peng S (2019) Salt-tolerant native plants have greater responses to other environments when compared to salt-tolerant invasive plants. Ecol Evol 9:7808–7818

Loch D, Lees T (2001) Halophytic native grasses for stabilization and rehabilitation of salt-affected sites in northern Australia. In: Adkins S, Bellairs S, Bell L (eds) Proceedings of the 4th Australian workshop on native seed biology for revegetation, Mildura Victoria, pp 235–246

Luna DF, Aguirre A, Pittaro G, Bustos D, Ciacci B, Taleisnik E (2016) Nutrient deficiency and hypoxia as constraints to *Panicum coloratum* growth in alkaline soils. Grass Forage Sci 72(4):640–653

Luna DF, Saavedra Pons AB, Bustos D, Taleisnik E (2018) Early responses to Fe-deficiency distinguish *Sorghum bicolor* genotypes with contrasting alkalinity tolerance. Environ Exp Bot 155:165–176

Maas EV, Hoffman GJ (1977) Crop salt tolerance: current assessment. J Irrig Drain Div ASCE 102:115–134

Marchesini VA, GiménezRl Nosetto MD, Jobbágy EG (2017) Ecohydrological transformation in the Dry Chaco and the risk of dryland salinity: following Australia's footsteps? Ecohydrology 10(4):e1822

Marinoni LD, Zabala JM, Taleisnik EL, Schrauf GE, Richard GA, Tomas PA, Giavedoni JA, Pensiero JF (2019) Wild halophytic species as forage sources: key aspects for plant breeding. Grass Forage Sci 74(3):321–344

Marschner H (1995) Mineral nutrition of higher plants, 2nd edn. Academic Press, London, pp 147–148

Miller GAD, Suzuki N, Sultan Ciftci Yilmaz, Mittler RON (2010) Reactive oxygen species homeostasis and signalling during drought and salinity stresses. Plant, Cell Environ 33(4):453–467

Munns R, Termaat A (1986) Whole-plant responses to salinity. Funct Plant Biol 13(1):143–160

Nosetto MD, Jobbágy EG, Tóth T, Jackson RB (2008) Regional patterns and controls of ecosystem salinization with grassland afforestation along a rainfall gradient. Global Biogeochem Cycles 22(2):1–12

Otondo J, Jacobo EJ, Taboada MA (2015) Mejora de propiedades físicas por el uso de especies megatérmicas en un suelo sódico templado. Ciencia del Suelo 33(1):119–130

Passioura J (2010) Scaling up: the essence of effective agricultural research. Funct Plant Biol 37(7):585–591

Pensiero JF, Zabala JM, Marinoni L, Richard G (2017) Recursos fitogenéticos forrajeros nativos y naturalizados (RFNyN) para suelos salinos de la región chaqueña de la Argentina. In: Taleisnik E, RS Lavado RS (eds.), Ambientes salinos y alcalinos de la Argentina. Recursos y aprovechamiento productivo. Orientación Gráfica Editora and Universidad Católica de Córdoba, Buenos Aires, pp 373–418

Perelman SB, León RJC, Oesterheld M (2001) Cross-scale vegetation patterns of Flooding Pampa grasslands. J Ecol 89(4):562–577

Pla I (1979) Suelos salinos y elementos secundarios. Revista Suelos Ecuatoriales 10(2):26–50

Pla I (2014) Advances in the prognosis of soil sodicity under dryland and irrigated conditions. Int Soil Water Conserv Res 2(4):50–63

Rajendran K, Tester M, Roy SJ (2009) Quantifying the three main components of salinity tolerance in cereals. Plant, Cell Environ 32(3):237–249

Rascovan N, Carbonetto B, Revale S, Reinert MD, Álvarez R, Godeas AM, Roxana Colombo R, Aguilar M, Novas MV, Iannone L, Zelada AM, Pardo A, Schrauf G, Mentaberry A, Vázquez MP (2013) The PAMPA datasets: a metagenomic survey of microbial communities in Argentinean pampean soils. Microbiome 1(1):21

Richards LA (1954) Diagnosis and improvement of saline and alkali Soils, Agriculture Handbook No. 60, United States Department of Agriculture, 160p

Rodríguez AA, Taleisnik EL (2012) Determination of reactive oxygen species in salt-stressed plant tissues. In: Sergey Shabala S, Cuin TA (eds) Plant salt tolerance. Humana Press, Totowa, NJ, pp 225–236

Rogers ME, Craig AD, Munns RE, Colmer TD, Nichols PGH, Malcolm CV, Barrett-Lennard EG, Brown AJ, Semple WS, Evans PM, Cowley K, Hughes SJ, Snowball R, Bennett SJ, Sweeney GC, Dear BS, Ewing MA (2005) The potential for developing fodder plants for the salt-affected areas of southern and eastern Australia: an overview. Aust J Exp Agric 45(6):301–329

Roy SJ, Negrão S, Tester M (2014) Salt resistant crop plants. Curr Opin Biotechnol 26:115–124

Sierra EM, Montecinos Urbina ER (1990) Floodings and droughts chronology of the Salado river depression. Revista de la Facultad de Agronomía 11(1):35–45

Snyman HA (1994) Evapotranspiration, water-use efficiency and quality of six dryland planted pasture species and natural vegetation, in a semi-arid rangeland. Afr J Range Forage Sci 11(3):82–88

Taleisnik E, Pérez HE (2016) *Chloris gayana*, una forrajera perenne con alta tolerancia a salinidad: 130-155. In: Toll Vera JR (ed) Grama Rhodes: Centenario de suliberación en Argentina, 1916–2016. Facultad de Agronomía y Zootecnia. Publicación Conmemorativa. Universidad Nacional de Tucumán, San Miguel de Tucumán, pp 130–155

Taleisnik E, Lavado RS (2017) Ambientes salinos y alcalinos en la Argentina. Recursos y aprovechamiento productivo. Orientación Gráfica Editora, Universidad Católica de Córdoba, Buenos Aires, p 624

Turner NC, Ward PR (2002) The role of agroforestry and perennial pasture in mitigating water logging and secondary salinity: summary. Agric Water Manag 53(1–3):271–275

Villagra P, Passera C, Greco S, Sartor C, Naranibar J, Meglioli P, Alvarez J, Allegretti L, Fernández N, Cony M, Kozub P, Vega Riveros C (2017) Uso de plantas nativas en la restauración y recuperación productiva de ambientes salinos de las zonas áridas de la región del Monte, Argentina. In: Taleisnik E, Lavado RS (eds) Ambientes salinos y alcalinos de la Argentina. Recursos y aprovechamiento productivo. Orientación Gráfica Editora, Universidad Católica de Córdoba, Buenos Aires, pp 419–444

Violle C, Navas ML, Vile D, Kazakou E, Fortunel C, Hummel I, Garnier E (2007) Let the concept of trait be functional! Oikos 116(5):882–892

Wu H, Zhu M, Shabala L, Zhou M, Shabala S (2015) K^+ retention in leaf mesophyll, an overlooked component of salinity tolerance mechanism: a case study for barley. J Integr Plant Biol 57(2):171–185

Zhang J, Nguyen HT, Blum A (1999) Genetic analysis of osmotic adjustment in crop plants. J Exp Bot 50(332):291–302

Zorin M, Loch D (2007) Development of new *Chloris gayana* cultivars with improved salt tolerance from 'Finecut' and 'Topcut'. In: Proceedings Sixth International Herbage Seed Conference. Gjennestad, Norway, pp 92–96

Chapter 4
Groundwater and Its Role in Maintaining the Ecological Functions of Ecosystems—A Review

Elisabet Verònica Wehncke and Néstor Alberto Mariano

Abstract Groundwater dependent ecosystems (GDEs) constitute one of the largest environments at the global level. In general, they are not easy to perceive, however in recent years, it has been found that they can contain high diversity of living forms with particular adaptive characteristics suggesting that the water quality is maintained and particular ecological functions providing numerous services to mankind. Due to the increasing need of natural elements and the intensive and unplanned land use in the last years, which has led to environmental degradation and diverse social conflicts, studies on groundwater and ecosystems, or the ecology of groundwater are gaining momentum. Thus, holistic perspectives are encouraged to properly understand GDEs's connections, their functional roles and visualize proper management perspectives under different scenarios of global change. In this chapter, the following questions are addressed: (i) why is there a need of an integrated view of groundwater? (ii) what are the ecological values and services of groundwater? (iii) what is known about the connection between ecosystems and groundwater? (iv) which will be the effects of decoupling the connectivity between surface water–groundwater ecosystems? (v) may their functions and services be assessed and valued? and (vi) what are the implications for management?

Keywords Aquifer · Biodiversity · Contamination · Ecosystem services · Geoethics · Groundwater management · Human impacts · Regional flows

E. V. Wehncke (✉)
Centro de Investigación en Biodiversidad y Conservación, Universidad Autónoma del Estado de Morelos, Avenida Universidad 1001, Col. Chamilpa, 62209 Cuernavaca, Morelos, Mexico
e-mail: lizwehncke@gmail.com

N. A. Mariano
Instituto de Ambiente de Montaña y Regiones Áridas, Universidad Nacional de Chilecito, 9 de Julio 22, 5360 Chilecito, La Rioja, Argentina
e-mail: ness07cba@gmail.com

M. M. Alconada-Magliano (ed.), *Intensified Land and Water Use*, Springer Earth System Sciences, https://doi.org/10.1007/978-3-030-65443-6_4

4.1 Introduction

Water can be in various spaces and states on Earth. Once it precipitates or melts from ice, it flows as surface water or through the air with the wind, it has numerous opportunities to penetrate the surface and be part of the soil and subsoil. After crossing the superficial zone of the soil and plant roots, water may or may not find barriers in its passage depending on the type of material found and eventually, it may accumulate, travel and completely saturate the subsoil. The portion that remains above that saturated zone and that is contained in the ground constitutes the vadose zone, which is separated from the saturated zone by the level of the water table. Water levels thus may vary according to available humidity or drought cycles, or anthropogenic water extraction intervention.

Groundwater present in geological materials may store large amounts of water depending on the space between pores, interstices and fissures of the geological material in question, whether fractured rocks such as basalt, granite, limestone, sand, gravel, chalk, or other sedimentary rocks etc. Water moves gravitationally filling the saturated zone that may vary from a few to tens of hundreds of meters in thickness; it may have a surface extent up to hundreds of squared kilometers overcoming regional and international boundaries (Cech 2010). In turn, springs may arise due to gravitational forces as described by Tóth (1962, 1995) which may be assisted by tectonic action or through fractures that allow groundwater that may be rich in calcium and bicarbonates to maintain a unique aquatic fauna of invertebrates and fish such in karstic limestone environments where groundwater dissolves calcite and flows through cave systems that may be several kilometers long and contain a particular fauna, many of them crustaceans, insects, worms, mites, gastropods, and fish that meet primordial functions of water purification through various processes such as denitrification (Griebler and Lueders 2009; Hahn and Fuchs 2009; Humphreys 2006) (Fig. 4.1). ***Stygofauna*** may live within fresh groundwater and within the pore spaces of limestone, calcrete or laterite, while larger animals may be found in cave waters and even found in wells, like the Mexican cavefish *Astyanax jordani*. Stygofaunal animals often have very low metabolism and feed on biofilm, plankton, bacteria, and plants found in streams. They usually measure less than 1 mm in length and are extremely sensitive to changes in water quality. They are highly specialized and have lived on the Earth for hundreds of millions of years. As to facilitate their study, Goonan et al. (2015) proposed a classification for stygofauna according to their degree in dependency on groundwater:

- *stygophilos*: organisms that spend their life cycle in both surface and groundwater;
- *stygoxenos*: surface species that occasionally move to groundwater, and
- *troglofauna*: species of arachnids, millipedes, beetles and crickets that live in caves in contact with groundwater and breathe air.

Fig. 4.1 The stygofauna are aquatic microorganisms living in groundwater, in saline habitats, freshwater and fissures. Examples of subsurface fauna: (1) *Lepidepecreum longicornis*; (2) Blind Diving Beetle larva, *Paroster* n. sp., Northern Goldfields, Western Australia; (3) Amphipod, *Melitidae gnathopod* & prey, Pilbara, Western Australia (Taken from Subterranean Ecology; http://www.subterraneanecology.com.au/); (4) Ozark cavefish, *Troglichthys rosae* (Taken from https://www.riverhillstraveler.com/wp-content/uploads/endangered-species.pdf); (5) *Phreatoicus typicus*, a 20 mm long hypogean phreatoicid from 20 m below ground surface in the Canterbury aquifer taken from Templeton, near Christchurch. G. Fenwick and GDF Wilson (2007); (6) Mites (blog.parksaustralia.gov.au; Julian Finn, Museum Victoria); (7) New Zealand phreatoicids, species of the endemic genus *Notamphisopus*, GDF Wilson; Like all New Zealand phreatoicids, species of this endemic genus are unpigmented, lack obvious eyes and are similar in general shape and body plan. The large first pereopods or legs show that this is a male specimen. (8) Caprellidae sp., Christian Gloor 2015; Wikimedia Commons, Creative Commons CC0 License

Many species of stygofauna, especially the obligate stygobites, are endemic to particular regions or even particular caves. This makes them focal points for conservation of groundwater related ecosystems; this will be further discussed in more detail.

In turn, circulating groundwater may be found connected to a lake or wetland, or provide water to streams and rivers supporting riparian vegetation. These diverse types of ecosystem constitute the surface expression of groundwater and may vary regarding their dependency on the unsaturated zone or on surface water (Foster et al. 2006a). GDEs comprise a diverse and complex group, generally characterized by being biologically rich in species and by having and maintaining a particular thermal and water chemistry, and specific ecological functions (Boulton and Hancock 2006). A good ecological status of GDEs is essential to maintain the provision of drinking water of good quality for the human being (Ducci et al. 2016; Foster et al. 2006a). In addition, the functional ecological integrity of both surface water and ecosystems depends on maintaining good connections in both groundwater quality and quantity (Boulton 2005).

Therefore, to protect groundwater in a comprehensive way it is necessary to know the biological and ecological status of the soil and subsoil, of rivers and streams (Silva et al. 2012). In the European water legislation, there has been a breakthrough as groundwater has begun to be considered not only as a resource but as a living ecosystem (Griebler et al. 2010). However, there is still a need to evaluate ecosystems more thoroughly, to define and extensively develop the ecological criteria that will help to understand the groundwater-ecosystem connection (www.kindraporject.eu: Gutjahr et al. 2014; Korbel and Hose 2011; Stein et al. 2012). Although there are several examples of approaches to conceptual frameworks that attempt to delineate an ecological assessment of groundwater systems (Eamus and Froend 2006; Griebler et al. 2010; Murray et al. 2006; Tomlinson and Boulton 2010), their development are in their first steps and commonly the connections between these systems are not very well understood. Australia has shown several attempts to incorporate ecological criteria in the management and administration of groundwater, establishing that the way to really ensure a benefit for present and future generations lies in the need to maintain or restore biodiversity and ecological processes in as many cases as possible (NSW-SGDEP 2002).

Rivers, besides being fed by precipitation, runoff may be derived from groundwater flows. Part of these flows may be inter-flows, that is, flows that move through the unsaturated zone without penetrating to the water table (Arthington 2012). The provision of groundwater flows to rivers is critical during periods of drought and often, is the only source of water that flows and maintains a riparian ecosystem, mainly in arid and semi-arid regions. Some of these rivers may re-infiltrate in different portions of their course and become intermittent or episodic (Boulton and Hancock 2006). Groundwater is almost always in contact with subsurface water that fills the interstices of sand and gravel in the bank of a river floor, this humid zone of saturated sediments (*hyporreic zone*) may extend a few meters below the river channel and several kilometers on either side, depending on the slopes forming habitats and shelters for life (*parafluvial zone*) (Fig. 4.2). Several components that might be a system in its own, interact with the *hyporreic* and *parafluvial zone* of a river, surface water, groundwater, alluvial aquifer, and riparian vegetation zone.

These systems operate in three spatial dimensions according to the dependence on groundwater: (i) the *sediment scale*, where chemical and microbial processes generate fine environmental gradients; (ii) the *scale of exchange between surface and groundwater* where the residence time of water in the hyporreic zone is variable and generate gradients along undulations of the river, its bars, banks and other segments; and (iii) the *basin size* that integrates gradients of the hyporreic zone and sediment retention along the river from the headwaters to downstream areas. Downstream, the composition of biological communities becomes more similar to that of rivers with permanent waters (Boulton and Hancock 2006). The exchange of surface and

Fig. 4.2 Simplified diagram of the hydrological compartments that may interact with the hyporreic zone of a river (E. Wehncke modified from Boulton et al. 1998, Arthington 2012)

groundwater along a river activates various ecological processes such as the movement of dissolved nitrogen and the dropping of benthic algal production (White et al. 1992). Besides, the degree of hydrological exchange between surface and groundwater may determine the distribution, as well as the activities of the decomposing microbes and invertebrates of organic material (Dent et al. 2000). Some fish species are also benefited by groundwater that arises forming springs in different portions of the stream, either because of the temperature it provides for egg laying stages or because of the oxygenation of the water in these sites that increase the availability of food (Dent et al. 2000). At evolutionary time scales, these ecosystems have provided refuge from changing environmental conditions on the Earth's surface; it is supposed that this fauna inhabited these environments for more than 400 million years (Finston et al. 2007). There is, therefore, at the scale of the basin a functional dependence of the riparian surface ecosystems to the upwelling or discharge of groundwater. Stanford and Ward (1993) proposed the concept of the Hyporreic Corridor, which suggests the existence of processes along a continuum of exchanges between groundwater and surface water, limited by geomorphic elements. In turn, there is a lateral exchange that connects riparian zones, paleo-channels and floodplains that may be very extensive. Thus, groundwater and the habitats that it traverses expand across the width and length of a river, maintaining various riparian ecosystems (Boulton et al. 1998).

These dynamic ecotones are areas of exchange of materials and energy become potential routes for the dispersal of wildlife and the transmission of nutrients, as well as contaminants. An important aspect both for the management and conservation of these systems is to understand the implications of anthropogenic alterations on groundwater dependent systems, for which it is extremely critical to know their connectivity and general functioning. Thus, the ecohydrological approach proposed by Tomlinson and Boulton (2010) offers a global vision that allows evaluating the ecological water needs of the systems.

Based on the above, the aims of this chapter are addressed in the following questions: (i) why is there a need for an integrated view of groundwater? (ii) what are their ecological values and services of groundwater? (iii) what is known about the connection between ecosystems and groundwater? (iv) which will be the effects of decoupling the connectivity between surface water–groundwater ecosystems? (v) may their functions and services be assessed and valued? and (vi) what are the implications for management? Questions whose answer were pursued following research advances on the topic through a methodology as that indicated in the Annex in this chapter.

4.2 Why Is There a Need for an Integrated View of Groundwater?

Groundwater is an ecosystem in its own and plays an integrating role in supporting other types of aquatic, terrestrial, and coastal ecosystems, and the landscapes associated with them, in both humid and arid regions. Therefore, it is a key factor difficult to isolate from other key ecosystems due to the multiple processes, connections and functions that they all perform in an integrated manner.

- The lack of control in the extraction and protection of groundwater has had negative impacts, mainly on aquatic flora and fauna due to changes in its flow systems. In some areas where groundwater is under intensive extraction (especially in arid or in highly populated regions) the ecological function of groundwater has largely been lost as a result of the drawdown of the groundwater level. In other cases, it is threatened by the deterioration of groundwater quality caused by diffuse contamination by agriculture (mainly due to excess use of nutrients and pesticides).
- Maintaining the connections between ecosystems and groundwater as healthy as possible is one of the key aspects to achieve the sustainability of these complex systems. Various sources of contamination on these connections through runoff from agricultural land, soil erosion, urban development, industries, and the introduction of exotic species threaten such sustainability. Joint efforts are required at various levels of organization, politics, and society to make visible and more compatible the use of groundwater and environmental conservation.

4.2.1 *The Aquifer as an Ecosystem*

In recent years, groundwater systems have finally been seen as true ecosystems (Hancock et al. 2005), and not only their enormous value as a resource has been appreciated, but also their relative susceptibility to contamination, which increase the need to learn more about its function and resilience capacity. Three components interact with each other in groundwater ecosystems, (i) the *matrix* with several types of pores according to the *sediment* or type of rock (porous, karstic, and fractured), (ii) *groundwater* occupying interstices, and (iii) the various *living organisms* that inhabit this space.

4.2.1.1 Main Characteristics of Life Forms in Groundwater Ecosystems

In comparison with taxonomic groups found in surface water ecosystems it has been found that aquifers are surprisingly rich in biodiversity with phylogenetically isolated species (Tione et al. 2014). At the scale of the landscape the subterranean biodiversity is very high with hundreds and even thousands of species comparable with the surface one (Hose and Lategan 2012; Humphreys 2008; Thulin and Hahn 2008). In contrast, at a local scale biodiversity is very low with approximately less than five species present (Hose and Lategan 2012). Both systems are very different: (i) the absence of light means that there are usually no primary producers in these sites (upper plants and algae) that can direct trophic webs, although a small amount of primary production can occur from chemo-autotrophic bacteria and protozoa that obtain their energy through chemical reactions with inorganic molecules such as hydrogen sulfate, sulfur and ammonium in an anaerobic environment or with very low levels of oxygen (Hose and Lategan 2012); (ii) bacteria and fungi at the base of the trophic chain of aquifers metabolize carbon that filters from the surface into subterranean ecosystems (Boulton 2000).Some invertebrates are adapted to live in saturated sediments very close to and below the surface where they graze the biofilm and the dissolved organic matter produced by microbial communities (Goonan et al. 2015). The subterranean fauna, known as stygofauna and detailed above, provides an important ecosystem service by avoiding the obstruction of channels and pores through its feeding and digging habits, and in this way also contribute to the water purification (Thulin and Hahn 2008). Normally they are blind, their bodies are translucent or whitish, elongated and flat, with low rates of reproduction and metabolism that makes them well adapted to low oxygen conditions; they are usually long-lived (Marmonier et al. 1993; Thulin and Hahn 2008; Tione et al. 2014). As it has been mentioned before, these communities are composed mainly of crustaceans, as well as tardigrades, oligochaetes, nematodes, and mites and in turn, they are composed of species with different states of adaptation to the subterranean environments (Gibert and Deharveng 2002). Several of the large taxonomic groups commonly found in surface water habitats also have a strong underground presence (Arnscheidt et al. 2012; Tione et al. 2014). Because physical and chemical conditions of groundwater ecosystems are relatively stable and predictable, it is expected that stygofauna are quite sensitive to environmental changes, which is why these microorganisms are good bioindicators of ecosystem state (Griebler et al. 2010).

4.2.1.2 Groundwater Microbes

Unlike the stygofauna they do not have important endemisms, but are widely distributed in both surface and groundwater (Danielopol and Griebler 2008). They are very sensitive to sources of human contamination, which can alter key functional biogeochemical processes such as denitrification, sulfate reduction, nitrification, and methane oxidation, so it is important to know the preliminary reference conditions of sites (Hemme et al. 2010). At a small scale, groundwater ecosystems can present a very wide diversity of habitats depending on the aquifer matrix, and porous spaces can vary in size, shape, water velocity, etc. (Dole-Olivier et al. 2009). Subterranean fauna is more susceptible to pollution than surface-dwelling species (Bright et al. 1998), possibly because they respond more directly to changes in nutrient contents in groundwater, as trophic networks are shorter than in surface environments due to the absence of primary producers. The arrival of nutrients can promote the occurrence of invertebrates through trophic links with bacteria.

Thus, invertebrates may reflect the small-scale variability of chemical components and be good indicators when there is an increase in the incorporation of nutrients and carbon (Tione et al. 2014). Interstitial organic matter can be incorporated into biofilms, which efficiently incorporate dissolved organic carbon from water actively converting it into particulate organic carbon, which can in turn be food for interstitial invertebrates. Thus, subterranean fauna can consume bacteria and pollutants and contribute to ecosystem services, particularly water purification (Boulton et al. 2008). Invertebrate communities can present a spatial distribution associated with different land uses in the overlying environment of an aquifer (such as livestock breeding, sanitation systems, poultry activity, horticultural production, park and urban sites), which condition the presence of several taxa.

4.2.2 Main Groundwater Dependent Ecosystems

A first step has been to conceptualize the presence of groundwater in the landscape, many questions add to the more completely understanding how ecosystems function as a whole. In some countries, the GDEs have been recategorized; this new classification includes aquifers and caverns, and ecosystems that depend on the surface expression of groundwater as different discharge zones. One way to classify ecosystems related to groundwater discharge could be by its morphological environment—aquatic, terrestrial, coastal, etc.—and the associated hierarchy of the Tóthian groundwater flow system—local, intermediate or regional. On this basis, different

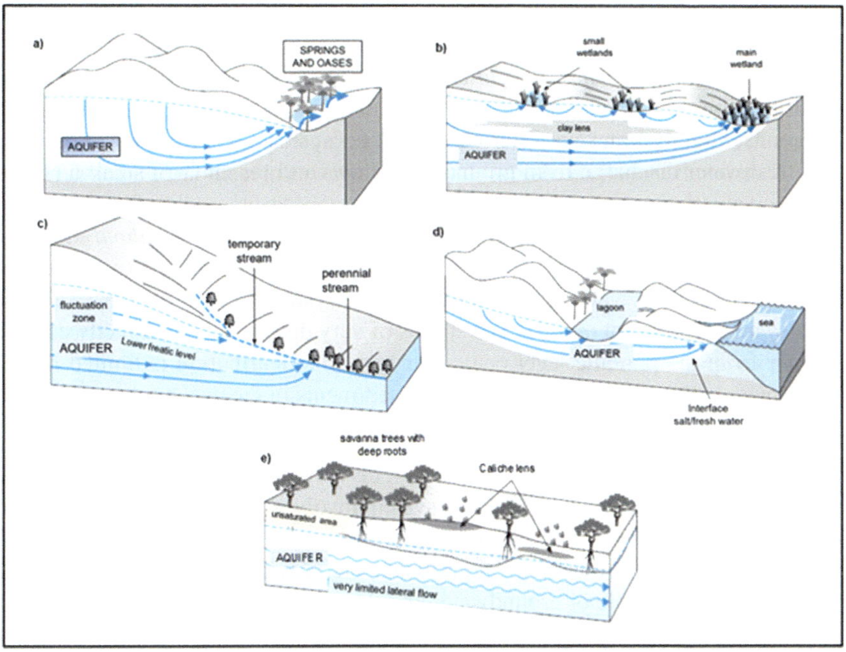

Fig. 4.3 Examples or main classes of ecosystems related to groundwater and its associated groundwater flow systems. **a** Wetlands of arid regions; **b** Wetlands of humid regions; **c** Aquatic ecosystems of riverbeds in humid regions; **d** Coastal lagoons and the entry of small amounts of seawater during exceptionally high tide; **e** Terrestrial ecosystem in arid regions (Modified from Foster et al. [2006a]. GW-MATE Core Group/World Bank/License: CC BY NC-SA 4.0)

classes may be recognized, such as those described by Foster et al. (2006a) and are listed below. Figure 4.3 shows some of the most important described by this author.

- *Natural discharge from relatively deep groundwater flow systems*, rising to form permanent well-defined springs with associated (and often unique) aquatic ecosystems. Marshes or oases in highlands, fed by surface water may become recharge for groundwater. Wetlands of arid regions: depend on the discharge of deep regional groundwater flows, sometimes with limited contemporary water recharge and groundwater with old age.
- *Wetland ecosystems related to discharge of local groundwater flow systems*, as seepages in land surface depressions. Wetlands of humid regions: individual ecosystems may depend upon or use groundwater from different depths flowing along several strata.
- *The discharge from extensive flow systems* (i.e., regional) provides (in part perennial and elsewhere ephemeral) flow during dry weather in the upper reaches of river systems that represent aquatic ecosystems. Aquatic ecosystems of riverbeds in humid regions: variable ecosystems along the upper reaches of rivers, partly

fueled by perennial groundwater discharge and partly by intermittent groundwater flows.

- The *discharge of groundwater flow systems into coastal lagoons*, which are critical to dilute salinity by marine influence and provide unique habitats. Coastal lagoons: slightly brackish water dependent ecosystem, generated by the mixing of freshwater discharge from groundwater flows and the entry of small amounts of seawater during exceptional high tides.
- Some extensive *semi-arid and humid terrestrial ecosystems without standing water*, but with very deep-rooted phreatophytic vegetation extracting moisture directly from the water table. Terrestrial ecosystem in arid regions: savanna ecosystem dependent on trees and shrubs of very deep roots that directly capture water from the phreatic level or its hair band (its distribution is limited by the thickness and degree of consolidation of sediments in the unsaturated zone).

Focusing on understanding how groundwater flow functions would give a better understanding of the requirements of ecosystems and biota in general which may be variable and in accordance with the flow system involved. The development of more detailed studies on these issues may reveal connectivity between biodiversity and groundwater along hydro-ecological gradients and provide a model to gain a fuller understanding of the role of groundwater and the ecological value of these systems.

4.3 What Are the Ecological Values and Services of Groundwater?

The daily life of the human being depends on many services given by countless ecosystems on the planet (Daily et al. 1997; MEA 2005). Groundwater provides ecosystem services with immense social impact and great economic value; however, the protection and management of these services require a quantitative and qualitative knowledge of the processes at different spatial and temporal scales, and an understanding of their resilience to the various anthropogenic impacts. The concept of ecosystem services points to a tendency of understanding beyond purely hydrogeological processes, focusing on the functions and processes that are connected to them. An approach leading to an understanding of ecological principles will allow predicting the responses of ecosystems as a whole and their services in the face of disturbances such as pollutants spills, climate changes, alterations to land use, among others. For this purpose, it is necessary to develop simple predictive tools so that decision-makers and environmental managers may adequately protect ecosystems and generate instruments for risk assessment (WLE 2015).

Groundwater is the largest source of drinking water in the planet and serves as a dissolving and cooling agent in industrial uses; in particular, it includes agrarian systems, which subtract groundwater for food production. Globally, 20% of irrigation water and 40% of the water used in industry is derived from groundwater (MEA 2005). Growth in industrialization, deposition of waste, and the pollution generated by the production and use of synthetic chemicals that are released into the environment, placing a very high risk to groundwater (Griebler and Avramov 2015). A calculation of the groundwater footprint showed that the extraction of groundwater from several large aquifers in the world significantly exceeds their annual renewal rate (Gleeson et al. 2012).

This suggests an important threat to the health of aquifers, their organisms and processes, and to many groundwater dependent ecosystems. Ecosystem services provided by groundwater have been divided into four main categories (Fig. 4.4), support services, provisioning services, regulation services, and cultural services (MEA 2005).

The presence of groundwater by itself is a support service and every aquatic and terrestrial ecosystem depends on its availability in good quantity and quality. Regulating services include water purification, biodegradation of pollutants, and the elimination of pathogenic microorganisms and viruses, which in turn contributes to the control of diseases. Cultural services include large bodies of water in caves and hot springs that are tourist attractions and of great spiritual importance to many indigenous communities. The conceptual framework of ecosystem services has been a very powerful weapon to raise awareness about their importance to humanity; however, the next step that is the protection and the adequate and sustainable management

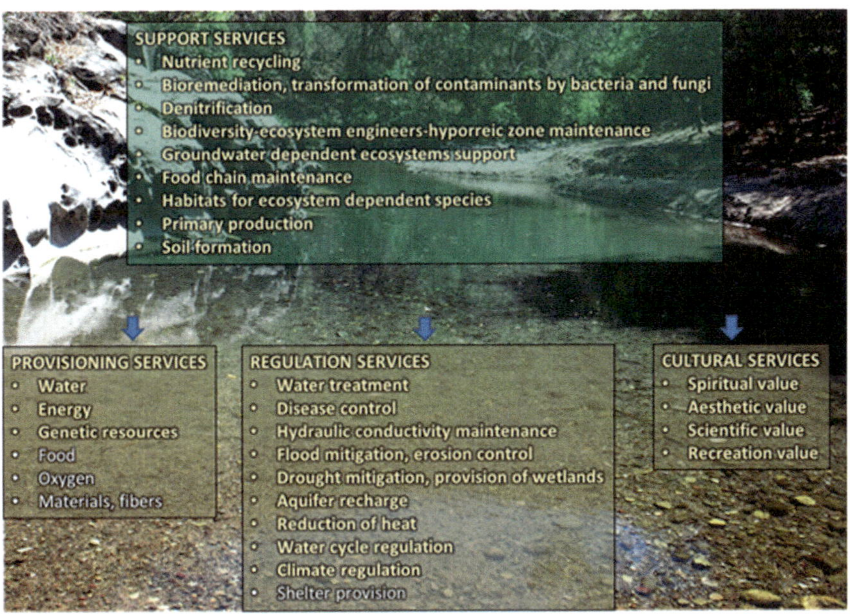

Fig. 4.4 Example of goods and services provided by groundwater according to the categories defined in the Millennium Ecosystem Assessment (MEA 2005). Goods and services that are directly related to groundwater ecosystems are highlighted in color (E. Wehncke modified from Griebler and Avramov [2015], and Goonan et al. [2015])

of the ecosystems, their organisms and the functions they represent, still awaits an efficient implementation.

This is coincident with other ecosystem services derived from subsurface aquatic ecosystems:

- Functions of karstification and drip filtration (Wilhartitz et al. 2009)
- Decontamination (Kinner et al. 2002)
- Nutrient remineralization, nutrient recycling at higher trophic levels, bioturbation and flow maintenance (Kinner et al. 1998, Mattison et al. 2002, 2005)
- Flood mitigation and maintenance of GDEs through infiltration and discharge at the local scale (Carol et al. 2009)
- Groundwater discharge provides the service of attenuating nitrogen loads derived from the earth through biogeochemical activity (Kroeger and Charette 2008, Griebler and Avramov 2015).

There are still many questions about subsurface ecosystem services, generally those provided by microbial activity and the impacts of irrigation on microbial function, as well as the impacts of seawater intrusion on the composition of the microbial community of functional groups involved in nitrogen recycling (Santoro 2009).

4.3.1 Degradation and Threats of Groundwater and Their Dependent Ecosystems

Aquifers store and transport water, in general, in an extremely slow way (approximately 1% of fresh water in world's aquifers is replaced by rain each year). However, they store huge amounts of fresh water involved over decades and millennia. For more than 8 thousand years man has managed to subtract groundwater, either for domestic use and/or for agriculture, and today, approximately one-third of the world's population extracts groundwater for different uses (Vörösmarty et al. 2005); however, approximately 90% of groundwater is inaccessible and difficult to extract with current technology (Cech 2010; Pearce 2007).

In the United States, approximately 85% of water goes to irrigate agriculture, and more than a third of this comes from groundwater (Postel and Richter 2003). In many places in the World, such as Australia, North Africa, the Middle East, South and Central Asia, Europe, and North America, excessive extraction of groundwater has dried up springs and altered river flow regimes (Konikow and Kendy 2005). Terrestrial and aquatic ecosystems vary in their degree of groundwater dependence in order to maintain their functions and their composition (Hatton and Evans 1998; Humphreys 2006), and therefore their vulnerabilities to changes in regimes and surface groundwater connectivity are also different. Human activities can alter groundwater levels by bringing them to very low levels and disconnecting them from surface water of rivers through excessive extraction for use in agriculture, domestic, industrial, or mining (Hancock 2002). Likewise, the recharge of this water may be altered by activities such as deforestation, afforestation, cultivation, urbanization, construction of dams, and changes in the course and flow of rivers (Foster et al. 2006a).

Quantifying the ecological responses of groundwater dependent ecosystems to these kinds of alterations is not an easy task; however, it is imperative to quantitatively evaluate the directions and implications of these responses to control the deterioration of water sources and be able to achieve an ecological and socially sustainable management.

To achieve such understanding, it is necessary to know and apply the ***theory of groundwater flow systems*** (Tóth 1962, 1995). This involves estimating flows horizontally and vertically, and the areas of recharge and discharge considering three systems of groundwater flow within the topographic and geological framework: local, intermediate and regional (Tóth 1962, 1995). In recharge areas, the natural chemical, biological and physical hydrological conditions influenced the soil cover and these contrast with the discharge areas. Water in these areas is controlled and manifested by downward and upward subsurface flow movement, oxidation and reduction conditions, acid and alkaline conditions, and cold and hot water temperatures, respectively. Groundwater movement under natural conditions is very slow from a few centimeters to a few hundred meters per year. In turn, changes in the environment in terms of quality and quantity of water or other related changes are evident after long periods of time, so these changes, whether natural or anthropogenic are very difficult to recognize and identify, and therefore, they have been insufficiently defined in the literature (Carrillo-Rivera et al. 2007).

Pollution, that is, alteration of the water quality that can cause harm related to human alterations, at times it may occur naturally (Foster et al. 2004). Aquifers are contaminated when the subsurface pollutant load generated by anthropogenic discharges and leachates (from urban, industrial, agricultural and mining activities) is not adequately controlled and (in certain components) it exceeds the natural attenuation capacity of the ground and the underlying strata (Foster et al. 2004). Normally, natural subsurface profiles actively degrade multiple water contaminants and have commonly been considered potentially effective for a safe final disposal of human excreta and domestic waste waters. This occurs as a result of biological degradation and various chemical reactions, but it is also due to the delay in the contaminant transport and the increase in the time available for other processes to eliminate contaminant products. However, when the unsaturated zone and water levels are shallow, their profiles are not equally effective in the removal of contaminants; therefore, the high probability of contamination of these unconfined aquifers is of concern (Foster et al. 2004).

The danger of groundwater contamination can be assessed, and actions can be taken to protect its quality. Such evaluation should become an essential component of ***good environmental practices***. The logical definition of the danger of contaminating groundwater starts from the analysis of the interaction between the vulnerability of an aquifer to pollution and the pollutant load that could occur to the subsurface environment as result of human activity on the soil surface (Foster et al. 2004). The vulnerability of groundwater may be assessed from the hydrogeological characteristics of the unsaturated zone or the overlying confining layers, and its functioning.

Likewise, water pollution can occur naturally (Foster et al. 2006b) and the ecological integrity of the various types of aquatic, terrestrial and coastal ecosystems can be naturally affected. Chemical reactions of rainwater in the soil/rock profile during infiltration and percolation give groundwater its essential mineral composition. During the journey, water absorbs carbon dioxide and the resulting weak acid dissolves soluble minerals in the soil. In humid climatic regions with regular recharge, groundwater continuously moves and consequently contact times can be relatively short, such that only readily soluble minerals dissolve (Foster et al. 2006b). Nine important chemical constituents (Na, Ca, Mg, K, HCO_3 Cl, SO_4, NO_3, and Si) make up 99% of solute content in natural groundwater (Foster et al. 2006b). Trace compounds represent only some 1% of the dissolved constituents present in natural groundwater, but sometimes they may make it inadequate for human consumption. However, at the same time, many of these trace elements are essential in small quantities for human and/or animal health, for example fluorine (F^-) and iodine (I^-) must be ingested with water or food. However, at higher concentration they may be harmful (Foster et al. 2006b). Other substances always harm health, even in very small concentrations (for example, arsenic (As) and uranium (U). Some elements such as As, F^-, and manganese (Mn) may generate known problems in groundwater ingestion. Others, such as nickel (Ni) and aluminum (Al) are of growing concern and may require further investigation under certain conditions. The concentration of some of these elements may be increased by various polluting activities on the soil surface. Thus, for management purposes it is important to differentiate impacts of anthropogenic origin from responses that occur naturally, for which detailed studies and groundwater monitoring must be carried out. The World Health Organization lists several other trace elements (including notably Ni, U, and Al) as potentially hazardous in drinking water (Table 4.1), and therefore it is necessary to carefully verify their occurrence particularly when extraction of groundwater begins in aquifers characterized by slow circulating water and/or by an important geothermal activity.

Arsenic is the most worrying trace element in groundwater since it is both toxic and carcinogenic at low concentrations. The range of hydrogeological conditions that facilitate its solubility in groundwater is only beginning to be known, but its mobility in shallow depths under strongly reducing conditions in geologically recent (Holocene) aquifers is worrisome because it seriously complicates the supply of low-cost water in deltas and alluvial zones, particularly in Southeast Asia (Foster et al. 2002, 2006b).

Groundwater is a vital source for human life and groundwater dependent ecosystems, as well as for economic and social development. Groundwater is also the most significant source of drinking water in natural emergency situations, such as flood and drought events. Groundwater bodies are naturally less vulnerable and more resilient to external influences than surface waters, which are highly vulnerable; and in many cases they provide a reliable and sustainable solution to water challenges (Carrillo-Rivera et al. 2013). During the sixth (2002–2007) and seventh (2008–2013) phases of UNESCO's International Hydrological Program (IHP), the Groundwater Emergency Situations Project (GWES) was implemented. The main objective was to consider the events of extreme natural disasters that could affect health and human

Table 4.1 Dissolved constituents in groundwater and their effect on human health (Taken from Foster et al. 2006b; GW-MATE, Briefing Note Series 14)

Trace elements (μg/l)				Major elements (mg/l)		
Measurement requires expensive equipment				Mainly simple and cheap to measure		
*V**	Li*	P*	Sr*	*Mg**	*Na**	HCO_3*
*Se**	*Ba**	*B*	*F**	*K**	Ca*	
As	*Cu**	Br		Si*	*SO_4**	
Cd	*Mn**	*Fe**			*Cl^-**	
*Co**	*U*	*Zn**			*NO_3**	
*Ni**	I*					
*Cr**						
Pb						
Al						

*Probably essential for human/animal health
As toxic or undesirable in excessive amounts (also probably essential where indicated)
B other elements
Al: aluminum; As: arsenic; B: boron; B: barium; Br: bromide; Ca: calcium, Cd: cadmium; Cl^-: chloride; Co: cobalt; Cr: chromium; Cu: copper, F: fluoride; Fe: iron; HCO_3: bicarbonate; I: iodide; K: potassium; Li: lithium; Mg: magnesium; Mn: manganese; Na: sodium; Ni: nickel; NO_3: nitrate; P: phosphorus; Pb: lead; Se: selenium; Si: silica; SO_4: sulfate; Sr: strontium; U: uranium; V: vanadium; Zn: zinc

life identifying means of potential safety, groundwater sources with low vulnerability to contamination that could replace contaminated domestic and public water supplements in emergency situations. For this purpose, several maps were developed, including a global map of vulnerability of groundwater to floods and droughts, at a scale of 1:25,000,000 (The Global Map of Groundwater Vulnerability to Floods and Droughts 2015). Experts from different countries contributed to this initiative and WHYMAP program that attempts to collect and visualize hydrogeological information at a global scale. The products generated provide information on the quality, quantity, and vulnerability of groundwater sources in the World. Possibly, visualizing water hidden under the earth through global maps contributes to the political discussion on groundwater issues, and serves as a means of communication among experts, decision-makers on issues related to groundwater, and the general public.

Vulnerability is an intrinsic (natural) property of a groundwater system, which depends on its sensitivity to natural and/or human impacts, and the ability of the system to cope with impacts in question (Vrba and Zaporozec 1994). The "*intrinsic vulnerability*" is a function of natural factors: the physical, biological, and chemical characteristics of the aquifer, of the overlying soil, as well as topography and climate. Groundwater vulnerability is based upon the assessment of several parameters that vary over regions as a function of the environment. However, the selection and evaluation of parameters used for this project varied according to the goals and scope of a vulnerability assessment. Aggregating several parameters to create a single index of vulnerability involved the steps of selection, scaling, rating and weighting. Thus, a final groundwater vulnerability class becomes an aggregate of individual parameters that have different units of measurements, so the final groundwater vulnerability score is considered dimensionless.

Several methods have been developed and applied in the evaluation of groundwater vulnerability to contamination which still wait to be fully in agreement with groundwater functioning as well as in proper field scale model testing and its validation which, ideally, should include three steps: calibration, validation, and post audit (NRC 1993). Further, most produced reports fail to comply with NRC (1993) recommendation of "*Those who generate vulnerability assessment must ensure that users are aware of the uncertainties associated with modeling schemes and data used.*" The DRASTIC is a numerical index approach (Aller et al. 1985) to which Holden et al. (1992) concluded that "the complex weighting and coding procedures used in the DRASTIC scoring are self-defeating." DRASTIC is an overlay and index method whose resulting value, unlike a concentration, cannot be measured in the field. This method is based on depth to groundwater table (D), net recharge (R) aquifer media (A), soil media (S), topography (T), impact of the unsaturated (vadose) zone (I) and hydraulic conductivity of the aquifer (C). The rating system SINTACS (Civita 1990), which is based on the same parameters as DRASTIC, but has a more complex structure. Its discretized input stage (based on grid squares) and its output (mapping and numerical tables) are both entirely computerized (Civita 1994). The European COST Action 620 project, which is focused on vulnerability and risk assessment for the Alpine aquifer system (Cichocki and Zojer 2007), assessed groundwater vulnerability based on parameters determined from a water-balance equation (combined with hazard maps). Geographical Information Systems (GIS) are also widely applied to present vulnerability scenarios on vulnerability maps, which conceptually add little, if any, to standard index mapping. Vrba and Zaporozec (1994), Witkowski et al. (2007), among others, have summarized a range of other methods and numerical models used for groundwater vulnerability assessment. In areas with deep aquifers isotope hydrology methods may complement the commonly used vulnerability parameters, since it helps to identify factors such as the groundwater residence time (age) and groundwater sources that are resilient to hazardous events (For more details visit the open access UNESCO: The Map of Global Groundwater Vulnerability to Floods and Droughts, scale 1: 25 000 000. © BGR Hannover/UNESCO Paris 2015).

4.4 What Is Known About the Connection Between Ecosystems and Groundwater?

- *Connection potential*: many wetlands are hydrologically and ecologically linked to adjacent groundwater flows, but the degree of interaction may vary greatly. The grade of wetlands dependence on groundwater may be complete or very limited, for example, in conditions of extreme dryness. In turn, some groundwater flows depend almost entirely on wetlands to gain recharge or in cases may not receive any recharge from wetlands.

- *Hydrological links*: knowing the way in which water enters and leaves the wetland, the so-called water transfer mechanisms, and quantifying their frequency is a prerequisite to assess the consequences for a wetland of any type of external hydrological impact. In addition to geographical (horizontal) analyzes of wetlands, by mapping open water bodies and zoning vegetation the understanding of the interactions with groundwater requires a three-dimensional hydrogeological view, that is, to look at the geological material and groundwater flow in vertical sections beneath the wetland and neighboring sides.
- *Determining the mechanisms of water transfer and its importance*: to help discover water transfer mechanisms Acreman (2004) developed a hydrological typology of wetlands based on landscape environment. There are three levels of evaluation that can contribute to the understanding and quantification of water transfer mechanisms: (i) *Office data*: spatial data of topographic maps, land and vegetation use, geological maps and aerial photographs or satellite images. Old photos are very useful to explain the hydrological links with wetlands. Geological maps may sometimes reveal the proximity of geological formations to wetlands; however, such maps may be unreliable as they usually fail to provide with permeability (or hydraulic conductivity) data of the strata; (ii) *Field visits* by a multidisciplinary team, including a hydrogeologist and a botanist/ecologist is recommended to explore the wetland after prolonged rainfall to see if springs or ephemeral courses of water might be discovered. Also, a visit after a prolonged dry period, when the vegetation profiles may indicate sites where the wetland depends solely on groundwater. It is necessary to drill shallow boreholes to investigate soil properties of a wetland, particularly to discover sites where groundwater is leaving or entering the wetland and those of permanent water register in the dry season; data from boreholes drilled at different depth to record the water-head, among other data, may allow to obtain zones of discharge or recharge. Finally, to conduct interviews with local people may provide useful information on the wetland behavior; (ii) *Data collection* in the field, such as piezometer levels, soil properties and hydraulic conductivity. Based on initial knowledge, monitoring programs should be established in the field to collect the necessary data over a time period. Data will serve as a basis for obtaining more detailed knowledge and the corresponding construction of numerical models.
- *Water-balance*: the calculation of a water-balance, which involves the quantification of water transfer rates, provides means of testing the degree of hydrological understanding achieved. The principle of balancing inputs, storage and outputs serves as proof that all water transfers have been considered and quantified. However, these balances only address global transfers in water volumes, and not hydrological processes per se; therefore, it may not give a full explanation of regime changes experienced by wetlands.
- *Uncertainty*: the water-balance cannot provide a definitive determination and prediction of wetland responses due to hydrological impacts, such as excessive groundwater extraction, because usually not all the hydrological components

in these natural systems have been identified and not related processes may be defined. It also fails to provide information on the rhythm, frequency and impact of hydrological events. To define hydrological properties detailed modeling is necessary considering an adequate and realistic conceptual model. Additionally, it is advisable that uncertainty (or level of confidence) should be calculated for any water transfer mechanism using methods from different disciplines (chemistry, biology, isotopes, hydraulic, etc.) in interdisciplinary mode.
- *Modeling*: sometimes complex models not necessarily guarantee an understanding of phenomena. After a correct conceptual understanding of water transfer mechanisms resorting to modeling implies to: (i) generate quantitative information about processes that drive water transfer mechanisms; (ii) understand temporal and spatial variability of processes; and (iii) predict what will happen in different possible climatic or water management situations (Secretariat of the Ramsar Convention 2010).

4.5 Which Will Be the Effects of Decoupling the Connectivity Between Surface Water–Groundwater Ecosystems? How to Recognize Changes?

4.5.1 Effects of Decoupling Ecosystems

Some of the outstanding effects of decoupling ecosystems to surface and/or groundwater could be listed as follows:

- Physical changes at the base of rivers, including banks.
- Chemical changes by blocking interstitial spaces of the hyporheic zone with contaminants or bacterial biofilms (Boulton and Hancock 2006).
- Preventing the exchange of ecological processes between these systems.
- Potential localized extinction of vulnerable species (Arthington and Balcombe 2011), mainly aquatic due to changes in the periodicity of drought events manifested as excessive groundwater extraction.
- Alteration of ecosystem processes such as litter decomposition and transformation of organic matter and nutrients, ultimately affecting the structure of trophic chains (Bunn et al. 2006). Various microbial nitrification and denitrification processes depend upon gradients in aerobic environments, therefore perturbations in surface-groundwater connectivity and alteration of groundwater regimes that alter redox gradients impede the microbial function of nitrate attenuation (Ayraud et al. 2006), and potentially reduce the ability of riparian and hyporheic areas to limit eutrophication (Rivett et al. 2008).
- Saline intrusion in aquifers mainly from coastal areas (Barlow and Reichard 2009).
- Decreases but also unexpected increases in groundwater levels may cause changes on ecosystems dependency, for example, species adapted to live in environments of intermittent water presence may be replaced by species adapted to permanent

waters when these levels increase (Bond et al. 2010), and these tend to be exotic species (Bunn and Arthington 2002).

4.5.2 How to Recognize the Ecological Responses of Groundwater Dependent Ecosystems?

It is difficult to obtain a reliable quantifiable response in the short time of the relationships between some attribute of the groundwater regime and some ecological value of an ecosystem (Boulton 2009). This is due to several reasons including, among others:

- That the timing of evaluation may not coincide with the ecological responses to be evaluated, as these may vary throughout the range of hydrological conditions.
- That there are multiple components of the groundwater regime that can interact and have a significant ecological response, rather than one in particular; as well as, that there are various environmental impacts that could have synergistic effects on one or several attributes of groundwater.
- That it is difficult to identify the time lapse between a disturbance in the groundwater regime and the corresponding response or ecological effect of ecosystems and vice versa (Tomlinson 2011).

Due to the complexity of these systems, being able to determine quantitatively the connections between superficial and groundwater flow regimes with ecosystems has been difficult (Lloyd et al. 2004), although alterations of ecosystems have repercussions in flow regimes and vice versa (Bunn and Arthington 2002). Research on functional responses of ecosystems should be planned as research programs rather than as isolated short-term investigations. Meanwhile, any water supply plan to maintain ecological values supported by groundwater needs some measure or contingency approach that groundwater managers can use for example, adaptable preliminary recommendations supported by a good monitoring program using databases on hydrogeological and ecological variables that can be analyzed through correlations and regressions. There are still many gaps in knowledge about the degree or specific characteristics of surface ecosystem dependence with groundwater, whether it is the time, volume and quality of water needed to maintain the ecological values of surface and subsurface ecosystems.

Currently, functions of GDE's ecological responses have been related to only some attributes of the regime such as depth of water table and decrease in groundwater level, also with indicators such as the presence of some specific vegetation type (Shafroth et al. 2000). More research is needed to relate groundwater requirements for ecosystems and other attributes such as: connections with surface water, its duration, as well as variability and periodicity in the magnitudes of these hydrological exchanges. Regarding the riparian vegetation response, it is necessary to incorporate:

- Knowledge about changes in species grouping and guilds.

- Soil studies, to understand in a more integrated manner the flow requirement of ecosystems (Merritt et al. 2010).
- Scale from field measurements on water requirements and use by vegetation, transpiration by certain tree species, and combine them with remote sensing studies at regional scale or a whole stand by vegetation types (Crosbie et al. 2008; Lubczynski 2009; Steward and Ahring 2008).

4.6 May Their Functions and Services Be Assessed and Valued?

4.6.1 Identifying Groundwater Ecosystems

Some authors have developed different approaches to formally delineate techniques and procedures that indicate the use of groundwater by ecosystems (Clifton et al. 2007). Here, it may be concluded the major actions to be related to:

- *Mapping tools:* of geological structures, aquifer depth; distributions, condition, and composition of the vegetation.
- *Water-balance techniques*: identification and quantification of groundwater use by measuring and estimating the components of the water cycle in an ecosystem.
- *Measurements of plants water potential*: pre-dawn measurements of water potentials in plants to identify groundwater use and depth.
- *Isotopic analysis in plants*: comparison and measurement of stable isotopes in water of plant xylem to identify groundwater intake (isotope signatures need to be established).
- *Ecosystem responses to environmental changes*: long-term monitoring of the composition and ecological functions of ecosystems in response to management intervention, climate and surface and groundwater conditions.
- *Surface groundwater hydraulics*: application of hydraulic principles and statistical analysis of flow hydrographs and sites measurements to derive the degree of interaction between groundwater and surface water characteristics; information to agree with chemical and isotopic data.
- *Physical properties of water*: measurements of water electrical conductivity and temperature change along the river wetland extension, or in time, to identify the contribution of groundwater.
- To identify *ascending groundwater flow* in river and wetland beds by means of shallow piezometric devices.
- *Chemical and physical analysis of surface and groundwater*: isotopy, temperature, pH, Eh, anions and major cations and trace elements. The mixing relationships among end-members show the contribution of groundwater.
- Observation of *mixing and dilution ratios* to evaluate groundwater contribution to a stream.

- *Modeling of water use by plants*: mathematical representations or models of the water balance in plants to estimate their water requirements and the possible groundwater uptake and likely response to decrease in groundwater level.
- *Groundwater modeling*: mathematical representations in two or three dimensions or models of the movement of water in saturated and unsaturated zones to assess the potential level of interaction between surface groundwater and between groundwater and terrestrial ecosystems.
- *Conceptual modeling:* use of expert knowledge of similar ecosystems, biophysical environments and relevant data to develop a conceptual model of the ecosystem and its interactions with groundwater.
- *Depth and morphology of roots*: evaluation of depth and morphology of root systems in plants and comparison with measured depths of the phreatic level to evaluate the potential of groundwater collection.
- *Analysis of aquatic ecology*: use of ecological techniques to identify diverse aquatic species including the stygofauna, with requirements of particular habitats that indicate the dependence on groundwater.

Finally, as above indicated, to evaluate groundwater input, it must be associated with the *hierarchy of the Tóthian groundwater flow system—local, intermediate, regional*—which may be traced with the joint analysis of physical-chemical-hydrological data of the ecosystem.

4.6.2 How to Evaluate the Status of Ecosystem Services with Measurable Indicators

According to Tomlinson (2011), there are several indicators that may be used to assess the state of ecosystems natural processes, for example:

- The fraction of *organic carbon* in sediments can be used as a proxy for potential denitrification, which occurs above a threshold of 3% (Smith et al. 2009).
- *Vegetation* plays an important role in maintaining subsurface ecosystem services, such as night water intake by roots from deep layers and their release in the root zone in the upper layers of the soil. Therefore, the increase of denitrification can be measured through the action of root exudates (Rivett et al. 2008), and the increase of mineralization of soil nutrients and the maintenance of mycorrhiza associations (Dawson 1993).
- Effects due to *land use and deforestation of riparian vegetation*. There will be implications for ecosystem services if vegetation is removed or altered in any way.

It is important to recognize that ecosystems provide services that are critical for the maintenance of groundwater systems in addition to recharging groundwater in wetland, river and riparian vegetation zones. In addition, changes or alterations due to climate change will affect both, ecosystems and all processes of connection with groundwater and consequently, direct and indirect regulation, and support of services provided.

4.7 What Are the Implications for Management?

The identification of ecosystem services is useful because it provides a common currency by means of which every person may understand the meaning of values and goods maintained by ecosystems, the benefits of maintaining these values, and what would be put at risk in case of not conserving them, even of completely losing those values. It also helps to decide which aspects of a flow regime are most important to maintain certain values such as water quality, fish, floodplain productivity, threatened species, etc. This process may also help bring together a group of local people to work together on these initiatives and help reinforce their sense of commitment and collaboration for the good of a greater cause in the planning and management of water of an area (Arthington 2012). In terms of water management and associated ecosystems some steps can be outlined such as:

- *Identification of values*: through a combination of information sources such as field studies, monitoring programs, and existing research projects, classification of habitats and species, literature reviews, historical information, and consultation with management and conservation agencies, research groups, and people who depend on the aquatic ecosystem directly. If there are different values these should be prioritized according to whether they are nationally or internationally recognized (e.g., lists of Ramsar sites), threatened, rare or endemic species, conservation areas declared for other values such as recreation, tourism, and fishing.
- *Agreement with the communities about desired future values*: develop scenarios of water flow requirements for the ecosystems in question. Even when the group of inhabitants has identified a specific value and accepted their condition, it is desirable that alternative scenarios be presented showing the risks of not providing the quantity or quality, or the timing when flows are required. Scientific participation should be able to estimate risks of certain results, such as, the loss of habitats and refuge for certain species, loss of floodplains, or the estuary productivity.
- *Determination of the acceptable risk level for each ecosystem by local people*: this will be done once risks are identified and quantified as precisely as possible, supported by governance and administration systems.

In Australia, quality monitoring evaluation programs have been implemented, which have been developed under a risk assessment framework regarding the condition of waters, by the Environmental Protection Authority (Goonan et al. 2012). This has provided good information to support decision-making regarding water quality through the sectors of government, industry and communities, and it is basically supported on the following:

- The definition of the most significant *aquifer systems* in the region in question

- The *identification of key environmental* values that are relevant to each aquifer
- The development of a conceptual understanding of the aquifer systems with knowledge on their environmental values, which incorporate *key processes* (geochemical and hydrological), the pressures that are acting on an aquifer, and how these are likely to affect the condition of the aquifer
- The development of a *group of indicators* that include measures of pressure, condition and management response for each aquifer
- The implementation of a *spatial and temporal sampling* design appropriate to each aquifer needs, and
- The development of a report that combines the information and requirements of the Environmental Authority or other relevant beneficiaries.

Finally, the challenges to detect impacts on groundwater flow systems and their associated ecosystems implies the need for knowledge about each flow system and how it is linked to a specific portion of the ecosystem, the precautionary principles that can be taken and the capacity to prevent any type of impact, some type of surveillance or long-term monitoring that should be followed, and the implementation of corrective measures to change negative trends on the degradation of the system (WLE 2015).

4.7.1 What Is a Sustainable Extraction of Groundwater as Related to Ecosystems?

A sustainable extraction of surface water would be an extraction pattern that reflects the natural hydrological regime composed of the magnitude, frequency, duration, periodicity, and rate of change (Hamstead 2009; Poff et al. 1997). Groundwater still lacks a more complete understanding to develop an expression of a sustainable extraction, since it is still very unclear how much water their dependent ecosystems require. In groundwater studies the hydrogeological concept of ***safe yield*** refers to the amount of groundwater that may be extracted without exhausting reserves (Kalf and Woolley 2005). This rather physical concept lacks consideration of the ecological value concept supported by groundwater or its interconnection with ecosystems, the actual size of the called "*reserve*" and the hydraulic interconnection between this and the various surface watersheds as well as other groundwater sources involved in the short and long terms, and apparently, lacks also consideration of any of the principles of environmental water provision. Thus, the global operational rule seems to be that water for the environment is considered after consumptive demands have been satisfied.

The management of surface water is inconceivable beyond the management of groundwater. Methods for determining ***environmental flows*** are fairly well developed for surface water (Dyson et al. 2003; Tharme 2003; Acreman and Dunbar 2004; Poff and Zimmerman 2010), although there is still a lack of knowledge about their effectiveness, and flow-ecology relationships in general to the scale necessary to achieve objectives of water planning and management (National Water Commission 2011). However, the recognition of the role played by groundwater in issues of environmental flows is generally inadequate and considered as an additional concept but not at all as a priority. The complexities of surface water dynamics that have not yet been resolved in terms of environmental flow, are multiplied when consideration is to be given to groundwater. This is perhaps, because the original and currently employed water balances usually work with long term and bulk values where chemistry and isotopes are not included in the equation. Groundwater flow systems are currently ignored. Groundwater flows are not always uni-directional, there are fissures, ducts, and pores where water can pass quickly or very slowly with gradients of variable magnitude and direction such as in seasonal climates. In addition to the temporal variability there is a spatial distribution in patches in flow rates (Stanford and Ward 1993; Boulton et al. 2010), which direct vertical and lateral redox gradients and a chemical and nutrient zoning is generated impacting microbial processes that provide ecosystem services. Key attributes of groundwater that can be recognized are: levels, flow, pressure, and water quality. Residence times are also key in biogeochemical functions (Boano et al. 2010).

4.7.2 How Much Water Does the Environment Need?

To integrate groundwater, it is necessary to consider water regimes necessary to maintain the ecological values of groundwater dependent ecosystems at low levels of risk (Tomlinson 2011). This implies:

- To identify ecological values (such as biodiversity and ecological processes).
- To understand the dynamics of the water regime that maintains these values.
- To evaluate the risks that these values may have due to disturbances to the groundwater regime and those due to changes and alterations in the surrounding environment (Carrillo-Rivera et al. 2007).

It is also necessary to know the management unit, the tributaries involved, their potential flow, their isotopic, and their chemical signatures; a similar knowledge should be required to define the dynamics of the groundwater flow systems involved. Surface water may face challenges in their management such as, the confused effects of multiple factors that can alter the aquatic ecosystem, the uncertainty in the magnitude, scale and variability of the responses, and the lack of integration in conceptual models of hydrological systems. However, a possibility of acquiring an understanding of groundwater flow to tackle similar uncertainties as in surface water might be reached when the different groundwater flow systems affecting an ecosystem or water bodies, are thoroughly defined by means of their hydraulic, chemical and isotopic characteristics in their geological framework of reference (Carrillo-Rivera et al. 2007).

4.8 Final Considerations

Groundwater and ecosystems: what is known and what remains to be known?

Thus, it might be stress that,

- Carefully planned adaptive monitoring and evaluation is crucial to inform decision-making.
- Integrated assessment of ecological quality and groundwater parameters is necessary, because the lack of evidence that environmental policies converge toward common objectives such as, guidelines that protect the quality of drinking water which do not necessarily contribute to the protection and maintenance of groundwater biodiversity and ecosystems.
- The training of skills in methodologies and techniques (water security) is mandatory for the scientific determination of ecological water requirements to establish a robust and accurate water provision program.
- Progress in water governance should consider protecting the full spectrum of values and benefits of groundwater.

Globally, there is a recently begun struggle with the reality of recognizing the ecological values of groundwater in water planning and management. There are several gaps in knowledge and a lot of uncertainty for what is important for the scientific integrity in decision-making in water and ecosystems management programs. Given this uncertainty, it should be recognized that the current management of groundwater for ecological purposes is fully experimental. Thus, the integrated evaluation of ecosystems water requirements and actual groundwater parameters is necessary, because it is not at all obvious that environmental policies converge toward common objectives i.e., conservation of ecosystems.

Likewise, training capacities in methodologies and techniques for groundwater flow determination are required to contribute to their scientific determination and link to the ecosystem, from where a more realistic groundwater requirements might be resulting. In the case of limited capacity for monitoring, the use of traditional knowledge in the dynamics of groundwater systems is recommended, which also contributes to better community-based groundwater management. It is important to understand the system from the point of view of social capital as well as its options for adaptation and capacity in situations of ecosystem services degradation. For example, it is important to know if populations have alternative water sources on which to depend; and/or if they may increase the potential of their groundwater sources through strategies such as infiltration, collection, and storage, as well as if they can develop alternative livelihoods with less or no dependence on groundwater. These capacities depend on the social capital knowledge (water capacity), which is critical for the future sustainability of social structures dependent on groundwater.

The scientific community must take a great responsibility, that is, to articulate knowledge and procedures in a language that is understandable to all also to develop and implement co-participatory management solutions with landowners and decision-makers. Regarding groundwater dependent ecosystems, a great challenge is to be able to select useful and sensitive parameters to identify natural reference conditions for individual units of groundwater ecosystems, identify bioindicators and finally determine the appropriate scale for their application.

Concluding, whether there is a plan to conciliate development needs of humanity and the maintenance of ecosystem services in a sustainable way, it is essential to promote holistic perspectives and to advance in the incorporation of an adequately understanding on the connections between groundwater flows and ecosystems and their functional roles, as to visualize proper management strategies under different scenarios of global change.

ANNEX

Results obtained from an exhaustive review of the existing scientific literature in the Web of Science databases (https://www.webofknowledge.com/) that address the topic of **groundwater** and **ecosystems** (Reyes et al. 2016), show the state of the art that existed at least, until the middle of the year 2018. From this review it was evident that in recent years, the number of publications has been increasing, and this was attributed to the increase in recognition of the ecological value of groundwater flows (Bertrand et al. 2012). The years 2015 and 2017, had the largest number of publications (32 and 33, respectively). This is also a result of the increase in the development of mathematical modeling and computational tools, as well as the development needs of water and riparian ecosystem management and conservation plans in a sustainable way.

The search and selection of articles was structured according to 25 research areas (Table 4.2). For the search we combined the words "Groundwater flow" and "Ecosystems" and "Landscapes".

After classifying and coding the information according to the year of publication, country/region and study area, among other characteristics we found a total of 320 publications as of August 26, 2018, of which 306 were used for the study (unpublished data).The oldest article dates from 1989 (Ryszkowski 1989), and the United States is the country with the largest number of publications (139 publications), followed by Australia (20 publications) and Sweden (18 publications). Regarding the number of publications found according to 25 different research areas, **Environmental Sciences Ecology** studies had 197 publications, followed by **Water Resources** studies with 93, **Geology** with 71 and **Freshwater Biology** with 51. Finally, to know how often certain keywords appear in publications, 564 different keywords were determined where **Groundwater** appears in 69 publications, followed by **Modeling** with 51, **Wetlands** with 36 and **Landscape** with 33.

Table 4.2 List of research areas selected for the bibliographic review

Agriculture	Genetics, heredity
Anthropology	Geography
Biochemistry, molecular biology	Geology
Biodiversity, conservation	Microbiology
Biotechnology, applied microbiology	Imaging science, photographic
Chemistry	Technology
Computer science	Marine freshwater biology
Energy fuels	Meteorology, atmospheric science
Engineering	Oceanography
Environmental sciences, ecology	Physical geography
Evolutionary biology	Plant sciences
Fisheries	Science technology other topics
Forestry	Water resources

References

Acreman MC (2004) Impact assessment of wetlands: focus on hydrological and hydrogeological issues. Phase 2 report. Environment Agency, Bristol (W6-091) and Centre for Ecology and Hydrology, Wallingford (C01996)

Acreman MC, Dunbar MJ (2004) Defining environmental river flow requirements? A review. Hydrol Earth Syst Sci Discuss 8:861–876

Aller L, Lehr JH, Petty R (1985) DRASTIC: a standardized system to evaluate ground water pollution potential using hydrogeological setting. In: Proceedings of the NWWA/API Conference on Petroleum Hydrocarbons and Organic Chemicals in Ground Water: Prevention, Detection and Restoration. Water Well Journal Publishing, Ohio, USA, pp 38–56

Arnscheidt J, Dooley J, Eriksson K, Hack C, Hahn HJ, Higgins T, Wood P (2012) Biogeography and ecology of Irish groundwater fauna: assessment of the distribution, structure and functioning of subterranean fauna within Irish groundwater systems (2007-W-MS-1-S1) STRIVE report. Environmental Protection Agency STRIVE, Irish

Arthington AH (2012) Environmental flows: saving rivers in the third millennium. University of California Press, Los Angeles

Arthington AH, Balcombe SR (2011) Extreme flow variability and the "boom and bust" ecology of fish in arid-zone floodplain rivers: a case history with implications for environmental flows, conservation and management. Ecohydrology 4:708–720

Ayraud V, Aquilina L, Pauwels H, Labasque T, Pierson-Wickmann AC, Aquilina AM, Gallat G (2006) Physical, biogeochemical and isotopic processes related to heterogeneity of a shallow crystalline rock aquifer. Biogeochemistry 81:331–347

Barlow PM, Reichard EG (2009) Saltwater intrusion in coastal regions of North America. Hydrogeol J 18:247–260

Bertrand G, Goldscheider N, Gobat JM, Hunkeler D (2012) From multi-scale conceptualization to a classification system for inland groundwater-dependent ecosystems. Hydrogeol J 20(1):5–25

Boano F, Demaria A, Revelli R, Ridolfi L (2010) Biogeochemical zonation due to intrameander hyporheic flow. Water Resour Res 46(2)

Bond N, McMaster D, Reich P, Thomson JR, Lake PS (2010) Modelling the impacts of flow regulation on fish distributions in naturally intermittent lowland streams: an approach for predicting restoration responses. Freshw Biol 55:1997–2010
Boulton AJ (2000) River ecosystem health down under: assessing ecological condition in riverine groundwater zones in Australia. Ecosyst Health 6:108–118
Boulton AJ (2005) Chances and challenges in the conservation of groundwaters and their dependent ecosystems. AquatConserv 15:319–323
Boulton AJ (2009) Recent progress in the conservation of groundwaters and their dependent ecosystems. AquatConserv 19:731–735
Boulton AJ, Hancock PJ (2006) Rivers as groundwater-dependent ecosystems: a review of degrees of dependency, riverine processes and management implications. Aust J Bot 54:133–144
Boulton AJ, Findlay S, Marmonier P, Stanley EH, Valett HM (1998) The functional significance of the hyporheic zone in streams and rivers. Annu Rev Ecol Syst 29:59–81
Boulton AJ, Fenwick GD, Hancock PJ, Harvey MS (2008) Biodiversity, functional roles and ecosystem services of groundwater invertebrates. Invertebr Syst 22:103–116
Boulton AJ, Datry T, Kasahara T, Mutz M, Stanford JA (2010) Ecology and management of the hyporheic zone: stream–groundwater interactions of running waters and their floodplains. J North Am Benthol Soc 29:26–40
Bright JC, Bidwell VJ, Christina R, Ward Jonet C (1998) Environmental performance indicators for groundwater. Lincoln University. https://hdl.handle.net/10182/3358
Bunn SE, Arthington AII (2002) Basic principles and ecological consequences of altered flow regimes for aquatic biodiversity. Environ Manage 30:492–507
Bunn SE, Thoms MC, Hamilton SK, Capon SJ (2006) Flow variability in dryland rivers: boom, bust and the bits in between. River Res Appl 22:179–186
Carol ES, Kruse EE, Pousa JL (2009) Eco-hydrological role of deep aquifers in the Salado sedimentary basin in the Province of Buenos Aires, Argentina. Environ Earth Sci 60:749–756
Carrillo-Rivera JJ, Cardona A, Huizar-Alvarez R, Graniel E (2007) Response of the interaction between groundwater and other components of the environment in Mexico. Environ Geol 55:303–319
Carrillo-Rivera JJ, Ouysse S, Hernández-García GJ (2013) Integrative approach for studying water sources and their vulnerability to climate change in semi-arid regions: Drâa Basin, Morocco. Int J Water Resour Arid Environ 3(1):26–36
Cech TV (2010) Principles of water resources: history, development, management, and policy. Wiley, USA
Cichocki G, Zojer H (2007) VURAAS: vulnerability and risk assessment for Alpine aquifer system. In: Witkowski A, Kowalczyk A, Vrba J (eds) Groundwater vulnerability assessment and mapping: selected papers on hydrogeology, Taylor & Francis, London, pp 191–197
Civita, M (1990) La valutazione della vulnerabilita degli acquiferi all'inquinamento (Assessment of aquifer vulnerability to contamination). Proc. 1st Conv. Naz 'Protezione e Gestione de lleacque sotterrance, Metodologie, Technologie e Obiettivi', Marano sul Panaro, vol 3, pp 39–86
Civita, M (1994) Data needs and presentation. In: Vrba J, Zaporozec A (eds) Guidebook on mapping groundwater vulnerability: international contributions to hydrogeology, vol. 16. Verlag Heinz Heise, Hanover, Germany, pp 57–73
Clifton C, Cossens B, McAuley C (2007) A framework for assessing the environmental water requirements of groundwater dependent ecosystems. Report 1 Assessment Toolbox, Resource and Environmental Management, Kent Town, South Australia. https://publications.csiro.au/rpr/download?pid=procite:f576497c-4cf3-42f9-a6b6-befa9048137e&dsid=DS1
Crosbie RS, Wilson B, Hughes JD, McCulloch C (2008) The upscaling of transpiration from individual trees to areal transpiration in tree belts. Plant Soil 305:25–34
Daily GC, Alexander S, Ehrlich PR, Goulder L, Lubchenco J, Matson PA, Mooney H, Postel S, Schneider SH, Tilman D et al (1997) Ecosystem services: benefits supplied to human societies by natural ecosystems. Issues in Ecol 2:116

Danielopol DL, Griebler C (2008) Changing paradigms in groundwater ecology—from the "living fossils" tradition to the "new groundwater ecology". Int Rev Hydrobiol 93:565–577
Dawson TE (1993) Hydraulic lift and water use by plants: implications for water-balance, performance and plant-plant interactions. Oecologia 95:565–574
Dent CL, Schade JJ, Grimm NB, Fisher SG (2000) Subsurface influences on surface biology. In: Jones J, Mulholland PJ (eds) Streams and ground waters. Academic Press, New York, pp 377–404
Dole-Olivier M-J, Malard F, Martin D, Lefébure T, Gibert J (2009) Relationships between environmental variables and groundwater biodiversity at the regional scale. Freshw Biol 54:797–813
Ducci D, de Melo MTC, Preziosi E, Sellerino M, Parrone D, Ribeiro L (2016) Combining natural background levels (NBLs) assessment with indicator kriging analysis to improve groundwater quality data interpretation and management. Sci Total Environ 569:569–584
Dyson M, Bergkamp G, Scanlon J (2003) Flow: the essentials of environmental flows. IUCN, Gland, Switzerland and Cambridge, UK
Eamus D, Froend R (2006) Groundwater-dependent ecosystems: the where, what and why of GDEs. Aust J Bot 54(2):91–96
Fenwick G, Wilson GDF (2007). Field guide to New Zealand phreatoicid isopods. NIWA Report, National Institute of Water & Atmospheric Research Ltd; Christchurch, New Zealand
Finston TL, Johnson MS, Humphreys WF, Eberhard SM, Halse SA (2007) Cryptic speciation in two widespread subterranean amphipod genera reflects historical drainage patterns in an ancient landscape. Mol Ecol 16:355–365
Foster S, Garduño H, Kemper K, Tuinhof A, Nanni M, Dumars C (2004) Groundwater quality protection defining strategy and setting priorities. Briefing note series 8, GW-MATE Global Water Partnership Associate Program, World Bank, Washington DC, EEUU. http://www.worldbank.org/gwmate
Foster S, Hirata R, Gomes D, D'Elia M, Paris M (2002) Groundwater quality protections: a guide for waters service companies, municipal authorities and environment agencies. GW-MATE Global Water Partnership Associate Program, World Bank, Washington DC, EEUU
Foster S, Koundouri P, Tuinhof A, Kemper K, Nanni M, Garduño H (2006a) Groundwater Dependent Ecosystems: the challenge of balanced assessment and adequate conservation. Briefing note series 15, GW-MATE Global Water Partnership Associate Program, World Bank, Washington DC, EEUU. http://www.worldbank.org/gwmate
Foster S, Kemper K, Tuinhof A, Koundouri P, Nanni M, Garduño H (2006b) Natural groundwater quality hazards avoiding problems and formulating mitigation strategies. Briefing note series 14, GW-MATE Global Water Partnership Associate Program, World Bank, Washington DC, EEUU. http://www.worldbank.org/gwmate
Gibert J, Deharveng L (2002) Subterranean ecosystems: a truncated functional biodiversity. Bioscience 52:473–481
Gleeson T, Wada Y, Bierkens MFP, van Beek LPH (2012) Water-balance of global aquifers revealed by groundwater footprint. Nature 488:197–200
Goonan P, Gaylard S, Jenkins C, Thomas S, Nelson M, Corbin T, Kleinig T, Hill R, Noble W, Solomon A (2012) The South Australian monitoring, evaluation and reporting program for aquatic ecosystems: context and overview. South Australian Environment Protection Authority, Adelaide
Goonan P, Jenkins C, Hill R, Kleinig T (2015) Subsurface groundwater ecosystems: a briefing report on the current knowledge, monitoring considerations and future plans for South Australia. Environment Protection Authority, Adelaide. https://www.epa.sa.gov.au
Griebler C, Avramov M (2015) Groundwater ecosystem services: a review. Freshw Sci 34:355–367
Griebler C, Lueders T (2009) Microbial biodiversity in groundwater ecosystems. Freshw Biol 54:649–677
Griebler C, Stein H, Kellermann C, Berkhoff S, Brielmann H, Schmid S, Selesi D, Steube C, Fuchs A, Hahn HJ (2010) Ecological assessment of groundwater ecosystems—vision or illusion? Ecol Eng 36:174–1190

Gutjahr S, Schmidt S, Hahn HJ (2014) A proposal for a groundwater habitat classification at local scale. Subterr Biol 14:25
Hahn HJ, Fuchs A (2009) Distribution patterns of groundwater communities across aquifer types in south-western Germany. Freshw Biol 54:848–860
Hamstead M (2009) Improving environmental sustainability in water planning. Waterlines Report, National Water Commission, Canberra
Hancock PJ (2002) Human impacts on the stream–groundwater exchange zone. Environ Manage 29:763–781
Hancock PJ, Boulton AJ, Humphreys WF (2005) Aquifers and hyporheic zones: towards an ecological understanding of groundwater. Hydrogeol J 13:98–111
Hatton T, Evans R (1998) Dependence of ecosystems on groundwater and its significance to Australia. Land and Water Resources Research and Development Corporation, Canberra
Hemme CL, Deng Y, Gentry TJ, Fields MW, Wu L, Barua S, Barry K, Tringe SG, Watson DB, He Z, Hazen TC (2010) Metagenomic insights into evolution of a heavy metal-contaminated groundwater microbial community. ISME J 4:660–672
Hose GC, Lategan MJ (2012) Sampling strategies for biological assessment of groundwater ecosystems, CRC CARE Technical Report no. 21, CRC for Contamination Assessment and Remediation of the Environment, Adelaide
Holden LR, Graham JA, Whitmore RW, Alexander WJ, Pratt RW, Liddle SF, Piper LL (1992) Results of the national alachlor well water survey. Environ Sci Technol 26(5):935–943
Humphreys WF (2006) Aquifers: the ultimate groundwater-dependent ecosystems. Aust J Bot 54:115–132
Humphreys WF (2008) Rising from down under: developments in subterranean biodiversity in Australia from a groundwater fauna perspective. Invertebr Syst 22:85–101
Kalf FRP, Woolley DR (2005) Applicability and methodology of determining sustainable yield in groundwater systems. Hydrogeol J 13:295–312
Kindra Project (2019) Knowledge inventory for hydrogeology research. www.kindraproject.eu. Accessed 13 Sept 2019
Kinner NE, Harvey RW, Blakeslee K, Novarino G, Meeker LD (1998) Size-selective predation on groundwater bacteria by nanoflagellates in an organic-contaminated aquifer. Appl Environ Microbiol 64:618–625
Kinner NE, Harvey RW, Shay DM, Metge DW, Warren A (2002) Field evidence for a protistan role in an organically-contaminated aquifer. Environ Sci Technol 36:4312–4318
Konikow LF, Kendy E (2005) Groundwater depletion: a global problem. Hydrogeol J 13:317–320
Kroeger KD, Charette MA (2008) Nitrogen biogeochemistry of submarine groundwater discharge Limnol. Oceanogr 53:1025–1039
Korbel KL, Hose GC (2011) A tiered framework for assessing groundwater ecosystem health. Hydrobiologia 661:329–349
Lloyd N, Quinn G, Thoms M, Arthington A, Gawne B, Humphries P, Walker K (2004) Does flow modification cause geomorphological and ecological response in rivers? A literature review from an Australian perspective. Technical Report 1, CRC for Freshwater Ecology, Canberra
Lubczynski MW (2009) The hydrogeological role of trees in water-limited environments. Hydrogeol J 17:247–259
Marmonier P, Vervier P, Giber J, Dole-Olivier M-J (1993) Biodiversity in ground waters. Trends Ecol Evol 8:392–395
Mattison RG, Taki H, Harayama S (2002) The bacterivorous soil flagellate heteromitaglobosa reduces bacterial clogging under denitrifying conditions in sand-filled aquifer columns. Appl Environ Microbiol 68:4539–4545
Mattison RG, Taki H, Harayama S (2005) The soil flagellate heteromitaglobosa accelerates bacterial degradation of alkylbenzenes through grazing and acetate excretion in batch culture. Microb Ecol 49:142–150
MEA-Millennium Ecosystem Assessment (2005) Ecosystems and human well being: wetland and water—synthesis. World Resources Institute, Washington, DC

Merritt DM, Scott ML, LeRoy Poff N, Auble GT, Lytle DA (2010) Theory, methods and tools for determining environmental flows for riparian vegetation: riparian vegetation-flow response guilds. Freshw Biol 55:206–225

Murray BR, Hose GC, Eamus D, Licari D (2006) Valuation of groundwater-dependent ecosystems: a functional methodology incorporating ecosystem services. Aust JBot 54(2):221–229

National Water Commission (2011) The national water initiative-securing Australia's water future: 2011 assessment. Canberra, NWC, 370 p https://www.agriculture.gov.au/sites/default/files/sitecollectiondocuments/water/nwi-assessment-2011.pdf

NRC (1993) Ground Water Vulnerability Assessment: Predicting Relative Contamination Potential Under Conditions of Uncertainty Committee for Assessing GroundWater Vulnerability, National Research Council. ISBN: 0-309-58508-2, 224 p

NSW-SGDEP (2002) The NSW State Groundwater Dependent Ecosystems Policy Department of Land Water Conservation NSW Government, 40 p. http://www.water.nsw.gov.au/__data/assets/pdf_file/0005/547844/groundwater_dependent_ecosystem_policy_300402.pdf

Pearce F (2007) When the rivers run dry: Water the defining crisis of the twenty-first century. Transworld Publishers/Eden Project Books, Boston, USA, p 302

Poff NL, Zimmerman JKH (2010) Ecological responses to altered flow regimes: a literature review to inform the science and management of environmental flows. Freshw Biol 55:194–205

Poff NL, Allan JD, Bain MB, Karr JR, Prestegaard KL, Richter BD, Sparks RE, Stromberg JC (1997) The natural flow regime. Bioscience 47:769–784

Postel S, Richter B (2003) Rivers for life: managing water for people and nature. Island Press, Washington, DC, p 220

Reyes AJ, Melhado J, Wehncke EV (2016) Agua subterránea y ecosistemas: Hacia una perspectiva holística de sus papeles funcionales qué sabemos y qué falta aún por conocer. Revista Latino-Americana de Hidrogeología Número Especial 10(1):345–351

Rivett MO, Buss SR, Morgan P, Smith JWN, Bemment CD (2008) Nitrate attenuation in groundwater: a review of biogeochemical controlling processes. Water Res 42:4215–4232

Ryszkowski L (1989) Control of energy and matter fluxes in agricultural landscapes. Agric Ecosyst Environ 27(1–4):107–118

Santoro AE (2009) Microbial nitrogen cycling at the saltwater–freshwater interface. Hydrogeol J 18:187–202

Secretariat of the Ramsar Convention (2010) Manejo de Aguas Subterráneas: Lineamientos para el manejo de las aguas subterráneas a fin de mantener las características ecológicas de los humedales Manuales Ramsar para el uso racional de los humedales, 4ª edición, V 11. Secretaría de la Convención de Ramsar, Gland, Suiza

Shafroth PB, Stromberg JC, Patten DT (2000) Woody riparian vegetation response to different alluvial water table regimes. West North Am Nat 60:66–76

Silva ACF, Tavares P, Shapouri M, Stigter TY, Monteiro JP, Machado M, Ribeiro L (2012) Estuarine biodiversity as an indicator of groundwater discharge. Estuar Coast Shelf Sci 97:38–43

Smith JWN, Surridge BWJ, Haxton TH, Lerner DN (2009) Pollutant attenuation at the groundwater–surface water interface: A classification scheme and statistical analysis using national-scale nitrate data. J Hydrol 369:392–402

Stanford J, Ward J (1993) An ecosystem perspective of alluvial rivers - connectivity and the hyporheic corridor. J North Am Benthol Soc 12:48–60

Stein H, Griebler C, Berkhoff SE, Matzke D, Fuchs A, Hahn HJ (2012) Stygoregions-A promising approach to a bioregional classification of groundwater systems. Sci Rep 2:673. https://doi.org/101038/srep00673

Steward DR, Ahring TS (2008) An analytic solution for groundwater uptake by phreatophytes spanning spatial scales from plant to field to regional. J Eng Math 64:85. https://doi.org/10.1007/s10665-008-9255-x

Tharme RE (2003) A global perspective on environmental flow assessment: emerging trends in the development and application of environmental flow methodologies for rivers. River Res Appl 19:397–441

The Global Map of Groundwater Vulnerability to Floods and Droughts (2015) The Map of Global Groundwater Vulnerability to Floods and Droughts at the scale of 1: 25 000 000 © BGR Hannover/UNESCO Paris International Hydrogeological Programme and German Federal Institute for Geosciences and Natural Resources (BGR), France https://www.whymap.org/whymap/EN/Maps_Data/Gwv/gwv_node_en.html

Thulin B, Hahn HJ (2008) Ecology and living conditions of groundwater fauna. Technical Report TR-08-06 Swedish Nuclear Fuel and Waste Management Co https://skb.se/upload/publications/pdf/TR-08-06.pdf

Tione ML, Blarasin M, Bedano J (2014) El acuífero como ecosistema: comunidades de invertebrados en aguas subterráneas y su relación con variables ambientales. Serie científica Katarumen, Ed UniRìo Editora, Río Cuarto, Argentina, 25 p

Tomlinson M (2011) Ecological water requirements of groundwater systems: a knowledge and policy review Waterlines Report 68, National Water Commission, Canberra, Australia, 144 p https://webarchive.nla.gov.au/awa/20160615081629/archive.nwc.gov.au/library/waterlines/68

Tomlinson M, Boulton AJ (2010) Ecology and management of subsurface groundwater dependent ecosystems in Australia—a review. Mar Freshw Res 61:936–949

Tóth J (1962) A theory of groundwater motion in small drainage basins in central Alberta, Canada. J Geophys Res 67:4375–4388

Tóth J (1995) Hydraulic continuity in large sedimentary basins. Hydrogeol J 3(4):4–16

Vörösmarty, C. J., Léveque, C., Revenga, C., Bos, R., Caudill, C., Chilton, J., ... & Reidy, C. A. (2005). Fresh water. Millenn Ecosyst Assess 1:165–207

Vrba J, Zaporozec A (1994) Guidebook on mapping groundwater vulnerability, international contributions to hydrogeology, vol 16. Verlag Heinz Heise, Hanover, Germany, p 131

White DS, Hendricks SP, Fortner SL (1992) Groundwater-surface water interactions and the distribution of aquatic macrophytes. In: Stanford JA, Simons JJ (eds) Proceedings of the First International Conference on Ground Water Ecology, American Water Resources Association, Bethesda MD, pp 247–256

Wilhartitz IC, Kirschner AKT, Stadler H, Herndl GJ, Dietzel M, Latal C, Mach RL, Farnleitner AH (2009) Heterotrophic prokaryotic production in ultraoligotrophic alpine karst aquifers and ecological implications. FEMS Microbiol Ecol 68:287–299

Witkowski A, Kowalczyk A, Vrba J (2007) Groundwater vulnerability assessment and mapping: selected papers on hydrogeology. Taylor & Francis, London, p 260

WLE—Water, Land and Ecosystems CRP (2015) Groundwater and ecosystem services: a framework for managing smallholder groundwater-dependent agrarian socio-ecologies—applying an ecosystem services and resilience approach (International Water Management Institute, Research Program on Water, Land and Ecosystems [WLE]), 29 p http://www.iwmi.cgiar.org/Publications/wle/corporate/groundwater_and_ecosystem_services_framework.pdf

Chapter 5
Groundwater Flows in Santa Fe Submeridional Lowlands. A Conceptual Model

Dora Cecilia Sosa, Eduardo Luis Díaz, and Silvana Luisa Castro

Abstract This chapter describes the functioning of groundwater flows in a large sedimentary flatland of 54,300 km^2, known as the "Bajos Submerionales" extended through the territory of the Chaco, Santa Fe, and Santiago del Estero provinces, in Argentina, applying the model proposed by Tóth (1963, 2000). This model allows us to understand the complex hydrogeochemical functioning of the region, characterized by the presence of waters of various types. Waters of sodium carbonate and bicarbonate from local flows, and sodium sulfate and chloride associated with intermediate and regional flows are described, with longer transit time and depth of travel. The presence of paleochannels represents special conditions of sandy sediments for fresh water storage. Therefore, this is the result of relationships between local flows interacting with deeper intermediate and regional flows that have their discharge in the region. Piezometric, hydrochemical and isotopic surveys have been made on surface water and groundwater for the whole basin that covers the three provinces, studying in detail an area which belongs to the province of Santa Fe. These concepts show the importance of vertical movements of groundwater in the region and the need to incorporate this criterion of water functioning in order to manage productive schemes that adapt to the alternating droughts and floods in the region.

Keywords Groundwater flows · "Bajos Submeridionales" · Paleochannels · Toth's theory · Lowlands · Argentina

5.1 Introduction

"Bajos Submeridionales" refers to an extensive region of approximately 54,300 km^2 located in the central north region of Argentina, within the Salado river mega alluvial

D. C. Sosa · S. L. Castro
Instituto Nacional Del Agua, INA, Patricio Cullen 6161 (3000), Santa Fe, Argentina
e-mail: sosa.dora@gmail.com

E. L. Díaz (✉)
Facultad de Ciencias Agropecuarias-Universidad Nacional de Entre Ríos, Ruta Provincial N° 11 Km 10,5 (3100) Oro Verde, Entre Ríos, Argentina

M. M. Alconada-Magliano (ed.), *Intensified Land and Water Use*, Springer Earth System Sciences, https://doi.org/10.1007/978-3-030-65443-6_5

fan which extends to the east of the Sierras Subandinas (i.e., mountain ranges), in the geomorphological province of the Gran Chaco Argentino(Padula and Mingramm 1968) (Fig. 5.1).

The basin has been identified in the region of the Chaco Wetlands (Canevari et al. 1998), within the project "Argentina Wetlands" (Kandus et al. 2008), which reinforces its essence of an ecosystem presenting characteristics and functioning closely linked to the presence of groundwater (Marchetti, personal communication).

Despite the apparent homogeneity that the "espartillares" (cordgrass, Sporobolus spartinus) transmit to any observer, they are far from being a homogeneous plant community. They have different characteristics according to their location in the flood gradient, rainfall, level, and hydrochemistry of groundwater.

The basin limits established by the National Water Resources Secretary have fixed what is known as the "Cuenca propia de Bajos Submeridionales," Basin N° 22 (Giraut et al. 2001), (Fig. 5.1). The Provincial Technical Committee of the "Bajos Submeridionales" has delimited in 2018, for the Province of Santa Fe, a region that exceeds the boundaries of Basin N° 22 and performs a regionalization according to its physiographic characteristics. In Fig. 5.2, four zones are delimited.

Furthermore, another boundary for the whole basin was proposed, based on the changes of runoff, as it can be observed in Fig. 5.3, in red lines. Usually, flat plain basins suffer changes in their hydrodynamics as a consequence of construction of artificial channels and roads. Additionally, these basins present large areas at risk of flooding or waterlogging. There can be observed the development of an important drainage network to mitigate hydric excesses (Fig. 5.3).

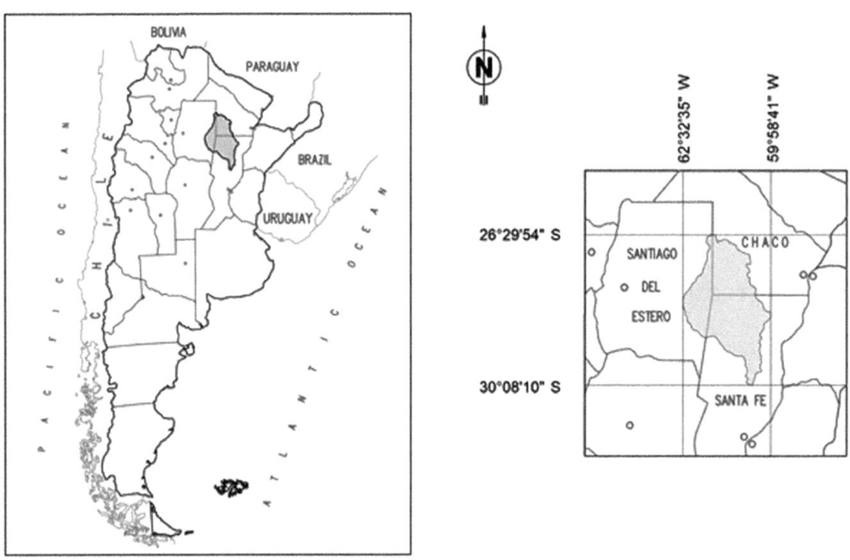

Fig. 5.1 Map of Argentina and location of the "Bajos Submeridionales" basin (based on Giraut et al. 2001)

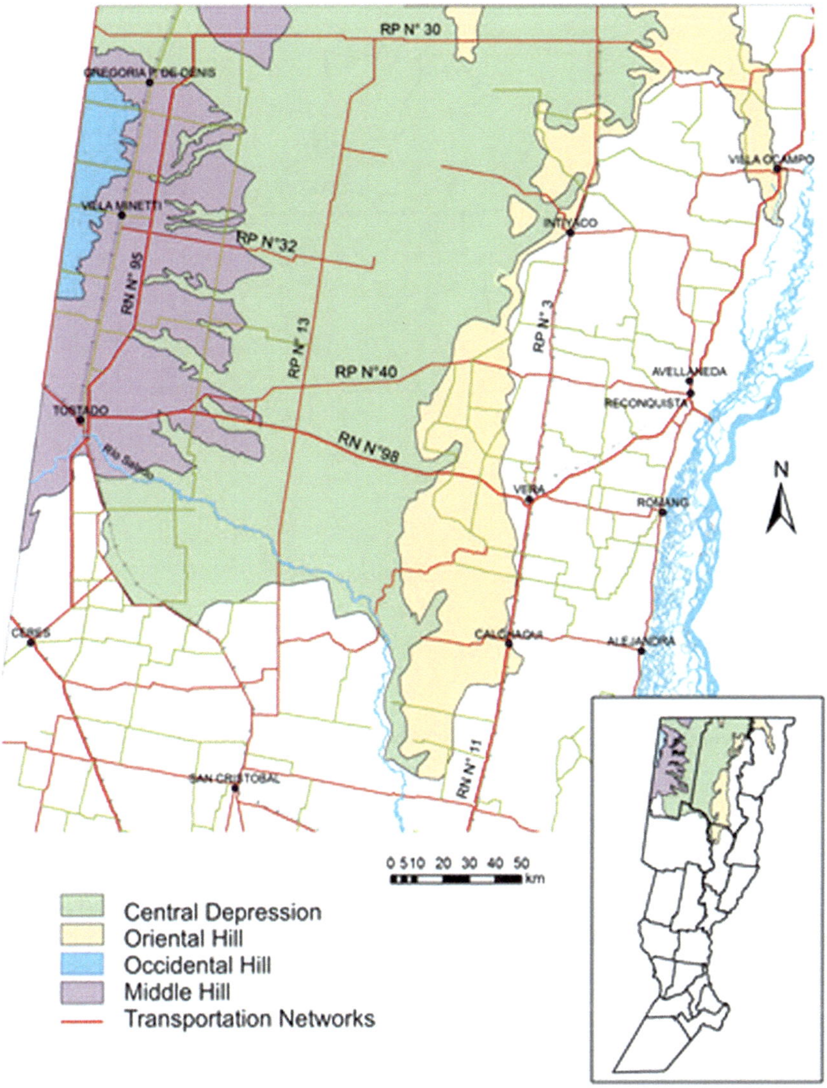

Fig. 5.2 Location of the region of the "Bajos Submeridionales" limits (adapted from the Provincial Technical Committee delimitatio 2018)

Cattle breeding is the main production activity; however, in wet years, in lands with low to deficient agricultural productive capacity, most of them are dedicated to agriculture.

In general, surface and groundwater contain a high salt content. The natural collector is the Arroyo Golondrinas stream, which at the same time discharges into the Calchaquí River, tributary of the Salado River (Fig. 5.3).

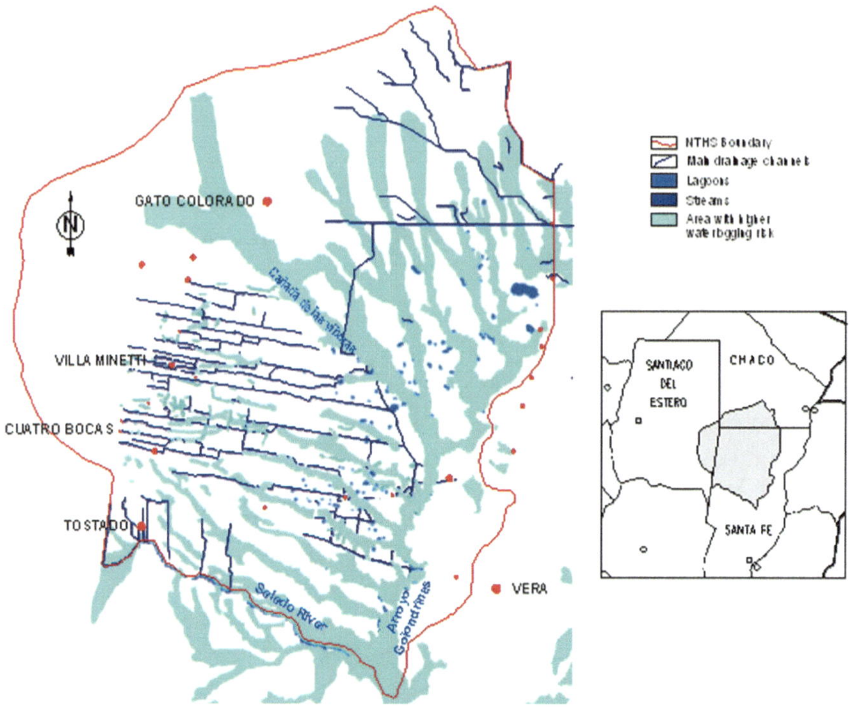

Fig. 5.3 Drainage network of the “Bajos Submeridionales” basin (adapted from Giacosa et al. 2017)

Fertonani and Prendes (1983, 1984) defined the studied area as a Non-Typical Hydrological System-NTHS (or SHNT for its Spanish acronym) by the behavior of surface water and where laminar flows prevail. These systems, with 5–30 cm·km^{-1} slopes (0.005%–0.030%), manifest extreme characteristics within the plain environments. The region is characterized by the alternation of droughts and floods, with serious social, economic, and productive consequences (Feldman 2017). In the Central Depression zone (Fig. 5.2), the softest northwest-southeast slopes of the entire area are located. There, during flood periods, water could remain from weeks to months. As the result of rainfall significant volumes of water in the form of a mantle can overflow the slight elevations that separate the depressions, in one direction or another, making it very difficult to delimit the surface flow divisions.

Consequently, the region has an important network of artificial channels (Fig. 5.3) built during wetter periods to drain surface water, thus reducing water retention times in shallow lakes and lowlands. In critical periods of drought it a system of side gates and spillway to control and storage superficial water has been implemented (only in some of them). It should be noted that for this and other regions with alternating droughts-floods periods, the construction of channels is not an adequate solution to control floods, intensifying the problems derived from the periods of drought, such as, for example, the increase in the salinization of soils that are already naturally affected by salts (Ameghino 1884; Maiola et al. 2013; Feldman 2017).

The aim of this chapter is to contribute to establish relationships between hydro-climatic pulses, geomorphology, soil, vegetation, and surface and groundwater flows that coexist, in the "Bajos Submeridionales" in Santa Fe Province, in order to understand the external elements of the landscape and propose agricultural and livestock management practices adapted to a droughts-floods dynamic. This would strengthen programs for the productive development of the province. Sosa (2012) identified groundwater flows in the "Bajos Submeridionales," applying the criteria of Engelen and Jones (1986) and Tóth (1963, 2009), who introduced the concept of flow systems and regional hydraulic continuity. Carrillo-Rivera (2000), cited by Alconada-Magliano et al. (2011), considered that flow systems can occupy a same aquifer unit, or by the contrary, that the same flow system can circulate through two or more aquifer units.

5.2 General Landscapes

Figure 5.4 shows different views of landscape characteristic of the "Bajos Submeridionales" region, and vegetation communities associated. Accumulation of water in the natural lowlands gives rise to large biodiversity reserves occupied by dense colonies of cordgrass, known as "espartillo," i.e., Sporobolus spartinus (Fig. 5.4a), in the Central Depression region (Figs. 5.2 and 5.4b). The palm areas associated with different environments are noteworthy, as shown in Fig. 5.4c. Also noted in the region are swampy low environments (Fig. 5.4d), which alternate with other higher environments (Fig. 5.4e) and with flat landscapes without trees with the presence of anthills almost one meter high (Fig. 5.4f). The different ecoregions and natural areas that characterize the province of Santa Fe in general and the "Bajos Submeridionales" region in particular, can be consulted in Biasatti et al. (2016).

5.3 Climate

Given the extension of the area, in the eastern central portion of the region, the climate can be classified, according to Thornthwaite (1948), as B´3 to B´4, where average annual rainfall ranges between 900 mm and 1000 mm with Summer–Autumn concentration. Climate in the western sector, according to Terré (personal communication), can be defined as Subtropical or Temperate warm Continental, with a wetter season (Summer) and a drier season (Winter), and a wide temperature variation between day and night.

Being precipitation determinant for the analysis of the hydric conditions of the region, and due to the low density and distribution of the meteorological stations within the basin, Scioli et al. (2020) complemented the rainfall information with the

Fig. 5.4 **a** Typical cordgrass ("espartillares," i.e., Sporobolus spartinus) landscape in the central depression, **b** Landscape of central depression, **c** Palms at the border between Chaco and Santa Fe provinces, **d** the Salado River backswamp, **e** Vegetation on the Occidental hilly area, **f** Anthills named as "Tacurúes," typical of the "Bajos Submeridionales" region

global database of the Climate Research Unit, CRU version TS 4.01 (Harris et al. 2020), corresponding to precipitation time series with grids every 0.5°. Figure 5.5 shows the distribution of mean monthly precipitation in the "Bajos Submeridionales basin" with respect to the Santa Fe basin, being similar in both. Figures 5.6 and 5.7, evapotranspiration and temperature, respectively, are presented for the same period considered for precipitation (Fig. 5.5) in the "Bajos Submeridionales" basin.

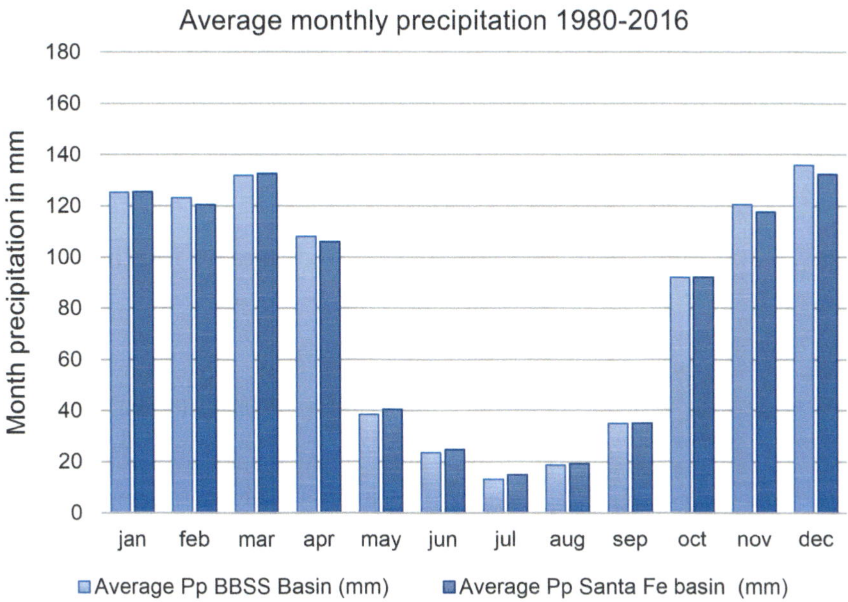

Fig. 5.5 Annual precipitation cycle 1980–2016 (adapted from Scioli et al. 2020)

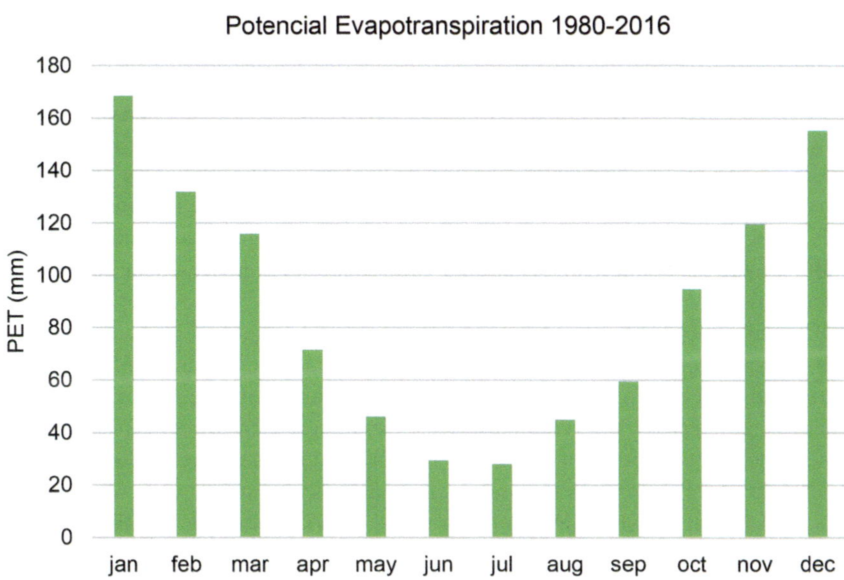

Fig. 5.6 Annual distribution of the weighted potential monthly evapotranspiration of the whole basin 1980–2016 (adapted from Scioli et al. 2020)

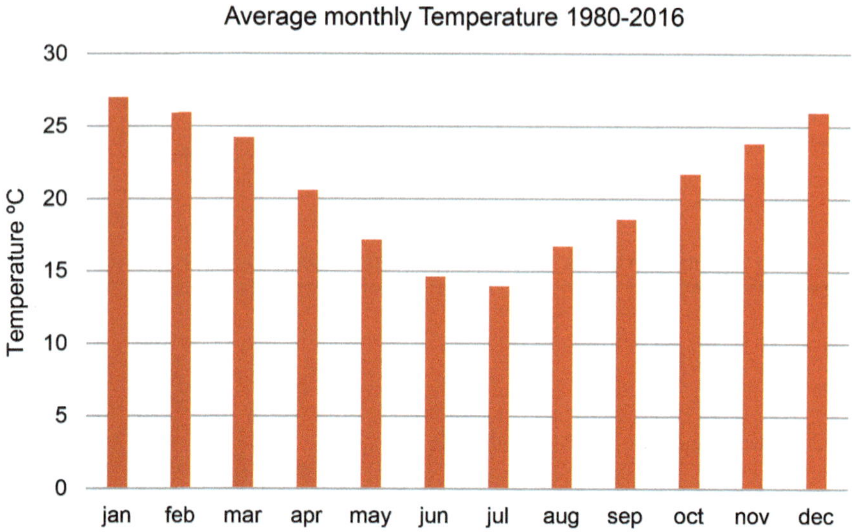

Fig. 5.7 Weighted monthly average temperature of the entire basin, 1980–2016 (adapted from Scioli et al. 2020)

5.4 Geomorphology

The basin is located within the geomorphological province of the Gran Chaco Argentino, which belongs to a wide plain with a strong climatic, geological and biogeographic identity, covering part of Bolivia, Paraguay, and Argentina (Iriondo 2010). In the basin, permanent swamps or wetlands, also temporary swamps or marshlands are common. Many of them are associated with the presence of materials of very low permeability, marshes with a significant percentage of colloids, with a silty-clay texture, and with an extremely low slope. This environment is typical of distal wetlands of alluvial fans (Thalmeier et al. 2019).

The Salado River alluvial mega-fan extends through the east of the foothills of the Sub-Andean Ranges ("Sierras Subandinas"), the distal part is characterized by large marshlands and wetlands related to a sub-humid to humid climate. There are shallow lakes interconnected by the Golondrinas Creek, numerous blowouts with permanent or semi-permanent bodies of water and paleobasins (Thalmeier et al. 2019).

Iriondo (2011) distinguished four geomorphological systems: one eolian system and three fluvial systems corresponding to the Salado, Bermejo, and Paraná rivers (Fig. 5.8).

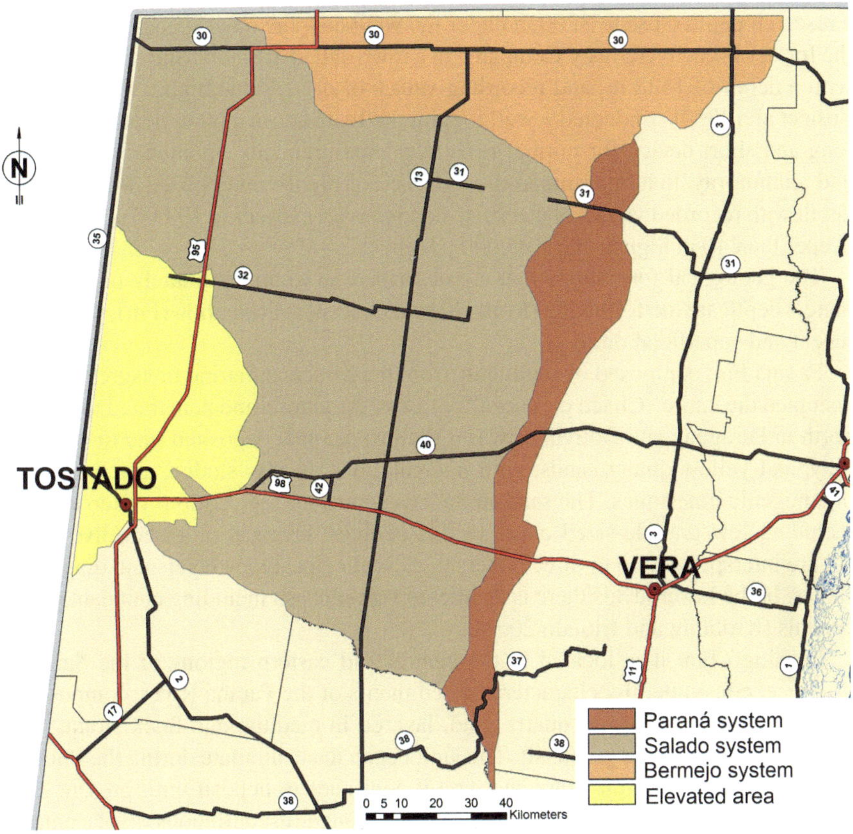

Fig. 5.8 Geomorphological systems (adapted from Iriondo [2011])

5.5 Geology

Thalmeier et al. (2019) described the Chaco Pampeana plain, as a landscape that includes a sedimentary sequence with a basement formed by large basins separated by ridges that record various geological events, occurring in different times, but with similar northeast-southwest orientation and consistent with other major features of Argentine geology (Peri 2013). In the Pliocene, the Chaco alluvial mega-fans began to develop with the rise of Sierras Subandinas (Iriondo 2010). The Salado River channel developed a large alluvial plain as well. Within the wide fan "Lomadas de Otumpa" (Rossello and Bordarampé 2005) is found, which constitute soft foothills of the topography spatially related to ancient sub-soil structures. Transverse paleobasins can be observed on them.

In the northwest boundary of the Santa Fe Province, although deep perforations are scarce, they provide relevant information; one was carried out for exploratory purposes for hydrocarbons (Las Mochas Locality in the 9 de Julio Department), with a research depth close to 3200 m under the wellhead, the other being carried out by the former French Rrailway Company, in a town called Tostado (Stappenbeck 1926) with a depth of 1428 m, and recording values of dry residue higher than 80 g l^{-1}. Sosa et al. (2000) conducted a well logging in Tostado, on a 60 m depth, including: long and short device for normal resistivity measurements, spontaneous potential, and gamma-ray, that determined the presence of highly mineralized water at 10 m depth with recorded values of electrical conductivity exceeding 10,000 $\mu S\ cm^{-1}$ and deeper than 40 m higher than 100,000 $\mu S\ cm^{-1}$.

The geological formations that are described up to approximately one hundred meters depth are the following (from oldest and deepest Formation (Fm), to the most recent and superficial one):

Paraná Fm: composed of sediments from the Miocene marine transgression, that occupied the entire "Chaco pampean" region of Argentina and penetrated toward the north in Paraguay and Bolivia. It was a shallow sea that deposited fine to very fine, gray, and yellow quartz sands, with intercalations of gray sludge, green clay, and organogenic limestones. The sand strata have abundant segregations of ferric oxide, including few granule-sized clasts to fine pebbles. Extracts of green olive clayey silt are interspersed, up to three meters thick, with calcareous layers and diagenized shales. In the lower levels there is an altered volcanic ash including abundant plaster crystals (Kröhling and Iriondo 2003).

Ituzaingó Fm: it is located in the central and eastern regions of the Santa Fe Province, constituted by characteristic sediments of the Paraná River, composed of fine yellowish and reddish quartz sand, layered in medium and thick strata, interstratified with gray and green silt. This unit began to accumulate during the Pliocene between 2 and 4 million years ago, and it continues to deposit until present days. Being the mechanism of riverbed wandering through discontinuous displacements, the formation presents a great spatial variability (Iriondo 2011).

Tezanos Pinto Fm: covering the central region of Santa Fe Province, it is the classic Pampa loess (Iriondo and Kröhling 1995). It extends through the northwest of Santa Fe Province in a strip of several kilometers wide along the border with Santiago del Estero Province, and continues north in the province of Chaco. It is 20–25 m thick and its deposition took place in the Last Glacial Maximum.

Fortín Tres Pozos Fm: It is located in the center of the territory of the Santa Fe and Chaco provinces; its sediments correspond to marsh deposits. Its thickness is between 15 and 25 m, from greenish gray to grayish brown in perforations and light brown in outcrops; its granulometry is clayey silt. The mineralogical composition is dominated by quartz and illite clay, with varying percentages of carbonates introduced by chemical precipitation.

San Guillermo Fm: Described by Iriondo and Kröhling (1995), it lies in erosional unconformity on the partially eroded soil of the Tezanos Pinto Fm. It is composed of gray slit with scarce clay and very fine sand (Kröhling and Orfeo 2002). This sedimentary unit was deposited in the Late Holocene (Iriondo 2011).

5.6 Surface Water Dynamics

The soils that characterize the "Bajos Submeridionales" Region developed under conditions of high hydro-halomorphism due to the influence of a water table in general saline and alkaline, in a particular geomorphology and geology (points 5.4 and 5.5). The prevailing soils are natric, mainly belonging to the Great Soil Group of the Natraqualf, and undifferentiated complexes of Natraquoll and Natralboll (classification criterion of Soil taxonomy, USDA-USA, in GeoINTA 2020). All these soils have very low infiltration, they facilitate the accumulation of surface water with formation of swamps, ponds, and lowland environments.

The "Bajos Submeridionales" basin of Santa Fe province is a wide plain with slopes between 30 cm Km^{-1} at the border with Santiago del Estero province (to the west), and 5 cm Km^{-1} in the Central Depression. The general slope is northwest-southeast, and it discharges its runoff through the Golondrina-Calchaquí system, reaching the Salado River (Fig. 5.3). Soils, mainly with silty-clay textures and high sodicity have low infiltration capacity, with lowlands filled up with sediments through precipitation (Giacosa et al. 2017).

Due to its morphologic, edaphic, and climatic features, there is no defined drainage network. The area has a strong intervention of constructed channels and roads, which connect shallow lakes and wetlands to shorten flooding permanence, and modify natural dynamics of inputs and runoff (Fig. 5.3). Given the very flat slope, the surface water flow is laminar, toward the lowlands and shallow lakes where the processes of vertical water transfer (precipitation, infiltration, evaporation) predominate over the horizontal runoff.

Storage features prevail in the area, like swamps, shallow lakes, and lowlands; thus, runoff is extremely slow. In normal and drought periods, laminar flow dynamics run toward swamps and ponds. Later these waters evaporate and infiltrate, and during these conditions vertical water movements predominate over runoff. In wetter periods, floods may last several weeks or months, and they are caused by the entry of excess water from the northwest of a few decimeters depth. The runoff has a predominant northwest-southeastern direction toward the Golondrinas-Calchaquí stream system, working as a tributary of the Salado River (Fig. 5.3).

To understand the importance of runoff coming from the "Bajos Submeridionales," it should be taken into account that 70% of the runoff from the Salado River at its mouth is originated in the Golondrinas-Calchaquí system (INA 2007). The very low elevations that separate depressions can be overflown, alternatively in one way or another by significant volumes of water as a result of rainfall and accumulation in local areas of the region. Therefore, the runoff dynamics makes the study of drainage and storage more complex, which is closely linked to the degree of mineralization of both surface and groundwater.

5.7 Hydrogeology

According to Fili and Tujchneider (1977), the regional hydrogeological column of Santa Fe Province is summarized from the oldest to the most modern strata, i.e., Basement rocks, Hipoparanian, Paranian, Lower, and Upper Epiparanian. The regional discharge level is given by the alluvial valley of the Paraná River.

The Epiparaian section is composed of fluvial sediments covered by finer sediments of eolian origin or others transported by rivers.

During the Tertiary period, the study area was the seabed of a large inland sea that gave rise to marine sediments of the Paraná Formation, followed by saltwater lakes. During the Quaternary period, this area partially emerged, allowing the sedimentation of river sands from the Bermejo, Salado, and Paraná systems (Iriondo, 2011).

The Basement: it is composed of igneous and metamorphic rocks of certain aquiferous characteristics. It was not detected in the perforations of Calchaquí (at a depth of 2,500 m) and Las Mochas (3,200 m depth) (Sosa 2012).

Groundwater has high mineral content and salts, especially sulfates, chlorides, and sodium. Salt concentration increases toward the north showing values higher than seawater content. These variations in saline content both vertically and horizontally are observed in distances measured in meters. As a consequence of the reduced movement and the presence of minerals in the sedimentary matrix, water acquires high salts concentrations with electrical conductivities from 1,000 to 30,000 μS cm^{-1} (Sosa et al. 2011). Waters with lower salt concentration, and consequently lower density, float over the more salty waters, generating a saline gradient from lower to higher concentration (Fig. 5.9).

Groundwater flow in a vertical direction has a greater magnitude than the horizontal one, giving place to a high hydrochemical variability, and tendency to storage. The first aquifer water level, during the excess periods, remains close to the surface, and they rise rapidly, due to rainfall; and during water deficit periods they drain more slowly, due to evapotranspiration and percolation (Sosa 2012).

During the 2008–2009 period, a survey and subsequent maps were made on water table depths and groundwater electrical conductivities (Figs. 5.10 and 5.11). In Fig. 5.10 it is observed that depths of the water table between 3 and 6 m prevail in the measured sites. Figure 5.11 shows four intervals of electrical conductivities where

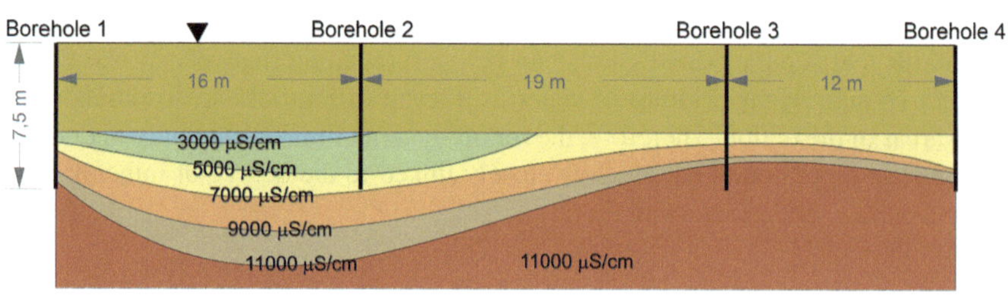

Fig. 5.9 Continuous profile of electrical conductivities (μS cm^{-1}) inferred depth from four boreholes in the northwest of Santa Fe Province

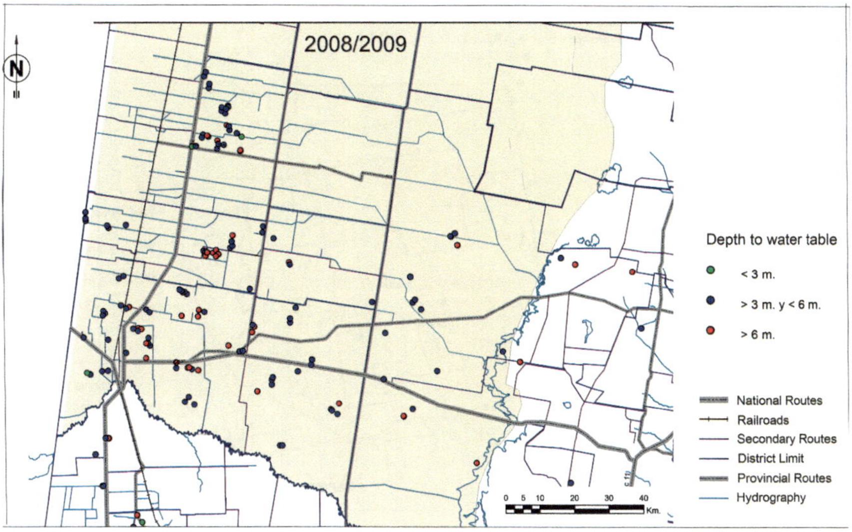

Fig. 5.10 Water table depth in meters, during 2008–2009 (Sosa 2012)

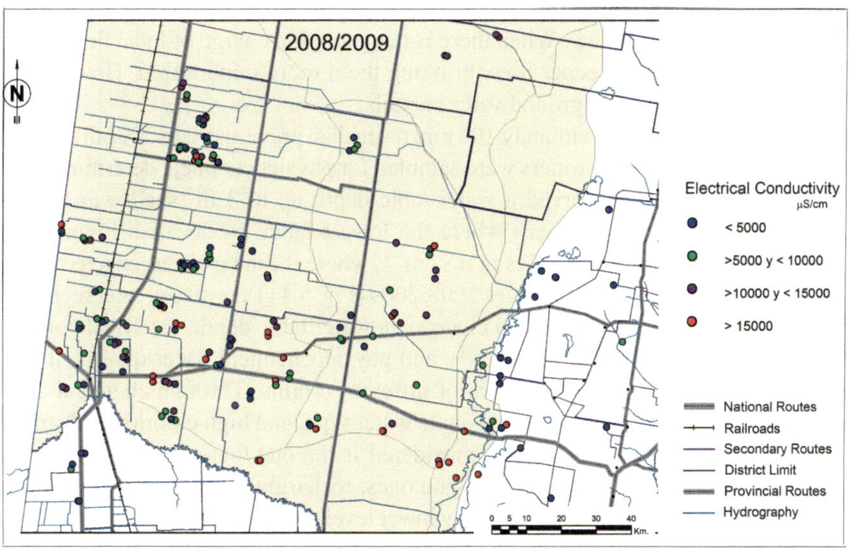

Fig. 5.11 Groundwater electrical conductivity in $\mu S\ cm^{-1}$ during 2008–2009 (Sosa 2012)

its spatial variability can be noted, although electrical conductivity values greater than 5000 $\mu S\ cm^{-1}$ prevail. As a consequence of low rainfall in previous years, the water table is deeper than in 2011 (Fig. 5.12) when a generalized ascent is observed in the entire region. This can be understood as the local flow effect. Also, there can

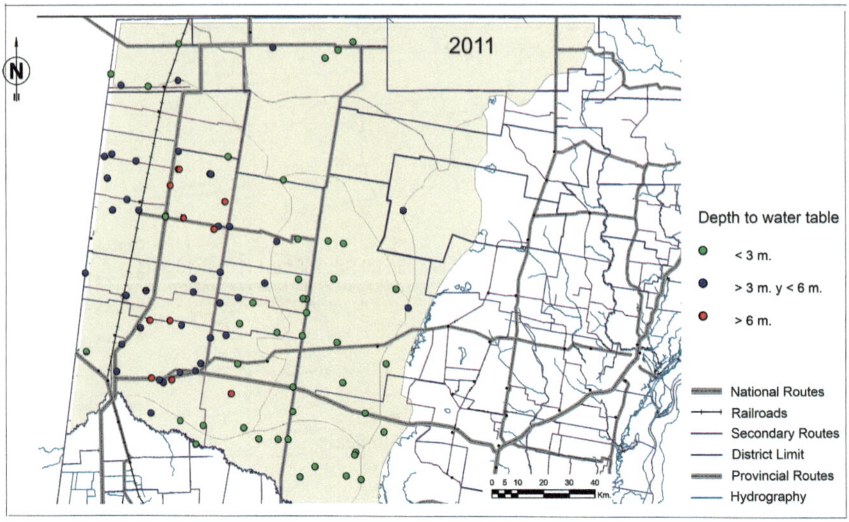

Fig. 5.12 Water table depth in meters, 2011 (Sosa 2012)

be observed a similar behavior in electrical conductivity values, which decrease as a result of greater local flow. When there is not enough recharge of local flow, available water is that from deeper flow, bringing them more mineralized. However, in all cases a high salinity of groundwater prevails.

In 2011, it rained approximately 100 mm more than previous years. Depth to water table was determined, and waters were sampled for physicochemical determinations. In Fig. 5.12 it can be observed a water table depth up to 3 m (shallower than in 2008/2009 Fig. 5.10) in the east where the lowest lands are located. Figure 5.13 shows the electrical conductivities in $\mu S\ cm^{-1}$, where it can be observed a reduction compared to electrical conductivities 2008/2009 (Fig. 5.11), due to a greater recharge.

Based on the results obtained in changes in water table depth, electrical conductivity (Figs. 5.10, 5.11, 5.12, and 5.13), and physicochemical water quality, it may imply the coexistence of water flows of different origins. Through chemical analysis results, it can be observed bicarbonate water type, and high chloride and sulfate contents. The regional flow system considered is the one that circulates in greater depth compared to local and intermediate ones, recharging in higher altitude areas and ending where discharge is favored by lower levels, thus several local flow systems can be found on it, and one or more of the intermediate type. From a chemical point of view, this water flow has a high content of dissolved salts. The above described is consistent with Toth's theory of flow systems (Tóth 2000).

During drought periods or extractions decreasing of hydraulic load occur, which could give rise to vertical upward flows of intermediate flows discharging in the upper aquifer, passing through a salinization process.

Within the regional scale, applying piezometric and hydrometric information of the main water courses in the region, Thalmeier et al. (2019) stated that flow patterns

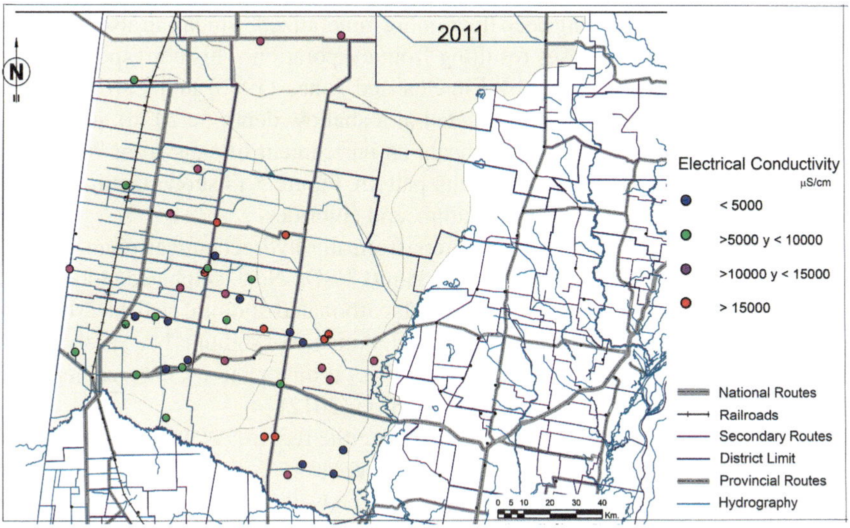

Fig. 5.13 Groundwater electrical conductivity in µS cm^{-1} in 2011 (Sosa 2012)

have a recharge at the foothill of Sierras Subandinas and discharge in the Golondrinas-Calchaquí system, and the Salado and Paraná rivers.

Based on the concept developed by Tóth (2000), water courses receive local groundwater flows. Intermediate flows drain into the Golondrinas-Calchaquí system and the Salado River, while the deepest ones would discharge into the Paraná River.

Thalmeier et al. (2019), with samples obtained from the entire basin, grouped the perforations according to their depth within the ranges < 15 m, 15-60 m, 60–150 m, and > 150 m and concluded that the hydrogeological characteristics of the deepest units (60-150 m and > 150 m) are variable. They also determined ascending (positive) vertical gradients, which suggests the potential contribution of lower to upper units.

5.8 Hydrogeochemistry

In the southern section of the "Bajos Submeridionales" basin, within the Santa Fe Province, waters were classified into four main groups based on anion and cation analysis. Groups were in decreasing order, sulfated-sodium; chlorinated-sodium, sulfate-chlorinated-sodium, and finally sodium bicarbonate (Sosa 2012).

Veizaga et al. (2019), with samples obtained from wells of the entire basin in 2017–2018, observed that the most numerous set belongs to perforations less than 30 m deep. Groundwater generally has a very variable degree of mineralization. The wide range of electrical conductivities recorded in shallow waters (450–50,000 µS cm^{-1}) can be the product of several factors related to: (i) waters with local recharges and low transit time, (ii) waters affected by evaporation processes, and (iii) finally, those discharges of ascending deep flows more mineralized, which can also be related to salt concentration processes resulting from evaporation. This corresponds to the hydraulic continuity described by Engelen and Jones (1986) and Carrillo-Rivera (2000). Very saline waters were detected at a shallow depth (< 20 m) and also in another set of deep samples that indicate an enrichment of salts from 75 m. The SO_4^{-2}, Na^+, and Cl^- ions follow a similar pattern. Figure 5.14 shows depth variation of electrical conductivity, sulfates, sodium, and chlorides.

Veizaga et al. (2019) classified waters from the hydrochemical point of view in five groups: chlorinated/sulfated-sodium(Cl/SO_4-Na), sulphated/chlorinated-sodium, chlorinated-sodium, chlorinated /bicarbonated-sodium, and bicarbonated /chlorinated-sodium, indicating the clear predominance of sodium (Na^+) as the dominant cation, and chloride (Cl^-) and sulfate (SO_4^{-2}) as major anions, in agreement with data for the Santa Fe Province, reported by Sosa (2012).

Cl/SO_4-Na type waters are found throughout the region. As we get closer to the border with Santa Fe Province, salt content increases. In the north-western Santa Fe Province, a set of samples with SO_4/Cl-Natype waters is recognized. The rest of the sampled surface waters corresponding to the Salado River and the Golondrina-Calchaquí stream system have an electrical conductivity between 540 and 16,000 $\mu S\ cm^{-1}$. It is the result of the different contribution of water from rainfall and that derived from what is mentioned in points 5.6 and 5.8, regarding the fact that the Golondrina-Calchaqui system functions as a discharge from the Sierras Subandinas (Thalmeier et al. 2019), and those are tributaries of the Salado River.

Gollán and Lachaga (1939) described salinity variation of the Salado River, typical of seasonal drier and wetter periods and by the discharge of background salty groundwater. Sosa (2012) and Veizaga et al. (2019) considered "background salty groundwater" as the regional or intermediate flows, discharging in downstream boundary conditions.

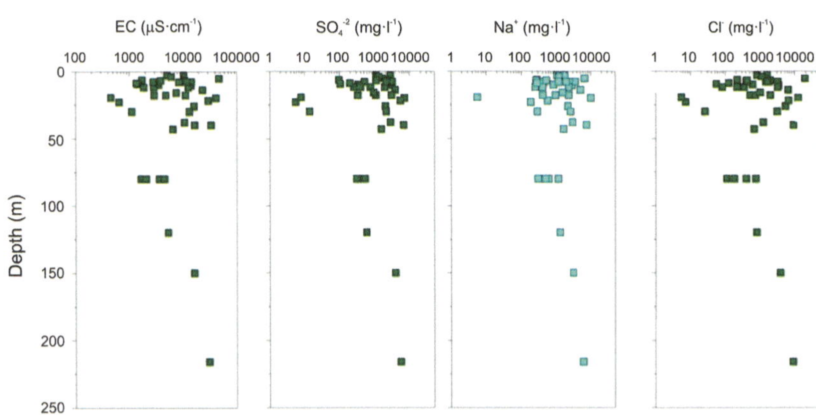

Fig. 5.14 Distribution of electrical conductivity (EC), sulfate, sodium, and chloride ion content with depth (modified from Veizaga et al. 2019)

The Salado River has mainly a chlorinated/sodic water type, with a considerable percentage of sulfates (Gollán and Lachaga 1939), coinciding with 2016/17 measurements published by Veizaga et al. (2019). Figure 5.15 shows some ionic relationships of interest of groundwater in the "Bajos Submeridionales" basin. The bulk of groundwater has a **rMg/rCa** between 0.1 and 2, an indicator of continental waters (Custodio and Llamas 1976). In boreholes deeper than 75 m, the ratio is greater than 1.5, whereas in those deeper than 220 m this relation exceeds 10. These high values indicate a mixture of fresh water with saline waters, from the Miocene marine entrance, stored in the Paraná Fm. On the other hand, the values of **rK/rNa** are consistent with those of the previous relationship; the deepest samples have values between 0.02 and 0.15, associated with marine waters. However, at depths below 50 m, values between 0.008 and 0.01 are found, which can be associated to both inland and marine waters or by a mixture of them.

rSO_4/rCl ratio was between 0.08 and 10, a range within which continental, marine waters, and mixtures are found. The sulfate ion, very abundant in the "Bajos Submeridionales," comes from the washing of lands formed in lacustrine-evaporitic environments from the oxidation of sulphides that are widely distributed in sedimentary rocks, from the decomposition of organic substances, etc. However, the dissolution of gypsum ($CaSO_4$, $2H_2O$) and anhydrite ($CaSO_4$) and other types of sulfates dispersed in the soil, would represent the quantitatively more significant contribution of this ion to groundwater (Custodio and Llamas 1976).

In Santa Fe Province, the first published classification of surface and groundwaters of 9 de Julio and Vera departments was carried out in the last century by Gollán and Lachaga (1939). These regional backgrounds were the only ones to date, with other studies found in more restricted areas (Bielsa and Fratti 1983), where the predominance of the Na ion is observed.

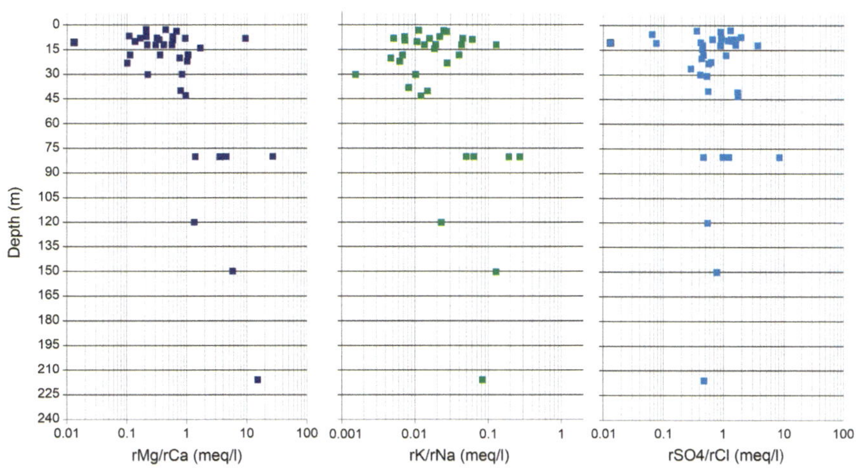

Fig. 5.15 Ionic relations as a function of depth (modified from Veizaga et al. 2019)

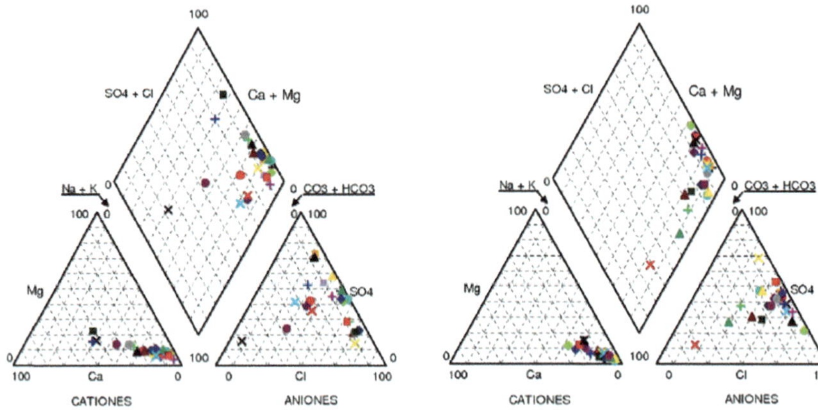

Fig. 5.16 Graphical representation of Piper of total sampled 50 water points (distributed in two graphics) at the first aquifer level from the "Bajos Submeridionales" (adapted from INA 2011)

INA (2011) sampled 50 water points at the first aquifer level, and the following determinations were made: Total solids, Total Alkalinity, Chlorides, Sulfates, Ammonia, pH, Nitrite, Arsenic, Nitrate, Electrical Conductivity, Calcium hardness, Sodium, and Potassium. The anions and cations diagrams of Piper (1944) from these samples are presented in two diagrams for better view in Fig. 5.16, where the predominance of sodium sulfate-chlorinated waters, or sodium chlorinated sulfates is observed. Approximately 80% of waters have sulfate and chloride as dominant anions.

In order to analyze the spatial distribution of the identified groups, Fig. 5.17 was done based in a Geographic Information System, where the diagrams of Stiff modified (Custodio and Llamas 1976) are presented. It can be clearly visualized that the *sodium cation* is the one that stands out in all water samples and *chloride and sulfate anions* are the predominant ones. The latter is the most problematic for cattle due to the laxative effect, when they exceed the limit value. An increase in salinities toward the east of 9 de Julio and the west of Ver*a* departments can be broadly observed.

5.9 Proposed Hydrogeological Conceptual Model

The "Bajos Submeridionales" basin was defined by Sosa et al. (2011) as "Non-typical Hydrogeological Systems" (NTHgS) due to the low hydraulic gradient, strong predominance of vertical over horizontal movements, storage tendency, and large hydrochemical variations over short distances, based on the concept developed by Fertonani and Prendes (1984) for these environment. These authors use the denomination of "Non-typical hydrological systems—NTHS" to identify plains with very

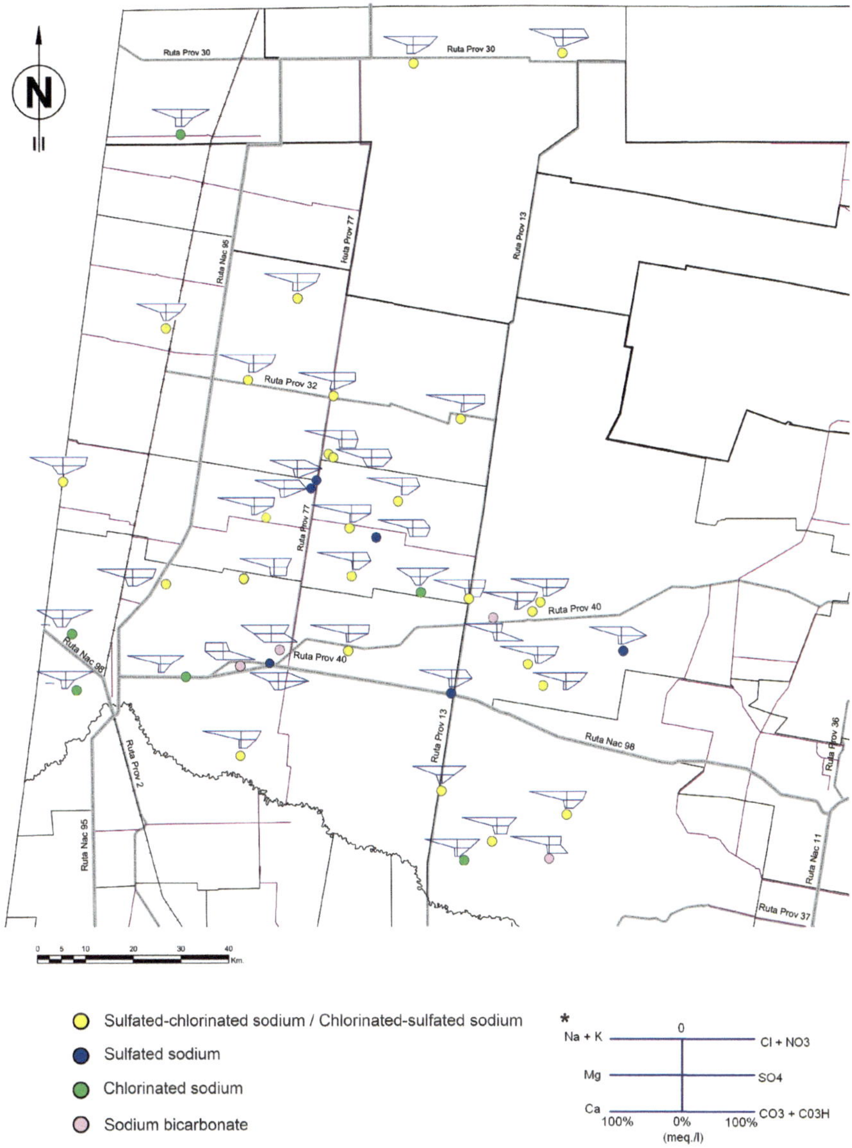

Fig. 5.17 Groundwater Stiff modified diagrams (adapted from Sosa 2012) *The meaning of the Stiff diagram is indicated

flat basin slopes, with limits difficult to define where surface waters have low kinetic energy and tend to be stored in lowlands.

To analyze the behavior of upper strata inside the same complex aquifer system, it is important to take into account the geomorphology of the region, where the work carried out by the river systems, which eventually resulted in numerous paleochannels, with silty sandy sediments, confer sectors of greater permeability and waters with lower electrical conductivity. Therefore, water coexists of different densities in an heterogeneous sedimentary system. These peculiarities gave rise to designate them as NTHgS. The salt variations found in the sediment-water system are also reflected in the resistivity contrasts from the application of geoelectric prospecting methods. Lower permeabilities of Quaternary sediments place those systems as aquitards. In addition, the system presents lower salt water stored in the more permeable sandy silt sediments, known as "freshwater pockets" and "paleochannels," and are the only ones that can be used for livestock production and human supply, in a vast sector of the Chaco Plain, and therefore they are of great impact on the productive resources of the region.

Recharge of local flows is caused by rainfall with a pulsating effect working as a vertical piston and diluting salt concentration inside the aquifer, in this way reservoir consists of waters of different qualities and densities, according to the Ghyben-Herzberg formula (Drabbe and Badon-Ghyben 1889; Herzberg 1901). Thus, there can be vertical upward or downward flow, depending on the hydraulic gradient that varies with climatic pulses.

The model proposed by Tóth (1963) is adopted as a conceptual hydrogeological model to understand the operation of the system (Fig. 5.18).

Local recharge occurs in sectors where the aquifer behaves as a free one in the western portion of the system. It is possible to infer, through the results of the chemical analysis, that local flows are associated with bicarbonate type waters, and intermediate flows may be associated with high chloride and sulfate contents (Tóth 1963).

Regional flow systems circulate in greater depth compared to local and intermediate flows, recharging in higher altitude (other regions) and discharging in areas with lower levels, so that several local flow systems can be found on one or several of the intermediate or regional type. Chemically, water from deeper flows has a high content of dissolved salts.

It is possible that in the Central Region of the "Bajos Submeridionales" in Santa Fe Province (see light green area, Central Depression in Fig. 5.2), part of the discharges of intermediate flows occur in the upper aquifer salinizing it, during periods with lower rainfall or pumping, due to vertical ascending flows.

When equilibrium conditions, resulting from water excesses, are modified, increases in groundwater levels are generated for extended periods, the interface between freshwater and saltwater shifts in depth. This mechanism can be compared as a "plug flow" (Sosa 2012).

To study the **interrelation of surface and groundwater flows**, major and minor ions were analyzed from samples extracted mainly from shallow lakes, and represented in a logarithmic diagram similar to that used by Schoeller-Berkaloff (Schöeller

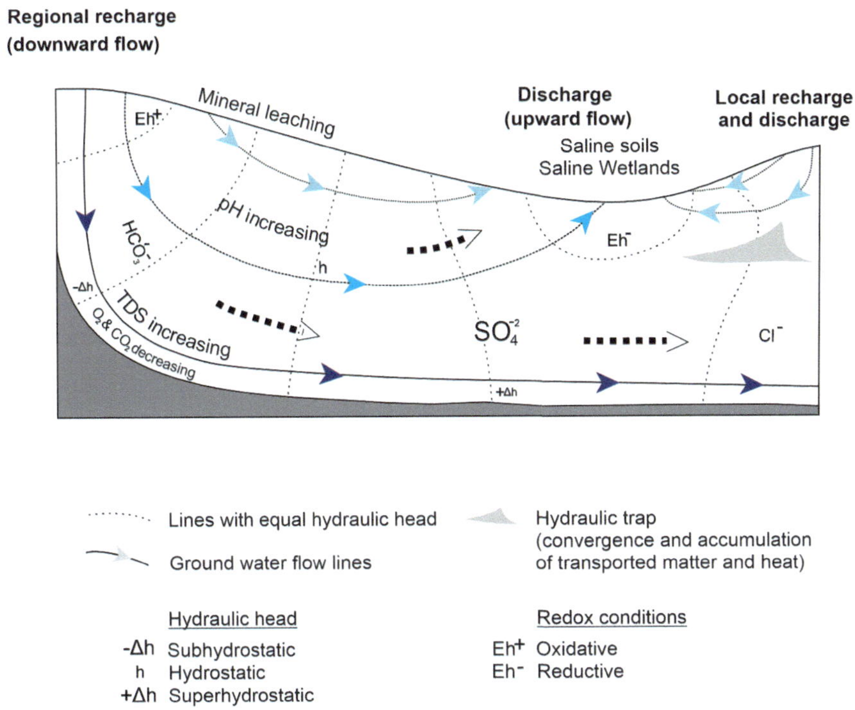

Fig. 5.18 Model of groundwater flows (modified from Tóth 1963)

1962) modified by Alconada-Magliano (2008), where the values in mg l^{-1} were represented on the vertical logarithmic axis (Fig. 5.19). La Loca shallow lake, although having lower electrical conductivity (539 $\mu S.cm^{-1}$) presents a higher concentration in minor elements Al, Si, Mn, Fe, Co, Ba, La, Ce, Pr, Nd, Sm, Eu, Gd, Tb, Dy, Ho, Er, Yb, Lu, Hf, and Pb, the same trend is observed in the components of all samples. La Tigra shallow lake, which has an electrical conductivity of 16,000 $\mu S.cm^{-1}$, has the highest concentrations for major ions corresponding to Cl^-, $SO_4{}^{2-}$, Ca, K, Na, Mg, and minor ones such as Li, Ni, Cu, and Mo. The trend of minor components, in lake samples, makes it possible to assume the presence of intermediate flows in these depressions that together with local flows and evaporation processes give rise to the physicochemical characteristics of waters, according to Sosa (2012).

Through a multivariate analysis on conglomerates of Euclidean distances (InfoStat® 2008, InfoStat 2008) using Dendrograms to compare chemical components in samples of water from wells and shallow lakes, it is possible to observe a strong bonding of chlorides, sodium, and sulfates, constituting the only differentiated family. Also applying the same statistical analysis of conglomerates to minor and major chemical components of shallow lakes, a strong relationship between the components is observed. Silica values represent a unique condition that does not

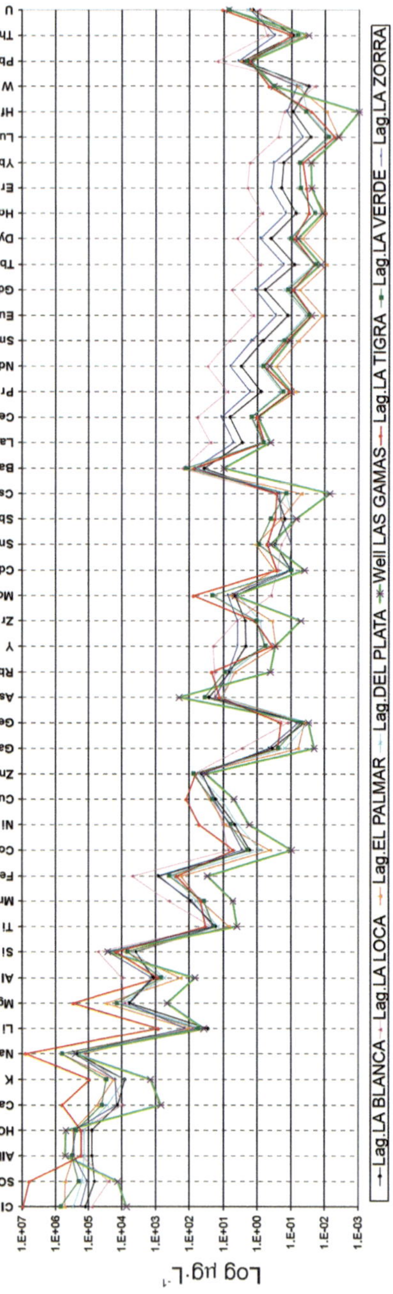

Fig. 5.19 Logarithmic diagram of major and minor ion concentrations in shallow lakes (Lag.) and one water well (Well) (Sosa 2012)

allow grouping with the rest of the variables, explained by its geothermal characteristic (Sosa 2012), and also concluded that water in shallow lakes has common elements with those found in the Puelches Fm. It is also observed that the water table in the "Bajos Submeridionales" is sodium chlorinated sulfate and few samples with bicarbonates presence, related to recharge of local flows.

An **isotopic** study of waters in the whole basin by Heredia Díaz et al. (2018), proposed a possible recharge zone of the hydrogeological regional system, a flow pattern, and the impact due to a possible geothermal behavior. Stable isotopes of water, deuterium (D or ^{2}H), and oxygen 18 (^{18}O), are a standard-use resource for characterizing hydrogeological systems. Based on the isotopic fractionation of the molecules due to their different molecular weight and the factors that influence it during the evaporation-sublimation, transport, and condensation of water throughout the hydrological cycle, characteristics of the recharge and circulation can be established, as well as flow and other features of hydrogeological systems (Kendall and McDonnell 1998).

The delta (δ) notation, δ2H or δ18O, indicates the isotopic deviation and is defined, δ = (—1) × 1000 (0/00) (1) where: and represent the 2H /1H or 18O /16O, the sample and the standard, respectively.

The relation δ18O δ2H allows identifying the possible origin of recharge of the underground flow or of the recharge water itself or isotopic enrichment processes (evaporation, geothermalism, etc.).

Samples obtained from boreholes of 15–70 m depth, in the northwest sector and most of them from the south and southeastern sectors, do not respond to conditions that could be explained by evaporative processes. This can be explained by a recharge in the subtropical foothills of the Sierras Subandinas, a circulation through the megafan of the Salado River, discharging intermediate flows in Santiago del Estero shallow lakes and in the Golondrinas-Calchaquí system (Heredia Díaz et al. 2018).

Groundwater samples in the northeast do not respond to an evaporitic scheme, but they do show some enrichment in ^{18}Heredia Díaz et al. (2018) concluded that shallow waters up to 15 m and some between 15–70 m, explain a local flow recharge in the north of the "Bajos Submeridionales." On the other hand, it is observed that those deeper than 70 m and some between 15–70 m, also present an enrichment in ^{18}O, which might be the result of an isotopic exchange from a possible deep geothermal system (regional/intermediate flow).

Heredia Díaz et al. (2018) analysed electrical conductivity and ^{18}O relationship (EC/^{18}O) which, allows to identify if, in addition to the evaporative-concentration, there is some other process that contributes to the enrichment of salts, such as hydrogeochemical evolution or the dissolution of evaporites, among others, and if the processes are common to other waters. They group them into three sets, two of which have a clear trend in the EC/^{18}O relationship.

These groups are:

Group 1 (Ev): constituted by all surface waters of the southeastern sector, with exception of La Tigra shallow lake, sampled in 2010/11 by Sosa (2012). It also includes samples of a waterlogged deflation zone and a shallow well during a wetter period, representatives of waters whose enrichment in salts is exclusively or predominantly due to evaporative concentration.

La Tigra shallow lake seems not to respond to an evaporative concentration of salts. A relevant methodological aspect is the hydroclimatic framework in which a study of this type is carried out in the BBSS. Because it is a system in which drier and wetter interannual cycles alternate, all with extreme years, and where "average" climate scenarios are not common. A paradigmatic case is the La Tigra shallow lake whose waters in the dry year 2010/11 (LT2016) clearly indicate a contribution of salts due to processes of acquisition of salts by transit, while in the wet year 2016/17 its enrichment only seems to be due to evapo concentration.

Group 2 (LNv): it is made up of all samples from boreholes between 15–70 m from the southeastern sector and some shallow ones from the south and one from the northeast. They are isotopically lighter, slightly evaporated or not evaporated waters, which suggest a fast local recharge.

Group 3 (OP): it is made up of all the waters of the northwest sector, almost all of the waters of the northeastern sector regardless of their depth, the shallow wells of the southeastern sector, and the saltwater shallow lakes of Santiago del Estero Province. The trend to salt enrichment may be mainly due to processes of acquisition by transit, hydrogeochemical evolution, and dissolution of evaporates rather than evaporation processes. The trend of the CE-δ18O relationship of this group indicates that in the salt enrichment of these waters compared to evapo concentration, the processes of acquisition of salts by transit prevail: hydrogeochemical evolution and dissolution of evaporites.

5.10 Final Considerations

The hydrogeological system studied has a high degree of complexity, due to the heterogeneity of the sedimentary column, transferring water with greater salinity to the upper level, caused by the alternative positive or negative piezometric gradient between bottom and surface aquifer (piston flow). So, vertical water movement is relevant and thus the "Bajos Submeridionales" can be considered a Non-typical Hydrogeological System, NTHgS. Hence, when occurring negative differences of piezometric loads, during several months, waters of greater salinity move toward the surface affecting soils, vegetation, and consequently, the productive capacity of the system.

Scale conditions of groundwater flow and the geomorphological characteristics of lowlands in floodplain, pose the existence of discharge zones with a long horizontal distance and a restricted presence of recharge zones inherent to local flows, which are strongly conditioned by the vertical circulation of water and turn out from the flow movements given by evaporation and hydroclimatic pulses.

Combined use of piezometric, hydrogeochemical, and isotopic measurements indicates a overall circulation associated with the alluvial mega-fan of the Salado River, with a general direction northwest-southeast, with hydraulic gradients between approximately 0.33-0.13 m Km^{-1} (0.033–0.013%), which recharges on the foothills of the Sierras Subandinas.

Water salinity is the result of mixing water from different origins, direct evaporation of free surface water and upflow of deeper salty water.

It is possible to infer, through the results of the chemical analysis, that local flows are associated with bicarbonate water type, and intermediate flows may be associated with high chloride and sulfate contents.

Finally, it is concluded that the conceptual flow model that can explain the functioning of groundwater in the "Bajos Submeridionales," and allows responding to the complex hydrogeochemical behaviors at a regional level, can be approached as the original model proposed by Tóth.

References

Alconada-Magliano M (2008) Procesos de inundación en el sector de médanos longitudinales al noroeste de la provincia de Buenos Aires, Argentina, su relación con la vegetación, suelo, agua y clima. Opciones de desarrollo. Unpublished doctoral thesis, for the degree of Postgraduate in Geography in Universidad Nacional Autónoma de México, México, 580 p, appendix 103p

Alconada-Magliano M, Fagundo-Castillo JR, Carrillo-Rivera JJ, Hernández PG (2011) Origin of flooding water through hydrogeochemical identification, the Buenos Aires plain Argentina. Environ Earth Sci 64(1):57–71

Ameghino F (1884) Las secas y las inundaciones en la provincia de Buenos Aires, 5° Edición. Ministerio de Asuntos Agrarios de la provincia de Buenos Aires, 62 p

Biasatti NR, Rozzatti JC, Fandiño B, Pautaso A, Mosso E, Marteleur G, Algarañaz N, Giraudo A, Chiarulli C, Romano M, Ramírez Llorens P, Vallejos L (2016) Las ecoregiones, su conservacion y las areas naturales protegidas de la provincia de Santa Fe. Ministerio de Medio Ambiente, 244 p. https://www.santafe.gov.ar/index.php/web/content/download/229660/1202209/file/LIBRO%20ECOREGIONES_web.pdf

Bielsa L, Fratti R (1983) Estudio de las condiciones químicas del agua superficial y freática existente en áreas de embalse para estimar la calidad del agua que operan las obras propuestas—Programa de desarrollo agropecuario para la región de los "Bajos submeridionales", provincias de Chaco, Santa Fe y Santiago del Estero. Consejo Federal de Inversiones, Buenos Aires

Canevari P, Blanco DE, Bucher E, Castro G, Davidson I (1998) Los humedales de la Argentina: clasificación, situación actual, conservación y legislación. Humedales para las Américas. Secretaría de Recursos Naturales y Desarrollo Sustentable de la República Argentina, Buenos Aires, Wetlands International, publicación N° 46, 208 p

Carrillo-Rivera JJ (2000) Application of the groundwater-balance equation to indicate interbasin and vertical flow in two semi-arid drainage basins. Hydrogeol J 8(5):503–520

Custodio E, Llamas MR (1976) Hidrología Subterránea. Ediciones Omega S.A

Drabbe J, Badon-Ghyben W (1889) Nota in verband met de voorgenomen putboring nabij Amsterdam, Tijdschrift van het Koninklijk Instituut van Ingenieurs, pp 8–22

Engelen GB, Jones GP (1986) Developments in the analysis of groundwater flow systems. IAHS 163:67–106

Feldman, I (2017) Recurrencia de sequías e inundaciones en llanuras argentinas.1. Revista de Divulgación Técnica Agropecuaria, Agroindustrial y Ambiental, Facultad de Ciencias Agrarias, Universidad Nacional de Lomas de Zamora, 4(1):56–66. http://www.produccion-animal.com.ar/inundacion/139-Feldman.pdf

Fertonani M, Prendes H (1983) Hidrología en áreas de llanura. Aspectos conceptuales teóricos y metodológicos. Actas del Coloquio de Olavarría de Hidrología de las Grandes Llanuras, Buenos Aires, Argentina, I, pp 118–156
Fertonani M, Prendes J (1984) Sistemas hidrológicos no típicos. Anales del Congreso del agua, Olavarría, pp 203–207
Fili M, Tujchneider O (1977) Características geohidrológicas regionales del subsuelo de la Provincia de Santa Fe (Argentina). Rev Asoc Cien Nat Litoral 8:105–113
GeoINTA (2020) Suelos de Santa Fe. Consulted June 2020 in http://visor.geointa.inta.gob.ar/?p=96
Giacosa R, Sosa D, Orue R, Castro S, Monteverde C, Macedo G, Maciel J; Nicuesa R, Tabernig D, Macedo S (2017) Relevamiento estado de obras de infraestructura hídrica-Bajos Submeridionales-área Provincia de Santa Fe. Unpublished final report, Instituto Nacional del Agua, Centro Regional Litoral, Consejo Federal de Inversiones, Gobierno de la Provincia de Santa Fe, 106 p
Giraut M, Laboranti E, Rey C, Fioriti M, Ludueña S (2001) Cuenca Propia de los Bajos Submeridionales. Creación de una unidad hídrica independiente. Seminario Internacional sobre manejo integral de cuencas hidrográficas, Rosario, Argentina, 8 pp
Gollán J, Lachaga D (1939) Aguas de la Provincia de Santa Fe. Ministerio de Instrucción Pública y Fomento, Instituto Experimental de Investigación y Fomento Agrícola Ganadero, Departamento de Química agrícola y Edafología, 385 p
Harris I, Osborn TJ, Jones P, Lister D (2020) Version 4 of the CRU TS monthly high-resolution gridded multivariate climate dataset. Sci Data 7(1). https://doi.org/10.1038/s41597-020-0453-3
Heredia Díaz J, Rodríguez L, Sosa D, Veizaga E, Thalmeier B, Moreno Merino L., ... & Castro S (2018) Indicios Isotópicos de circulación regional en el sistema hidrogeológico de los Bajos Submeridionales. XIV Congreso Latinoamericano de Hidrogeología, Salta, Argentina, 8 p. https://www.researchgate.net/publication/333024625_INDICIOS_ISOTOPICOS_DE_CIRCULACION_REGIONAL_EN_EL_SISTEMA_HIDROGEOLOGICO_DE_LOS_BAJOS_SUBMERIDIONALES
Herzberg A (1901) Die Wasserversorgung einiger Nordsee bader, J. Gasbeleuchtung und Wasserversorgung 44, pp 815–819, 842–844
INA (2007) Influencia de los cambios físicos y climáticos en el régimen de escurrimiento del Río Salado-tramo inferior. Ministerio de Asuntos Hidricos de la Provincia de Santa Fe,INA-CRL, UNL-FICH, INTA-EEA Rafaela. Giacosa R, Hammerly R, Zucarelli V, Morresi V, Sosa D, Bernal G, Picatto H, Paoli C, Monteverde C, Genesio M, Giorgi R, Tosolini R, Sapino V, Leon C, Chiavassa A, Perezlindo S, 286 p, 6 appendix, 7 maps
INA (2011) Comportamiento de las aguas subterráneas en el noroeste de la provincia de Santa Fe. Producción ganadera. CFI, Consejo Federal de Inversiones, 125 p. http://biblioteca.cfi.org.ar/wp-content/uploads/sites/2/2011/01/49231.pdf
InfoStat® (2008) InfoStat version 2008. Grupo InfoStat, FCA, Universidad Nacional de Córdoba, Argentina
InfoStat (2008) InfoStat, 2008 version. Manual del Usuario. Grupo InfoStat, FCA, Universidad Nacional de Córdoba. 1st edn. Editorial Brujas, Argentina, 334 p
Iriondo MH (2010) Geología del Cuaternario de Argentina. Museo Provincial de Ciencias Naturales Florentino Ameghino, 359 p. http://www.museoameghino.gob.ar/archivos/repositorios/138_descarga_95_version_digital_del_libro_geologia_del_cuaternario_en_argentina.pdf
Iriondo MH (2011) Aguas Subterráneas de la Provincia de Santa Fe. Primera Edición, Museo Florentino Ameghino, Santa Fe, ISBN:978-987-24637-2-4, 264 p
Iriondo MH, Kröhling D (1995) El sistema eólico Pampeano Museo Provincial de Ciencias Naturales Florentino Ameghino. Comun 5(1):5–45
Kandus P, Minotti P, Malvárez A (2008) Distribution of wetlands in Argentina estimated from soil charts. Acta Sci Biol Sci 30:403–409. https://doi.org/10.4025/actascibiolsci.v30i4.5870
Kendall C, McDonnell, JJ, (1998) Isotope Tracers in Catchment Hydrology, Elsevier, Amsterdam, 839 p
Kröhling DM, Iriondo MH (2003) El loess de la Pampa Norte en el Bloque de San Guillermo. Rev Asoc Argentina Sedimentol 10(2):137–150

Kröhling, D, Orfeo O (2002). Sedimentología de unidades loéssicas (Pleistoceno Tardío-Holoceno) del centro-sur de Santa Fe. AAS Revista, 9:135–154. ISSN:0328, 1159 p https://www.redalyc.org/pdf/3817/381747743003.pdf
Maiola O, Gabellone N, Hernández M (2013) (eds). Inundaciones en la región pampeana. Editorial Universidad Nacional La Plata, Argentina, 266 p. http://www.bibhuma.fahce.unlp.edu.ar/Record/aBIBLO38767
Padula E, Mingramm A (1968) Estratigrafía, distribución y cuadro geotectónico-sedimentario del "Triásico" en el subsuelo de la llanura Chaco-Paranense. Actas 3° Jornadas Geológicas Argentinas1, pp 291–331
Peri VG (2013) Caracterización morfotectónica de las Lomadas de Otumpa (Gran Chaco, Santiago del Estero y Chaco): influencias en el control del drenaje. Unpublished doctoral thesis, Facultad de Ciencias Exactas y Naturales, Universidad de Buenos Aires, 321 p
Piper AM (1944) A graphic procedure in the geochemical interpretation of water analyses. Trans Am Geophys Union 25:914–928
Rossello EA, Bordarampé CP (2005) Las Lomadas de Otumpa: nuevas evidencias cartográficas de deformación neotectónica en el Gran Chaco (Santiago del Estero, Argentina). Actas del 16° Congreso Geológico Argentino (La Plata, Buenos Aires), CD-ROM, Ponencia N° 210, 6 p
Schöeller H (1962) Les Eaux Souterraines. Masson et Cie, Paris, 1, 642 p
Scioli C, Espinosa E, Rodríguez L (2020) Chapter 1. Características climáticas. Manual de Buenas Prácticas para el Uso del Agua en Ganadería Bajos Submeridionales Santafesinos (BBSS). Unpublished report
Sosa DC, Genesio M, Diaz E (2000) Estimación de la conductividad eléctrica del agua de formación en sedimentos loéssicos y arenososl en la localidad de Tostado (Argentina). Actualidad de técnicas geofísicas aplicadas en hidrogeología. Editorial: Instituto Tecnológico Geominero de España. Editors Olmo Alarcón M, López Geta JA, ISSN: 84-740-388-4, pp 347–352
Sosa DC, Díaz E, Castro S, Genesio M(2011) Sistemas hidrogeológicos no típicos en la llanura chaqueña y las obras de recarga. VII Congreso de Hidrogeología de Salta
Sosa DC (2012) El Agua, Excesos y Déficits, en la Producción Agrícola de Secano y Pecuaria Dentro de la Cuenca Inferior del río Salado. Unpublished doctoral thesis, Universidad de la Coruña, Spain, 234 p. http://hdl.handle.net/2183/10158
Stappenbeck R (1926) Geología y aguas subterráneas de La Pampa. Unpublished manuscript, Obras Sanitarias de la Nación—Departamento técnico-Investigaciones técnicas. Biblioteca del Servicio Geológico Minero Argentino, SEGEMAR, Buenos Aires
Thalmeier M, Rodríguez L, Heredia J, Sosa D, Veizaga E, Díaz E, Marchetti Z, Castro S, Moreno-Merino L (2019) Patrones de Flujo Regionales del Gran Chaco Argentino: Región de los Bajos Submeridionales. VIII Seminario Hispano-Latinoamericano sobre temas actuales de la hidrología subterránea, Salta, Argentina
Thornthwaite, C. W. 1948 – An approach howard a racional classification of climate. Reprinted from the Geogrphical Review, XXXVIII: 55–94 WADSWORTH, F. 20
Tóth J (1963) Theoretical analysis of groundwater flow in small drainage basins. J Geophys Res 68(16):4795–4812
Tóth J (2000) Las aguas subterráneas como agente geológico: causas procesos y manifestaciones. Bol Geol Min 111(4):9–26
Tóth J (2009) Gravitational systems of groundwater flow: theory, evaluation, utilization. Cambridge University Press, Cambridge, UK, 297 p. ISBN: 978-0-521-88638-3
Veizaga E, Rodríguez L, Heredia J, Sosa DC, Moreno-Merino L, Thalmeier B, Díaz E, Marchetti Z (2019) Contribución a la Caracterización Hidrogeoquímica Regional de los Bajos Submeridionales en el Chaco Argentino. XIV Congreso Latinoamericano de Hidrogeología, Salta, Argentina

Chapter 6
Hydrogeochemical Characterization of Groundwater and Its Interaction with Other Components of the Environment in Mexico

Rafael Huizar-Álvarez and José Joel Carrillo-Rivera

Abstract Personal health and that of animals are often associated with the chemical composition of the groundwater they ingest. This primary source of water supply may affect the health status when significant changes in the concentration of some trace elements dissolved in drinking water are present. Indeed, adverse health effects occur due to chronic exposure to a high level of trace elements in drinking water. For example, groundwater consumption rich in arsenic or fluoride is causing severe and harmful health effects in broad sectors of the population in several countries. In Mexico, the quality of the drinking water supply is at risk due to water of an undesirable composition that rises to the extraction level of wells. This water inflow is with natural mineralization rich in certain trace elements that have been increasing with extraction time as well as with the obtained quantity; in other cases, there is a pollution effect by local inhabitants. The interest of this chapter is twofold: firstly, is to present different regions of Mexico with environmental and health responses related to groundwater consumption. The second is to emphasize the need to study the chemical evolution of groundwater based on the dynamic concept of the Tóthian groundwater flow systems.

Keywords Trace elements · Human health · Groundwater quality · Tóthian groundwater flow systems

---

Rafael Huizar-Álvarez—deceased.

R. Huizar-Álvarez (✉) · J. J. Carrillo-Rivera
Instituto de Geología, Universidad Nacional Autónoma de México, Circuito de la Investigación Científica S/N, Ciudad Universitaria, Ciudad de México 04510, México
e-mail: huizar@unam.mx

J. J. Carrillo-Rivera
e-mail: joeljcr@igg.unam.mx

M. M. Alconada-Magliano (ed.), *Intensified Land and Water Use*, Springer Earth System Sciences, https://doi.org/10.1007/978-3-030-65443-6_6

6.1 Introduction

Water is essential for human life through it the organisms consume and incorporate the nutrients they require to live. Nutrients are composed of different chemical elements, the biotic and abiotic processes move the elements from rocks, to soils, and water where they are incorporated into plants and animals and become parts of food chains. Likewise, these processes release the micronutrients that plants uptake from the soils, through acidifying or releasing organic acids, to the immediate root environment (rhizosphere) to liberate or chelate micronutrients and other trace elements into the proximal pore water (moisture) so they are available to plants, the utilization of a chemical element is solely dependent on its uptake into the living organism. Over the time, organisms have developed mechanisms for the uptake and utilization of elements that are more or less specific for each. Additionally, those elements that are not essential or even harmful to the organism are excluded usually in an efficient manner.

Elements are broadly classified as essential or toxic depending on their impact on human and animal health. However, elements, which are considered as being truly beneficial to human and animal health, may also lead to debilitating diseases if ingested in large doses, or if consumed in very lower doses than those required by the organisms. (e.g.,) the iron deficiency leads to nutritional deficiency that induce anaemia whereas high dietary iron intake or high iron stored are increased risk of coronary heart disease (Prashanth et al. 2015).

To understand the full context of the relationship between trace elements and the environment, it is important to obtain more interdisciplinary knowledge, and providing greater understanding of the mechanisms involved in both nutritional deficiency (need) of major elements, minor elements and the toxicity of trace elements The nutritional deficiencies can arise from lack of the essential elements in the drinking water or food. This can be due to lack of these elements in the soil where the food is grown or can be due to the eating habits of people. Toxicity might occur when high concentrations of metals in soil and drinking water lead to high exposure to metals. The potential risk of developing toxicity depends on the bioavailability of the specific trace element.

Groundwater is the main source of water supply in the world for personal consumption and for almost all economic activities. However, inadequate management can induce the concentration of certain dissolved ions to increase above the allowable limit, making water not recommended for consumption (World Health Organization 2010). This may occur when it is extracted by wells especially in an intensive manner. If ingested, in the long term, health problems might arise in the population consuming it. Groundwater may have a high content of trace elements for two reasons (i) dissolves the geological environment in which flows, and (ii) is contaminated by sewage infiltration.

There is technical and scientific information from different fields of knowledge relating to human health problems caused by the intake of water containing trace elements that exceed international and local regulations. In this chapter, a review and analysis of national studies regarding the hydrogeochemical characterization of groundwater and its environmental effects is established and confronted with international principles, particularly on some chemical elements and compounds in groundwater (mainly for $As^{3,5-}$, B^{3+}, $Cr^{3,6+}$, Cu^{2}, F^{-}, Fe^{2+}, $Mn^{2,3+}$, Hg^{2+}, Ni^{2+}, Na^{+}, Pb^{2+}, NO_3^{-}, SO_4^{2-}). When a population ingests groundwater under specific circumstances, with a high content of one more of these elements the risk of their health to be affected is to be taken with great concern.

In the interest of knowing, in which countries, the hydrogeochemical process of water is studied and analysed from the perspective of the theory of the Tóthian *groundwater flow systems* (TGFS), which is the approach proposed in this chapter. The TGFS also explains how the study of groundwater based on this theory assist in proposing a natural mitigation procedure for fluoride content in groundwater (Chebotarev 1955; Freeze and Cherry 1979; Tóth 1963, 1978, 1986, 1988; Edmunds et al. 2002; Carrillo-Rivera et al. 2008).

6.2 Geological Framework, Hydrogeology, and Health

It could be thought that there is no relationship between the geological framework and hydrogeology with human and animal health. However, knowing that rocks and minerals contain most chemical elements that occur in nature; it happens that several of these ions are essential in certain amounts for the health of plants, humans, and animals. The weathering of rocks and the biotic and abiotic processes form the soils where the plants and animals grow and feed; so, from soil water and solid phase of soil they consume different chemical elements. Soils may differ widely in their total concentrations of both macro and trace elements due to variations in the mineralogy of the geological parent material on which the soil has formed, even without inputs from environmental pollution or agricultural activity. The health problems of human beings, plants, and animals depend largely on natural factors, specifically on the sources of origin (rock, soil, air, and water). The chemical elements are acquired by the human body via food, water, and air. Excessive intake, or lack of some inorganic chemical elements (ions) mostly known as trace metals affects the health of people, animals, flora, and fauna (Selinus et al. 2005) ions are ingested and incorporated by the living beings in dissolved form, through food to consumed in water; during digestion the ions are dissolved and subsequently incorporated to the fluid system; they can also be ingested as supplements or airway, for example, derived from industrial processes or miner powders, or volcanic ash. Although there is sufficient historical data showing the effects on human and animal health by geological factors, which currently exist worldwide health problems associated with geological and hydrogeological factors. There is still little interest in increasing awareness of the relationships between these factors and the health of living beings.

World health problems caused by the production and handling of metals exist since ancient societies; the use of heavy metals by the population is today evident by its toxicity effects. Unlike the past, the relationship between various diseases and trace metals are well known today; lead, silver, cooper, iron have been widely used by the society since ancient times about 5000-year ago (BD), "e.g.," specific eras were named as copper, bronze and iron ages (Davies et al. 2005; Hong et al. 1994; Nriagu 1998). Poisoning with lead, silver, gold, and antimony have been documented in various Chinese dynasties 1000 (BD); Lead poisoning is reported in texts of the Assyrians (1550–600) (BD), and in the Egyptian medical papyrus Nriagu (1983). Different diseases of epidemic significance of lung, saturnism, sterility, mental disability, and still-births among others endured the Roman civilization, due to different uses made of lead and mercury (Nriagu 1983). Also in Central and South America (Silver and Rothman 1995) as well as in Egypt in the sixteenth century were used to treat syphilis (Fergusson 1990). Regarding arsenic this author mentioned that Greek, Roman, Arabic, and Peruvian societies used arsenic for therapeutic purposes, in fact it is used in homeopathic treatments.

From the industrial revolution, the population has been exposed to high levels of chemical elements and compounds as well as the intake in different forms (drinking, inhaling, by skin contact). The great technological development achieved in the different production processes determined that currently the bioavailability of different inorganic chemical elements is much greater than before, since all industrial processes generate and discharge liquids and solid waste in varying quantities and contents of the following ions: *antimony, arsenic, copper, chromium, fluoride, iron, mercury, silver, lead, selenium, and others*, which may damage the health of the population.

6.3 Contributions to Environmental Medicine by Earth Sciences

Most public health problems are associated with diseases caused by pathogens because they manifest themselves very fast; however, inorganic poisons also affect public health but takes longer time to be manifest and are more dangerous; these poisons include; *arsenic, chromium, fluoride, mercury, nickel, lead, silver, cadmium, and selenium.* Currently, great concern exists in keeping safe environmental levels of several trace metals and their compounds in various regions of the world especially in mining and transformation industrial regions (steel and petrochemical) among others, where serious health problems are presented but unfortunately little is known; when it is known solutions do not become seriously proposed due to the cover-up of Governments. Therefore, the geological framework occupied a substantial place in the nature and occurrence of trace elements, to assess its content and their bioavailability in the rocks and soil and relating its effect on the health of the population and animals. For this reason, it is possible to speak of *medical geology* (Davies et al. 2005, Dissanayake and Chandrajith 2009). The hydrogeology explains the water-rock interaction, the dissolution of trace elements in water from where their impact on human health may be established. Considering that human beings incorporate (ingest) the chemical elements in mainly dissolved form. Then, the role of hydrogeology in the context of health is fundamental, which together with other sciences intend to understand, mitigate, and solve health problems and assist in protecting the environment.

This indicates that addressing the problem of human and animal health from an *environmental perspective* should also be considered, in addition to medical geology and hydrogeology, geography, pedology, atmospheric and hydrogeochemical science, which also contribute knowledge to the health problem of living beings. It can be considered that these disciplines complement each other with environmental medicine. Then, in this chapter it is preferred to use a more integral term, *Earth Sciences*. According to Jefferson et al. (2005), *Environmental medicine* is the study of how the environment affects health, including the practice of how to minimize/or prevent adverse effects. This author specifies that the term environmental medicine is a synonym for the term *environmental health,* but the latter is often confused with "health of the environment", so in this chapter the previous term will be addressed.

6.4 Biological Importance of Trace Elements

In environmental health it becomes necessary to consider natural heavy metals and, of course, their physiological functions in the health and life of living beings. It is to consider that a deficiency, excess, or prolonged exposure to some ions cause health problems in organisms. This leads to wondering at what time an element is or is no longer essential to the life of living beings? According to Lindh (2005) an *element is essential* when: (i) is present in living organisms at relatively constant concentration, (ii) when it causes similar structural and physiological abnormalities in several species if it is removed from their organism, (iii) those anomalies are prevented or heal by supplementation of the same element. In this regard, Metz (1998) proposes, that it is essential for an organism when: (i) to reduce his exposure below a certain limit causes a consistent significant decrease of a physiological function, or when (ii) form an integral part of an organic structure that performs a vital function in the organism.

Essentiality of Elements. The concept of essentiality has the practical consequence that it is necessary to supply an organism with adequate amounts of the concerned elements. An immediate question raised by this consequence is *how much is adequate*?. Considering that fulfill a vital function in the organism, its deficiency can be supplemented with add-ons. However, for now, both the number of elements and their concentration in living beings still depend on experimental techniques to determine their essentiality, therefore, as experimental techniques improve, additional essential elements can be considered, and have greater accuracy of the concentrations required by the organism.

It is known to be approximately 11 known elements that are present in all biological systems in a constant and predominant manner (Table 6.1a). The human body is formed about 99.9% of the 11 elements, but only 4 of them—hydrogen, carbon, nitrogen, and oxygen—account for 99% of the atoms, or just over 96% of the body mass. These 4 elements, the major elements, comprise the bulk of living organisms. The remaining 7 elements are the minor elements—sodium, magnesium, phosphorus,

Table 6.1 (**a**) Major and minor elements in the human body by order of abundance and (**b**) Certain essential trace elements in the human body by their abundance in the human body (Modified from Lindh 2005; Hollabaungh 2007; Liy 2000)

Table 6.1 (a)	
Element	Element
Oxygen	Magnesium
Carbon	Potassium
Hydrogen	Sulfur
Nitrogen	Sodium
Calcium	Chlorine
Phoaphorus	
Table 6.1 (b)	
Element	Element
Silicon	Manganese
Iron	Nickel
Fluorine	Selenium
Zinc	Vanadium
Bromine	Chromium
Copper	Molybdenum
Arsenic	Cobalt
Tin	Lithium
Iodine	Tungsten

sulfur, chloride, potassium, and calcium. Comprise 3.78% of the body mass. One group of elements has still to be defined: the trace elements (Dissanayake and Chandrajith 2009). The entire group of noble gases is excluded from consideration because their chemical properties make them unlikely to fulfill any biological function.

According to Lindh (2005) of the 90 natural elements of the periodic Table 73 are trace elements. Of the 73, 18 are considered essential or possibly essential trace elements (Table 6.1b).

From Table 6.1b, there are still problems to prove the essentiality of some trace elements, since they are associated with health problems that should appear when completely removed from the diet of an organism, (e.g.,) growth retardation or loss of hair. The condition should be alleviated by supplementing the trace element and reverse the deficiency state. A first technical problem is that it is not possible to completely eliminate every element in food. It is not even possible because analytical techniques are still inadequate. Another problem is that to assess its essentiality, there is no well-founded hypothesis about its possible biological function. Removing the essential element of the diet may result in altering the absorption patterns of other trace elements, causing the results to be ambiguous. The results have been obtained in plants and animals. However, for obvious reasons in the case of humans, knowledge of essential trace elements is less advanced. According to Lindh (2005) and Michabata et al. (2002), in the essentiality of 12 of the trace elements of Table 6.1b, there is general agreement although perhaps not for all biological species.

All of the above shows the goodness of nature provided inputs that make possible for all forms of life on this planet through water, air, minerals, and soil. In this way it was possible the formation and evolution of the large and very diverse ecosystems in which man dwells. That in turn became its biggest predator. Modification and transformation of nature has not ceased since the emergence of the first sedentary villages when the exchange started and later the trade of natural goods and man-made products. Thus according to Smith (1990), and Harvey (1994) human activity produces social space, destroys it, transforms it, and constantly reorganized the creative destruction of the territory and its reconstruction in a different way that has led to serious environmental affectations in the natural assets, in the case of water and groundwater in particular, have affected its dynamics, its chemical and bacteriological quality in various regions of the world.

Increasing numbers of countries in the world recognize the need to manage natural goods (water, soil, vegetation, and air) in a more sustainable way. It is necessary to think about the pressures on those assets and adopt policies that help make sustainability a reality. Regarding groundwater, natural processes should be determined with systemic approaches that help to understand its operation and, at the same time, to mitigate the risk of contamination, to know which environments are most vulnerable and how they should be managed to conserve them for future usage. One characteristic of groundwater is that pollution usually takes a long time to appear on a water source, often decades or hundreds of years. In consequence, it is technically difficult and expensive to clean groundwater once it is contaminated, in many cases, deterioration has been proved to be already irreversible.

6.5 Environmental Impacts of Groundwater in Mexico

Despite its importance, groundwater is used inappropriately in Mexico; its chemical quality is threatened by pollution and intensive extraction; further water quality is not considered in groundwater or surface water evaluation. In Northern Mexico, groundwater constitutes 99% of the water supply, whereas depending on the geological conditions, in temperate areas could be about 70 and 50% in warm regions. Groundwater represents more than 80% of the supply required for all uses in Mexico (Comisión Nacional del Agua, (Conagua) 2017). Sustainable development must keep a balance between ecosystems and groundwater-surface water functioning. The knowledge of groundwater functioning as the axis of development has been ignored in Mexico, thus avoiding giving its inhabitants the inalienable right to a healthy environment, in which its quality depends directly on water. In this sense there are many recognized environmental responses related to a lack of knowledge on how groundwater functions and its inappropriate usage, these responses have been characterized and classified into two forms of impact on the environment (Carrillo-Rivera et al. 2008), (i) one whose environmental causes can be argued as invisible since population does not conceive them as cause and effect and (ii) other environmental impacts referred to as visible, are mostly observed and recognized by the population. Both are the result of an administration devoid of scientific technical foresight on environmental responses to groundwater extraction, without recognizing how this is linked with other components of the environment, and therefore suffers from the lack of a comprehensive understanding of the dynamics of water. The continuation of this vision in groundwater management will have most serious consequences and large geographic environmental impacts.

6.5.1 Change of Chemical Groundwater Quality

To fight against pollution any program must begin by clearly knowing the problem to be solved; however, this depends on the quantity and quality of the available information and how such information is interpreted. In Mexico there is some public data on water quality since 1990. However, the way in which the Government quantifies it varies, causing great difficulty in knowing how the chemical quality of groundwater has evolved and even what its exact original state was. In addition, there is a lack of visibility on the cause that raised the problem, in both quantity and quality terms, such information is hardly available to society. Also, the amount of evaluated parameters is negligible and monitoring has not been constant over time and space. Water supply of 95% of Mexico's population comes from a groundwater source whose natural mineralization of certain elements has been increasingly surpassing the official regulations having anthropogenic contamination effects. In both situations, the change in water quality consumed by the population is related to the way the water extraction is administered, in particular, pumping time of the well and the extraction yield. In other words, ignoring the integral functioning of groundwater prevents efficient well operation programs consisting of establishing the extraction flow and the respective pumping periods, thus avoiding extracting groundwater with a greater mineral content, not suitable for human consumption (Edmunds et al. 2002; Cardona et al. 2004; Huizar-Alvarez et al. 2004; 2016b; Carrillo-Rivera et al. 2002, 2007, 2008; Cardona et al. 2018; Ortega-Guerrero 2017). Since 1981, the water authority in Mexico (Conagua), reports that aquifers "over-exploitation" induces saline water intrusion decreasing the quantity of available groundwater and deteriorating its chemical quality, suggested by the content of Total Dissolved Solids (salinity) avoiding considering the presence of trace element in the water; However, in most cases the generated information is scarce, deficient and instead of reflect a solution of problem and advances in management it manifests failure and growth of the problems. In 2017, there were 18 coastal aquifers with what is termed as sea water intrusion, 32 aquifers with brackish water associated with saline soils and the easy dissolved evaporitic minerals. Also, Government recognizes that there are aquifers recharged intentionally with wastewater, this provides a variety of pollutants in the water supply; however, finding and applying solutions to these problems and their monitoring is poor (SEMARNAT 2002; De Alba 2004; Jiménez-Cisneros 2007; Carrillo-Rivera et al. 2008).

As a result of the large groundwater volumes required to be used in Mexico, drinking water quality is at an increased risk of unwanted water contamination rising toward the well extraction level. According to the parameters set out in the drinking water standard, NOM-1247-SSA-1996, SSA (Secretaria de Salud) (1996) water quality represent a health risk for people in various regions of Mexico with the presence of arsenic, cadmium, which are carcinogenic; chromium and manganese produce a neurotoxic effect; fluoride causes deterioration to the bone and teeth system; sodium causes high arterial hypertension; nitrate and chloride are indicators of microbial contamination in well water; lead affect central and peripheral nervous system as well as sulphate, organic compounds, aromatic hydrocarbons, solvents, among others. According to (Conagua 2006), in some sectors of most urban centers of the country the presence of wastewater contamination in drinking water of these cities is known.

In Fig. 6.1, several regions of environmental impact in Mexico related to excessive groundwater extraction are shown. In addition to Fig. 6.1, an extensive summary of Arsenic, Fluoride, Iron-Manganese, and Pesticides in groundwater in Mexico that are documented in the country is presented. Respect to other trace metals, and major inorganic ions only the areas where they have been reported by unpublished documents are indicated, such as, the mining and hydrocarbon-related areas.

6.5.2 *The Case of Arsenic in Groundwater, Mexico*

For several decades the consumption of water containing arsenic above 10 micrograms per liter ($\mu g\ l^{-1}$), is a public health problem worldwide. Arsenic affects various organs of the human body. The most affected are: the skin, liver, vascular system, bladder, lung, developing foetus, nervous and endocrine systems. It causes irreversible damage that incapacitate a human for work, and even may cause death. Some diseases are: palmoplantar keratoderma, sensory peripheral neuropathy, peripheral vascular diseases, and cancer. Among the most affected countries are Argentina, Chile, China, Bangladesh, Bolivia, USA, and countries of the European Union (Spain, Greece, Italy) as well as India, Nepal, Peru, Pakistan, Rumania, and of course Mexico, among others (Bhattacharya et al. 1997; World Health Organization 2010).

6.5.2.1 Arsenic Levels in Drinking Water in Mexico

Arsenicosis is a chronic disease In Mexico that in different degrees suffer more than two million inhabitants (Alarcón et al. 2012). Nationally, the Laguna region is the emblematic case of poisoning by the intake of groundwater with high arsenic content. Since 1960, serious cases of arsenicosis have been documented in that region, mainly the towns of San Pedro, Tlahualilo and Fco I Madero, where water intake may have up to 0.740 mg l^{-1} of total arsenic (Del Razo et al. 1990), and timely data reports up to 0.865 mg l^{-1}. Almost half of the national territory consumes water whose arsenic

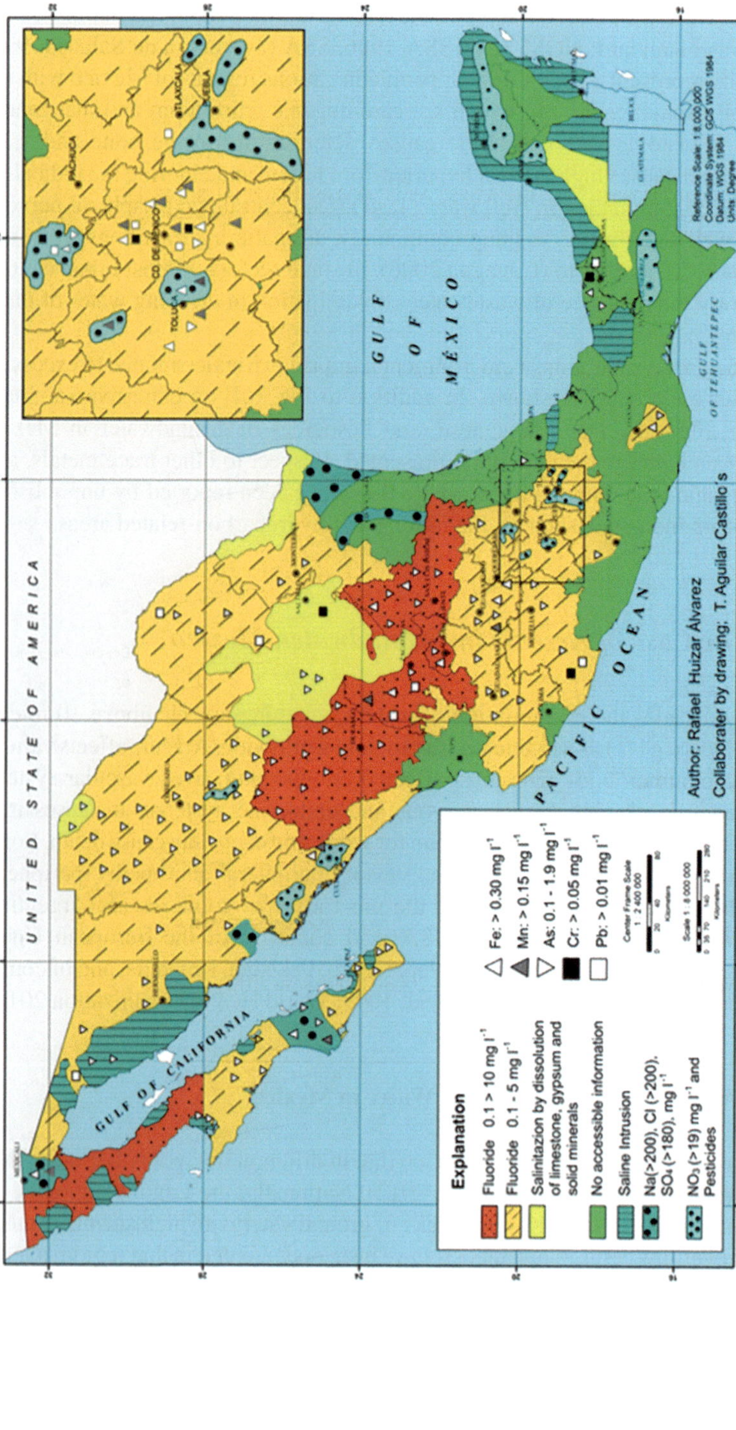

Fig. 6.1 Regions distribution in Mexico where high content of different ions present in groundwater are causing serious damage to the population's health.

content is equal to or greater than the permissible limit. In the following states: Aguascalientes, Baja California, Baja California Sur, Coahuila, Chihuahua, Durango, Guanajuato, Guerrero, Hidalgo, Jalisco, Morelos, Nuevo León, San Luis Potosí, Sinaloa, Sonora, and Zacatecas, there are many localities affected by arsenicosis at a different stage. Arsenic content in groundwater varies from 0.03 to 0.740 mg l^{-1} (Del Razo et al. 1990; Armienta et al. 1997, 2008; Arreguín et al. 2000; Planer et al. 2001; Carrillo-Rivera et al. 2008; Esteller et al. 2012; Irigoyen-Camacho et al. 2013, 2016). Out of these localities the cities of Juárez, Meoqui, Camargo, Delicias, and Chihuahua, have arsenic content in water rangings from 0.02 to 0.255 mg l^{-1}; Caborca, Sonora, 0.24 to 0.321 mg l^{-1}; Aguascalientes 0.24 mg l^{-1}; Rio Verde SLP, 0.54 mg l^{-1}; Guadiana Valley, Durango 0.24 mg l^{-1}; Guanajuato 0.12 mg l^{-1}; Hidalgo 1.31 mg l^{-1}; La Laguna region where it varies from 0.23 to 0.865 mg l^{-1}, and Zacatecas with 0.3 mg l^{-1}. There are also many rural population sufferings. The above-mentioned arsenic content is higher or slightly less than 25 μgl^{-1} which is the limit sanctioned by the Mexican standard NOM 127 SSA-1996 and greater than 10 μgl^{-1}, limit proposed by the WHO (2001). Groundwater consumed by the population of the mentioned areas is extracted in wells drilled in different lithology, mainly in felsic and intermediate volcanic rocks, sedimentary rocks encasing veins mineralized ridges with high presence of mineral sulphides, clastic rocks, and hydrothermal fluid zones.

Mining areas are an important source of arsenic to groundwater; in Mexico there is a very large number of mining sites; however, the arsenic content in the water that the population associated with the mining activity drinks has only been documented in some areas of the following states where the average values of arsenic content are Aguascalientes 0.23 mg l^{-1}; Baja California Sur 0.41 mg l^{-1}; Chihuahua 0.65 mg l^{-1}; Sonora 0.30 mg l^{-1}; Guanajuato 0.26 mg l^{-1}; Zimapan up to 1.31 mg l^{-1}; and San Luis Potosí 6.76 mg l^{-1} (Del Razo et al. 1993; Carrillo-Chavez and Drever 1998; Conagua 2000; Armienta et al. 1997; Niparajá 2005; Mejía-González et al. 2014; Ortega-Guerrero 2017). For the rest of the territory there is only unpublished information that reports a high content of arsenics in the water consumed by the population greater than that allowed by the standard. Also, in Mexico there are numerous sites where thermal groundwater discharges as well as hydrothermal areas, the latter (Los Humeros Oriental, Los Azufres and La Primavera) are used to generate electricity; arsenic content in water of these three sites are: 5 mg l^{-1}; 3.9–24 mg l^{-1} and up to 20 mg l^{-1}, respectively. Regarding the arsenic content in the water consumed by the inhabitants in the vicinity of these hydrothermal deposits it reaches 2.2 mg l^{-1} (Birkle and Merkel 2000; Venegas et al. 1991; Sánchez-Díaz 2007).

In Mexico, the groundwater chemical deterioration from the presence of arsenic and fluoride is a serious problem; unfortunately there are increasingly new areas affected and with the potential to be due to different causes, some of these are: (i) there is no preventive assessment of the chemical quality of water at the regional and local level, (ii) lack of constant updating of hydrogeological studies, (iii) the poor operational programming of extraction wells where the chemical evolution of water vs extracted volume and pumping period is considered. This requires water management institutions in different communities to be formed together with different parts of the society, to construct an efficient team to implement solutions providing the population with water with chemical quality without presenting a risk to health.

6.5.2.2 Sources of Arsenic

Arsenic is a mineral found in the form of granular massif frequently with some antimony, traces of iron, silver, and bismuth. It occurs associated with silver, nickel, cobalt in metallic ores, and is also found in other minerals in whose composition it is part, mainly in mineral sulphides, sulphates, mineral oxides, and silicates. Among the most common minerals, arsenic is found in pyrite, arsenolite, arsenopyrite, galena, blend, and rejalgar. In rock minerals: igneous, volcanic, mineralized reefs (silicates, biotite, amphibole, pyroxene, feldspar, and olivine); in sedimentary, and clastic rocks (mineral oxides: oxyhydroxide, hematite, magnetite, manganese oxides) and in hydrothermal system fluids (Dana and Ford 1986; Smedley and Kinniburgh 2005).

In nature inorganic arsenic occurs in the form of oxianions, such as trivalent arsenite (As(III)) H_3AsO_3, or pentavalent arsenate (As(V)) $H_3AsO_4^-$. Under oxidizing conditions at pH less than 6.9, arsenic is present as $H_3AsO_4^-$ mainly. While under reducing conditions and with pH lower than 9.2 arsenite species H_3AsO_3 predominates. The existence of one or the other species of arsenic in aqueous medium depends fundamentally on the conditions of the redox potential (Eh) and pH of the water. Then the mobilization of arsenic in water is especially favored by pH, depending on the relative oxidizing and reducing conditions (Smedley and Kinniburgh 2002). From the dissolution, adsorption, desorption from the rocks is incorporated into the groundwater, also by ion exchange of minerals is the mechanism responsible, mainly of metal oxides and also by evaporation. In natural conditions values vary a lot from environment to environment (Smedley and Kinniburgh 2005; Ortega-Guerrero 2017). The arsenic concentration in groundwater is controlled by water-rock interaction within the related flow system as it is released from the solid phase, and with transport phenomena and dispersion in specific geological environments. Sulphate-reducing acid conditions favor the precipitation of pyrite, oropimente, or other sulphide mineral containing co-precipitated arsenic; so, if there is a very high concentration of free sulphate, a high content of arsenic in water cannot be expect (Bhattacharya et al. 1997; Nickson et al. 2000; Ravenscroft et al. 2001; Smedley and Kinniburgh 2002; 2005; Alarcón-Herrera et al. 2012; BGS-DPHE 2001; Mejía-González et al. 2014; Razo et al. 1990; Ortega-Guerrero 2017; Navarro et al. 2017).

There is a group of authors who have documented the arsenic response in human health. Most authors agree that aquifers with arsenic producing problems are mainly those consisting of young intergranular (alluvial) sediments that are most vulnerable to the development of groundwater with high arsenic content at regional scale. They agree that the main sources of arsenic are those aquifers, because there sediments and clays present the most favorable hydrogeochemical conditions to release it through the processes of oxide-redox, dissolution, desorption, and adsorption, which control the arsenic behavior. These reactions occur near the water surface and are considered important in chemistry controls of trace elements in groundwater. Iron, aluminium, and manganese oxides are relatively abundant in most sediments and are commonly produced by wear of primary minerals in freshwater sediments. These oxides are the most abundant in the granular environment and are the best arsenic adsorbents (Manning and Goldberg 1997).

Smedley and Kinniburgh (2005) accept that geochemical processes that release arsenic into groundwater so far considered, in themselves, are not sufficient to explain the distribution of high concentration of arsenic in groundwater in many parts of the world. They consider that the lithology of most aquifers from which drinking water is extracted have several hundred million years of age and contain groundwater that may be several thousand years old. Therefore, fresh water has flowed through them in a great volume throughout the aquifer history. In young aquifers that have groundwater actively flowing occur a similar situation (i.e.), a great volume of water has been flowing through them. Conversely, in alluvial and delta aquifers constituted by relatively recent sediments where the hydraulic gradient is small, groundwater flows slowly becoming relatively old water. This allows for a longer water-rock interaction time resulting in geochemical processes acting longer and therefore the arsenic content in the water increases. In turn, they questionably consider that the high content of arsenic in groundwater presented at regional level is associated with local flow systems, where geochemical and hydrogeological conditions which release and retain arsenic, respectively, are combined; however, time for the required reactions to take place are missing (i.e.), proper pH and Eh conditions.

6.5.2.3 Groundwater Flow System and Arsenic

From the point of view of the geological and hydrogeological context of the entire Mexican territory, groundwater extracted anywhere in the country flows through at least two of the following types of rocks: volcanic, igneous, metamorphic, sedimentary, mineralized reefs, which constitute the diversity of the country's relief with its different elevations, which may be interdigitated or interstratified with clastic materials ranging in sizes from clay to conglomerate that are forming the filling of valleys, and plains. Based on the above, the groundwater recharge, throughflow, and discharge areas may be located at different elevations and different lithological relief. Discharge areas occur in the topographically lower sector; implying the presence of a groundwater flow system; that is, even if the lithology has a small hydraulic conductivity value suggesting that groundwater might not flow; however, discharge is always attesting; such movement is supported by its chemical evolution of the flowing water which according to Chebotarev (1955), Freeze and Cherry (1979), Tóth and Corbet (1986) the chemical composition of groundwater evolves from the recharging site to the discharge site. For this reason the water extracted in plains and valleys may have high content of Total Dissolved Solids and some trace elements. This allows to say that the high arsenic content or any other ion in the extracted groundwater in a plain does not come exclusively from the clastic material where the water is extracted. Since the rock that is the source of the classic sediments will be located some distance outside the plain, also could provide arsenic. Then, a percentage of arsenic is dissolved in the groundwater and increases from the recharge zone to the discharge zone. In situations like this, it is important to have the geological and hydrogeological model and the constructive design of the wells and thus know what the lithology which could mainly supply water to the well and what is its chemical composition. There are occasions, when wells drilled in alluvial material provide water that comes mostly from fractured rocks that are underlying or interdigitate with them.

Considering all the above, it is necessary to note that the incorporation of arsenic into groundwater has not been analysed within a totally integral and systemic point of view. That is, considering from the recharge to the discharge zone, under the perspective of the groundwater flow systems in its geological context (Tóth 1999) which reflects that the ions are incorporated and increasing their concentration in the flow system without reaching saturation, since during the longitudinal and vertical travel of the water the process of water-rock interaction results in the geochemical reactions that determine the sub-saturation in respect of that water mineral. However, they lead to increase in total dissolved solids concentration among these also some trace elements are included (Chevotarev 1955; Freeze and Cherry 1979; Tóth 1963, 1986; Carrillo-Rivera et al. 1996; Edmunds et al. 2002; Ortega-Guerrero 2009; Huizar-Alvarez et al. 2016a; Cardona et al. 2006, 2018).

6.5.3 Groundwater Fluoride, the Mexico Case

Fluorine (F^-) is considered essential for human health when ingested in small quantities; it is obtained in the form of fluoride (F^-) through drinking groundwater, which is the main route of exposure of human beings to this element. Excessive intake of fluoride can cause dental and skeletal fluorosis. It is estimated that some 200 million people in 25 countries are facing the problem of fluorosis due to excess of fluoride in the water sources; some of these countries are: Algeria, Australia, China, USA, Egypt, India, Kenya, Japan, Libya, Morocco, Mexico, New Zealand, Pakistan, and Uganda, among others (Bansiwal et al. 2009). In areas with endemic dental fluorosis drinking water comes from relatively deep wells. However, bottled drinking water and processed foods have been also a dangerous source of fluoride in the last 20 years (Betancourt-Linares et al. 2013; Smyth 1992; Armienta and Segovia 2008; Carrillo-Rivera et al. 1996, 2002; Huizar-Alvarez et al. 2016a; Cardona et al. 2006, 2018). In relation to the safety of the water, the WHO fluoride reference value is 1.5 mg l^{-1}. The levels greater than 1.5 mg l^{-1} are associated with dental fluorosis, where enamel loss (fluoride stings) and interaction with the bone appears; above 10 mg l^{-1}, paralyzing skeletal fluorosis appears (Irigoyen et al. 2013). On the contrary, if adequate amounts of fluoride are used, they could prevent tooth decay. Thus, since the 1950s, salt fluoridation has been promoted as a public health measure, disregarding the fluoride content in the water supply. Despite more than 50 years of research, it has been difficult to determine an optimal water fluoride concentration WHO (2015) recommended that it be 0.5–1.2 mg l^{-1}, while (USEPA 2011) proposes 0.7–1.2 mg l^{-1}, depending on the weather conditions and the amount of water consumed per day in the region. On the other hand Irigoyen et al. (2016) documented presence of dental fluorosis from moderate to severe by the intake of water containing 0.9–1.0 mg l^{-1} of fluoride. Regrettably, relevant diets are not given consideration.

6.5.3.1 Fluoride Levels in Drinking Water in Mexico

Dental fluorosis in Mexico is a public health problem, there are no figures available on the amount of population affected by fluorosis in Mexico; however, the National Dental Caries Survey documents a wide prevalence variation of fluorosis in different regions from (3.2 to 88.8%) (Bentancour-Linares et al. 2013). The more pronounced prevalence are the North, Northwest, Northeast, and Central states, where the natural fluoride groundwater concentration varies from 0.001 mg l^{-1}, to 20 mg l^{-1}, featuring high dental and skeletal fluorosis impact (Loyola-Rodríguez et al. 1998; Díaz-Barriga et al. 1997; Del Río 2001; Bonilla et al. 2002; Hurtado and Gardea-Torresdey 2004; Secretaria de Salud, 2004; Galicia et al. 2009; Carrillo-Rivera et al. 2008; Ortega-Guerrero 2009; García-Pérez et al. 2013; Varela-González et al. 2013; Cardona et al. 2018).

The most population affected by fluorosis live in localities in the states of Baja California Norte, Durango, Aguascalientes, Zacatecas, Jalisco, San Luis Potosí, Chihuahua, Coahuila, Guanajuato, Sonora, Puebla, Sinaloa, Michoacán, Queretaro, Mexico, Morelos, Nuevo León, Hidalgo, Guerrero and Oaxaca (Fig. 6.1) as well as in Mexico City; in the three latest States the prevalence of fluorosis is still low. In urban areas of Hermosillo, Sonora, fluoride concentration varies between 1.5 and 2.8 mg l^{-1}; in individual sources a concentration of up to 7.8 mg l^{-1}has been recorded (Valenzuela-Vázquez et al. 2006; Irigoyen et al. 2013). In Abasolo, Guanajuato State, an average content of 0.9–4.5 mg l^{-1} is recorded with a point high value of 16 mg l^{-1}. In Durango City, it is estimated that 95% of the residents are exposed to fluoride concentration from 0.5 to 10.2 mg l^{-1} (Fawell J et al. 2005; Hurtado and Gardea-Torresdey 2004; Carrillo-Rivera et al. 2008), reported in Meoqui, Chihuahua a concentration between 4.8 and 5.9 mg l^{-1}, in Los Altos de Jalisco region, in more than 40% of the municipalities the fluoride content in groundwater is > 1.8 mg l^{-1}; the town of Teocaltiche recorded up to 18.6 mg l^{-1}; Michoacan up to 16 mg l^{-1}; in Puente Grande, Zacatecas the fluoride concentration in drinking water reaches 13.2 mg l^{-1}.

This shows the Mexican government's little interest in providing good chemical quality water to the population, since NOM-127-SSA-1996, sanctions water quality for human intake but there is no a monitor system being implemented, there is not a permanent monitoring for at least the different ions which are hazard to the of health human beings from where the most appropriate corrective measures might be proposed. Of course there is insufficient support for research in this area of environmental and human health. As a result of government disinterest, the number of states and population affected by dental fluorosis instead of decreasing has increased a very worrying situation. Bentancour-Linares et al. (2013), in their dental study report that a third of the 32 states of Mexico are locations with high ICF (Community Fluorosis Index), while in the localities of the remaining states the prevalence found dental fluorosis was low and mild. They could not specify the main source of fluoride, but they considered that areas with high fluorosis are associated with high fluoride content in groundwater. This shows the need to carry out simultaneously dental and hydrogeochemical studies at a regional level, that could reveal the number of population exposed to water consumption with high fluoride, also to identify when the fluoride source is of a different intake; for example, when iodized and fluoridated salt are used?, or consumption of drinks with high fluoride content are the cause of the disease? For this reason to monitor the total consumption of fluoride in the child population and have update information on the main pathologies of the oral cavity it is necessary to make decisions based on scientific evidence to propose the most appropriate preventive program.

6.5.3.2 Fluoride Sources

Fluoride is one of the most volatile and characteristic element of volcanic and igneous rocks of felsic type; its average concentration in the earth's crust is 625 mg l^{-1} (Bailey 1977); the main controls of its concentration on the rocks are: the type of magmatic differentiation, and the temperature of the magma since the felsic rocks contain fluoride in greater quantity than mafic ones. In groundwater its molecules are shaped in highly variable concentrations, the latter depending on the type of mineral origin, the residence time of water in the rocks, pH, and temperature depending on the depth of movement (Nordstrom and Jenne, 1977).The ionic strength of the F^- also influences the solubility of the mineral candidate to disband in the ion exchange reactions (Apambire et al. 1997). The minerals that commonly provide fluoride are moderately soluble and release F^- to water slowly. Among the most common minerals which provide F^- to groundwater are fluorite CaF_2, fluorapatite (Ca_5 $(PO_4)_3$ F), amphibole, phlogopite, biotite, cryolite (Na_3AlF_6), villiaumite (NaF), topaz (Al_2 (SiO_4) F_2), triofluorite, hieratite, and some clays, where F^- replaces (OH^-) within the mineral structure (Hem 1985; Gaus et al. 2002; Ozsvath, 2009; Rao 2003; Handa 1975; Wenzel and Blum 1992). Although the weathering rate of micas and amphiboles is low, the F^- them is released from the hydroxyl position this increases the levels of dissolved F^- (Boul et al. 1981; Edmunds and Smedley 2005; Chae et al. 2007). These minerals are abundant in some volcanic, intrusive, metamorphic, crystalline, and sediments derived from the erosion of these igneous rocks.

The F^- content in groundwater may be increased to saturation level with respect to fluorite and precipitate, for example, when the water is close to the saturation level and its temperature decreases (producing saturation). Once the fluorite solubility limit ($Ca^{2+} + 2F^- \leftrightarrow CaF_2$) has been reached, there is an inverse relationship between Ca^{2+} concentration with respect to F^-, and positive between HCO_3^- and F^-. This indicates that if the content of Ca^{2+} in the rocks is lower than that of Na^+, the solubility of F^- in groundwater is favored since its solubility inside the magma is mainly associated with the Na^+ and K^+ cations (Handa 1975; Bardsen et al. 1996; Dhiman and Keshari 2006; Chae et al. 2007). Sánchez-Díaz (2007) reports that rocks of acidic composition contribute a greater quantity of F^- to groundwater in respect to intermediate and mafic rocks.

During dissolution of the constitutive regulatory minerals from rocks Na^+ content in the groundwater increases due to ion exchange, motivating a Ca^{2+} concentration decrease. Simultaneously, the content of F^- in groundwater tends to increase because there is a sodium-bicarbonate hydrochemical facies (NaHCO3). Saxena and Ahmed (2001) indicate that in the water-rock interaction, the water rich in $NaHCO_3$ accelerates the dissolution of minerals, releasing F^- into the water over time. Ortega-Guerrero (2009), documents the F^-enrichment due to the dissolution of fluorinated minerals with high content of Na^+ and HCO_3^- is product of albite (sodium feldspar) dissolution of volcanic rocks. Undoubtedly the chemical composition of groundwater is related to the chemical composition of the rocks it flows through. Martini (1984) and Gunnar et al. (2005) document that for a particular flow system (local, intermediate, or regional) the concentration of fluoride increases from the recharge zone reaching maximum values in the discharge zone. In addition, a differentiation of hydrochemical facies from $CaHCO_3$ to $NaHCO_3$, respectively, is also observed.

6.5.3.3 Attenuation of Fluoride Content in Groundwater

There are different procedures to decrease the fluoride content in water among others there are: (i) ion exchange resins, (ii) reverse osmosis, (iii) nanofiltration, and (iv) activated carbon. Each and every one of these processes is able to reduce the fluoride content to a limit sanctioned by health standards (López Paraguay 2013). However, attenuation of fluoride content in groundwater might be proposed through natural means (Carrillo-Rivera et al. 1996); this proposal is based on knowing the groundwater chemical evolution and temperature during extraction, in several pumping time intervals, and the safe intake of the population. In this regard Carrillo-Rivera et al. (1996, 2002, 2008), Varsányi and Kovács (1997), Huizar-Alvarez et al. (1998, 2004), Cardona et al. (2018), indicate that, by recording the temperature of the water in the extraction well in which the chemical composition of groundwater represents a mixture of intermediate and regional water flows, the concentration of fluoride close to the maximum allowed by the standard for drinking water (1.5 mg l^{-1}) may be defined by the relationship between fluoride content and temperature. Fluoride may be also controlled naturally if, for example, an enriched flow in fluoride is forced to travel through a lithological unit consisting of limestone debris or directly by limestone, conditions that control the presence of fluoride in the extracted water. Two important factors to consider at the well location are: the *geological framework* and the *variations in fluoride* concentration in groundwater with *depth and temperature*. Water management includes considering optimal pumping rates, especially where there is the possibility of mixing groundwater low in fluoride with fluoride-rich groundwater, e.g., old groundwater, or thermal, where the fluoride content will be increasing at high pumping rates. This allows to propose hydrogeological and geochemical control mechanisms for this and other trace elements that are a health risk, and program design pumping to regulate the percentage of different groundwater flows that supply, and thus mitigate and avoid strong environmental problems.

6.5.4 *Iron and Manganese in Groundwater, Mexico Case*

Iron and manganese are essential elements for living beings, but their intake after a certain concentration poses a health risk, so groundwater must have no more than the concentration recommended by the health institutions. Both elements are soluble in groundwater and oxidize by chlorination or in contact with water and air oxygen precipitates resulting in a dark colored water unpleasant to the population. Although manganese is not very toxic, its increased accumulation in tissue may cause toxic effects on the brain and lungs. It also causes neurotoxicity that is associated with motor and cognitive disorders known as manganism. The mechanisms underlying this toxic condition remain unknown, since clinical signs and symptoms are similar, but not identical, to those of Parkinson's disease (Ramírez and Azcona-Cruz 2017). The long-term effects of manganese compounds in people especially in the elderly should be controlled due to their known effects on catecholamine in the brain and their relationship to Parkinson's disease (Nordberg and Cherian 2005). In Mexico, the number of areas affected by the presence of iron and manganese in drinking water is unknown and in turn the health authority has implemented partial measures to mitigate this problem. In some cities water treatment plans have been installed, but the problem persists. At national level there are several regions where iron and manganese content in groundwater are higher than the limit sanctioned by the Mexican standard NOM-127-SSA-96. Some of these places are in cities as well as in agricultural or mining fields; among the first are: Culiacán, Camargo, Guaymas, Mexico (Texcoco, Iztapalapa, Tlahuac), Obregón, Novojoa and Zihuatanejo; the mining areas are: Molango, Zimapan, Veracruz, Zacatecas. In all these places, iron and manganese content have been recorded ranging from (0.01 to 2.3 mg l^{-1}) and (0.0001 to 3.72 mg l^{-1}), respectively (Lesser Illades and Sánchez-Díaz 1986; SEDESOL-INE 1994; Cardona and Hernández 1995; Cardona et al. 2004).

Iron toxicity may occur at high levels of intake. The average lethal dose of iron is 200–250 mg kg^{-1} of body weight. However, even an oral intake as low as 40 mg kg^{-1} of body weight has been lethal. Epidemiologic observations have linked high dietary iron intakes or high iron stores with increased risk of coronary heart disease (Salonen et al. 1992). A toxic potential of iron arises from its pro-oxidative effects, which yield reactive oxygen that attacks polyunsaturated membrane lipids, proteins, and nucleic acids. The iron overload disease, hereditary hemochromatosis, is caused by defect in the regulation of iron uptake, which leads to very high circulating transferring clinical signs that appear when body iron accumulates to about tenfold excess of normal, these include hepatic cirrhosis, diabetes, heart failure, arthritis (De Valk and Marx 1999; Schumann 2001).

6.5.5 Source of Iron and Manganese

Iron and manganese usually occur together but iron is more abundant. Both metals are part of some minerals that constitute igneous and volcanic rocks reach in olivine, amphibole, micas, and volcanic glass; as well as in lacustrine, alluvial sediments, and in soil where goethite and hematite are an important source of iron. These trace metals are widely used in the industry, manganese is particular to the steel industry, in the production of fertilizers, chlorine and medications among others. These products may then become possible sources of iron and manganese affecting groundwater quality. These elements come into solution in water when released by oxidation-reduction processes under pH control, and by dissolution. The reduction of goethite is an important source of iron to groundwater, in turn iron reacts with hydrogen sulphide to form sulphides (pyrite) that precipitate as part of the sediments form ferric hydroxide that may precipitate or be in suspension. Reduction of manganese oxide is the main source of manganese dissolved; as it does not oxidize so easily it may last sometime in the water and circulate for long distances. In Mexico City region, manganese increases in the direction of groundwater flow (Cardona and Hernández 1995; Lesser Illades and Sánchez-Díaz 1986; Edmunds et al. 2002; Huizar-Alvares et al. 1998; Stumm and Morgan 1981).

In the eastern, southeaster and northeaster areas of Mexico City the concentration of iron and manganese in groundwater are high (0.69-2.4 mg l^{-1}) and (0.43-3.72 mg l^{-1}), respectively, their origin is not yet fully identified, since they may come from: (i) the basaltic rock forming Sierra Santa Catarina; (ii) the lacustrine sediments where goethite and volcanic glass abound (Cardona and Hernández 1995; Díaz-Rodríguez et al. 1998; Huizar-Alvares et al. 1998; Edmunds et al. 2002); (iii) it is also highly possible that there is a source of anthropogenic origin, because in this area there are two ancient garbage dumps, buried Santa Cruz Meyehualco and San Lorenzo Tezonco. In addition, for over 32 years in the Sierra Santa Catarina consisting completely fractured and pyroclastic basaltic rocks might be artificially recharged with partially treated wastewater; therefore, that water must contribute with iron and manganese to groundwater. Likewise, organic matter that transports that wastewater must activate oxidation-reduction processes that contribute in their dissolution from the basaltic rock. This scenario makes it difficult to determine the main source of iron and manganese for groundwater in this sector of Mexico City (Cardona and Hernández 1995; Pitre 1994; Mazari and Mackay 1993; Macías and Mazari 2018; Huizar-Alvares et al. 1998). Likewise, Cardona and Hernández (1995), Edmunds et al. (2002), Huizar-Alvarez et al. (1998, 2016b) document the presence of an intermediate type groundwater flow system in the Santa Catarina well site, whose water due to heavy extraction is mixed with water from a regional flow that rises vertically in the area. The latter contains manganese and iron, which is manifested in the area of Iztapalapa and Texcoco. Therefore, it might suggest that it is a mixture of water induced by pumping involving two components: (i) downward flow of water from the aquitard and (ii) upward flow of regional water causing a mixture of water with high iron and manganese content. In fact Pitre (1994) and Huizar-Alvarez et al.

(1998), document groundwater contamination associated with surface wastewater and the presence of oxidation-reduction conditions that induce the dispersion and precipitation dissolution of iron and manganese.

In agricultural areas, groundwater may have a significant iron and manganese content due to: (i) the presence of natural sources of these ions contributed by water from intermediate and regional groundwater flow systems associated with a large water level drawdown in agricultural regions is not ruled out (ii) they come from soil that is being contaminated, since, the contamination of soil also occurs through the use of fertilizers and pesticides containing metal ions. In addition to these man made activities such as industrial and domestic discharges are also contributing toward the contamination to the soil (Gupta and Gupta 2005). In this case, the uncontrolled groundwater extraction causes the decrease of the water level that in turn induces the infiltration of return irrigation water, which transport varying amounts of agrochemicals within these the iron and manganese used in agricultural practice, which are incorporated into groundwater affecting further their chemical composition. This is very feasible due to the lack of iron and manganese micronutrients in agricultural lands is a common problem, which is solved by applying it through agrochemicals. However, the efficiency of this practice still has presented technical deficiency, a cause may be the interaction of the chemical with the soil whose chemical characteristics decrease its availability for cultivation. The efficiency of micronutrients also depends in part on how the product raffles the chemical mechanisms in the soil and in the rhizome, which block their assimilation by roots; this is specific for elements such as iron, manganese and zinc (Carpintero 2017). Then, using these trace elements in an uncontrolled manner likely to decrease its deficiency in agricultural soil, may lead to contamination of soil and groundwater. Some documented areas under such contamination conditions are: Mexicali; Mezquital, Hgo; Salamanca; Sinaloa; San Luis Potosí; Santo Domingo; Toluca; where iron and manganese content vary from (0.01 a 2.3 mg l^{-1}) (SEDESOL-INE 1994; Cardona and Hernández 1995; Cardona et al. 2004).

6.6 Final Considerations

In many countries the chemical quality of groundwater is increasing public health awareness requiring major and efficient attention from governments. In this chapter, based on a review and interpretation of numerous national and international studies concerning the deterioration of the obtained chemical quality of groundwater the *State of the Art in Mexico* is presented regarding major causes of the chemical deterioration of groundwater which is affecting human health and ecosystems. It was found to be necessary to generate information allowing to specify the geochemical processes that control the presence of trace metals in groundwater in different geological contexts to those of clastic materials. There are places where these processes have been mostly documented and the problem of human health is considered that the granular media is the greatest contribution of heavy metals to groundwater. In almost all countries that have studied trace elements in groundwater, the evaluation lacks an integral method (systemic view), from the perspective of groundwater flow systems (GFS), theory proposed by Tóth (1963, 1999), which allows the chemical evolution of the flow systems to be evaluated from recharge zone to the discharge zone. As it has been documented by Cardona and Hernández (1995), Cardona and Carrillo-Rivera (2006), Cardona et al. (2018), Carrillo-Rivera et al. (1996, 2007, 2008), Varsányi and Kovacs (1997), Edmunds et al. (2002, 2005), Huizar-Alvarez et al. (2014, 2016b), Ortega-Guerrero (2009), Sánchez-Díaz (2007) a high content of trace elements in groundwater may be of natural origin or by an anthropogenic contribution, so that in contrast to the second case, in the first is not appropriate to talk about pollution since nature does not contaminated itself.

Less than 40% of the states of the Mexican Republic have specific studies relating to the knowledge of how groundwater functions; these might allow us to understand the current situation of the chemical quality of groundwater in the north-central part of the country. In general, the answer of water extraction by boreholes usually induces a change in the chemical composition of the extracted water. Such response has been documented under particular hydrogeological conditions, in most of the cases the water quality degradation depends on time, flow rate of extraction and hydrogeological framework; this allows to positively propose the possibility of defining the particular conditions and propose controls on the quality of the extracted groundwater.

References

Alarcón-Herrera MT, Llorens E, Poch M (2012) Remoción de Arsénico del Agua para Consumo Humano en Latinoamérica. López Carrasco (ed). Centro de Investigación en Materiales Avanzados, Chih, México

Apambire WB, Boyle DR, Michel FA (1997) Geochemistry, genesis and health implications of fluoriferous groundwater in the upper regions of Ghana. Environ Geol 33(1):13–24

Armienta MA, Rodriguez R, Aguayo A, Ceniceros N, Villaseñor G, Cruz O (1997) Arsenic contamination of groundwater at Zimapán, México. Hydrogeol J 5:39–46

Armienta MA, Segovia N (2008) Arsenic and Fluoride in the groundwater of Mexico. Environ Geochem Health 30:345–353

Arreguín Cortés F, Chávez GR, Soto NP (2000) Una revisión de la presencia de arsénico en el agua subterránea en México. Comisión Nacional del Agua (CONAGUA), México

Bailey JC (1977) Fluorine in granitic rocks and melts: a review. Chem Geol 19:1–42

Bansiwal A, Thakre D, Labhshetwar N, Meshram S, Rayalu S (2009) Fluoride removal using lanthanum incorporated chitosan beads. Colloids Surf, B 74:216–224

Bardsen A, Bjorvatn K, Selving A (1996) Variability in fluoride content of subsurface water reservoir. Acta Odontologica Scandinamica 54(6):343–347

Betancourt-Lineares A, Irigoyen-Camacho ME, Mejía-González A, Zepeda-Zepeda M, Sánchez-Pérez L (2013) Prevalencia de fluorosis dental en localidades mexicanas ubicadas en 27 estados y el D.F. A seis años de la publicación de la Norma Oficial Mexicana para la fluoruración de la sal. Rev Investigación Clínica l 65:237–247

Bhattacharya P, Chatterjee D, Jacks G (1997) Occurrence of arsenic-contaminated groundwater in alluvial aquifers from the Delta Plain, Eastern India: options for a safe drinking water supply. Water Res Dev 13:79–92

Birkle P, Merkel B (2000) Environmental impact by spill of geothermal fluids at the geothermal field Los Azufres, Michoacán, Mexico. Water Air Soil Pollut 124:371–410

Bonilla A, Trejo-Vazquéz R, Márquez-Algara C (2002) Analysis of published health risks caused by exposure to fluorides in Aguascalientes, Mexico. Rev Internacional de Contaminación Ambiental 18(4):171–177

Buol SW, Hole FD, McCracken RJ (1981) Génesis y clasificación de suelos, Trillas (ed). México

BGS–DPHE (2001) Arsenic contamination of groundwater in Bangladesh. In: Kinniburgh DG, Smedley PL (eds) Vol 1: Summary. BGS Technical Report WC/00/19, British Geological Survey

Cardona A, Hernández N (1995) Modelo geoquímico conceptual de la evolución del agua subterránea en el valle de México. Ing Hidráulica en México X 3:71–90

Cardona A, Carrillo-Rivera JJ (2006) Hidrogeoquímica de sistemas de flujo intermedio que circulan por sedimentos continentales derivados de rocas riolíticas. Ing. Hidráulica en México XXI 3:69–86

Cardona A, Carrillo-Rivera JJ, Huizar-Alvarez R, Graniel-Castro E (2004) Salinization in coastal aquifers of arid zones: an example from Santo Domingo, Baja California Sur, Mexico. Environ Geol 45:350–366

Cardona A, Banning A, Carrillo–Rivera JJ, Aguillón–Robles A, Rüde T, Aceves de Alba1 J (2018) Natural controls validation for handling elevated fluoride concentrations in extraction activated Tóthian groundwater flow systems: San Luis Potosí, Mexico. Environ Earth Sci 77:121

Carpintero JM (2017). La eficiencia del fertilizante: corrección de carencias de hierro, zinc y manganeso. Grupo Fertiberia, www.fertiberia.com ›blog›

Carrillo-Chaves Drever JI (1998) Environmental assessment of the potential for arsenic leaching into groundwater from mine wastes in Baja California Sur, Mexico. Geofísica Intern 37:35–39

Carrillo-Rivera JJ, Cardona A, Moss D (1996) The importance of the vertical component of groundwater flow: a hydrogeochemical approach in the valley of San Luis Potosı, Mexico. J Hydrol 85:23–44

Carrillo-Rivera JJ, Cardona A, Edmunds WE (2002) Use of extraction regime and knowledge of hydrogeological conditions to control high-fluoride concentration in abstracted groundwater: San Luis Potosí basin. Mex J Hydrol 261:24–47

Carrillo-Rivera JJ, Varsanyi I, Kovács LO, Cardona A (2007) Tracing groundwater flow systems with hydrogeochemistry in contrasting geological environments. Water Air Soil Pollut 184:77–103

Carrillo-Rivera JJ, Cardona A, Huizar-Alvarez A, Graniel E (2008) Response of the interaction between groundwater and other components of the environment in Mexico. Environ Geol 55:303–319

Chae GT, Yun ST, Mayer B, Kim K, Kim SY, Kwon JS, Koh YK (2007) Fluorine geochemistry in bedrock groundwater of South Korea. Science Total Environ 385(1–3):272–283

Chebotarev II (1955) Metamorphism of natural water in the crust of weathering. Geochim Cosmochim Acta 8:22–48, 137–170, 198–212

Comisión Nacional del Agua (Conagua) (2000) Hydrogeological and hydrogeochemical exploration at the Mesilla Bolson, Conejos-Médanos Zone). Junta Municipal de Agua y Saneamiento de Ciudad Juárez

Comisión Nacional del Agua (2006) Estadística del agua en México. México

Comisión Nacional del Agua (2017) Estadística del agua en México. México

Dana Edward, Ford S, William E (1986) Tratado de mineralogía. CECSA (ed). México, XIII Printing

Davies BE, Bowman Ch, Davies TC, Selinus O (2005) Medical Geology: Perspectives and Prospects. In: Selinus O (ed in chief) Essential of medical geology impacts the natural environment on public health. Chapter 1. Elsevier Academic Press, Burlington MA 01803

De Alba A (2004) Suicidio o Renacimiento, (Metrópoli y Naturaleza). Ed. Indesol- Grupades, AC, Plaza y Valdés. México pp 342

De Valk B, Marx JJ (1999) Iron, Atherosclerosis, and ischemic heart disease. Arch Intern Med 159:1542–1548

Del Razo LM, Arellano MA, Cebrián ME (1990) The oxidation states of arsenic in well water from a chronic arsenicism area of northern Mexico. Environ Pollut 64:143–153

Del Razo LM, Corona JC, García-Vargas G, Albores A, Cebrián ME (1993) Fluoride levels in well water from a chronic arsenicism area of northern Mexico. Environ Pollut 80:91–94

Del Río Herrera S (2001) Prevalencia y grados de fluorosis en una muestra de población escolar de educación primaria de Ciudad de Tampico Tamaulipas, México. prodigy.com/dientitos/fluorosis.hml

Dhiman SD, Keshari AK (2006) Hydrogeochemical evaluation of high-fluoride groundwaters: a case study from Mehsana district, Gujarat, India. Hydrol Sci J 51(6):1149–1162

Díaz-Barriga F, Leyva R, Quistian J, Loyola-Rodríguez JP, Pozos A, Grimaldo M (1997) Endemic fluorosis in San Luis Potosi, Mexico. Sources of fluoride exposure. Fluoride 30:219–222

Díaz-Rodríguez A, Lozano Santacruz R, Dávila Alcocer V, Vallejo E, Girón P (1998) Physical, chemical, and mineralogical properties of Mexico City sediments: a geotechnical perspective. Can Geotech J 35:600–610

Dissanayake CB, Chandrajith R (2009) Introduction to medical geology: focus on tropical environments. Ed. André Freiwald. Springer—Verlag Berlin Heidelberg ISBN 978-3-642-00484-1

Edmunds WM, Carrillo-Rivera JJ, Cardona BA (2002) Geochemical evolution of groundwater beneath Mexico City. J Hydrol 258:1–24

Edmunds WM, Smedley PL (2005) Fluoride in natural waters. In Selinus O. (ed in chief), Essentials of medical geology impacts the natural environment on public health. Chapter 12. Elsevier Academic Press, Burlington, MA 01803

Esteller MV, Rodríguez R, Cardona A, Padilla-Sánchez L (2012) Evaluation of hydrochemical changes due to intensive aquifer exploitation: case studies from Mexico. Enviro Monit Assess 184:5725–5741

Fawell J, Bailey K, Chilton J, Dahi E, Fewtrell L, Magara Y (2005) Fluoride in drinking water. World Health Organization IWA Publishing, London

Fergusson JE (1990) The heavy elements: chemistry environmental impact and health effects. Pergamon Press, NY

Freeze RA, Cherry JA (1979) Groundwater. Prentice-Hall Ed. Inc. Englewood Cliffs NJ 07632

Galicia Chacón LF, Juárez López ML, Molina Frechero N (2009) Prevalencia de fluorosis dental y consumo de fluoruros ocultos en escolares del municipio de Nezahualcóyotl. Gaceta Médica Mexicana 145(4):263–267

García-Pérez A, Irigoyen-Camacho ME, Borges-Yáñez A (2013) Fluorosis and dental caries in Mexican schoolchildren residing in areas with different water fluoride concentrations and receiving fluoridated salt. Caries Res 47:299–308

Gaus I, Shand IN, Gale A, Williams T, Eastwood C (2002) Geochemical modeling of fluoride concentration changes during aquifer storage and recovery (ASR) in the chalk aquifer in Wessex England. England Geol Hydrogeol 35(2):203–208

Gunnar Jacks A, Prosun A, Bhattacharya K, Vikas P, Singh Chaudhary (2005) Controls on the genesis of some high-fluoride groundwater. India Applied Geochem 20:221–228

Gupta UC, Gupta SC (2005) Future trends and requirements in micronutrient research. Commun Soil Sci Plant Anal 36:33–45

Handa BK (1975) Geochemistry and genesis of fluoride containing groundwater in India. Groundwater 13:275–281

Harvey D (1994) La construcción social del espacio y del tiempo. Una teoría relacional. (Trad. Perla Zusman) Conf. Simposio de Geografía Socioeconómica. Geographical Review of Japan. Universidad de Nagoya 67 (Ser. B) (2):126–135

Hem JD (1985) The study and interpretation of the chemical characteristics of natural water. 3rd ed. U. S. Geological Survey Water-Supply Paper 2254

Hollabaungh, CL (2007) Modification of Goldschmidt's geochemical classification of the elements to include arsenic, mercury, and lead as biophile elements. In: Sarkar, Datta and Hannigan (eds) Concepts and applications in environmental geochemistry. Chapter 1. Elsevier Ltd. ISSN 147-8177/https://doi.org/10.1016/s1474 (07) 05004-8

Hong S, Candelone JO, Patterson CC, Boutron CF (1994) Greenland ice evidence of hemisphere lead pollution two millennia ago by Grek and Roman civilizations. Science 265:1841–1843

Huizar-Alvarez R, Carrillo-Rivera JJ, Cardona BA, Hergt T, Juárez Sánchez F (1998) Definición del control de la calidad del agua subterránea extraída de la subcuenca de México, Delegación Ixtapalapa. Concejo de Estudios Restauración y Valoración Ambiental. Gob. Cd, México. Aire-Agua 1:409–550

Huizar-Alvarez R, Carrillo-Rivera JJ, Angeles-Serrano G, Hergt T, Cardona BA (2004) Chemical response to groundwater extraction southeast of Mexico City. Hydrogeol J 12:436–450

Huízar-Álvarez R, Varela-González GG, Espinoza Jaramillo MM (2014) Sistemas de flujo subterráneo y contenido de fluoruro en el agua de Tenextepango, Morelos, México. Revista Mexicana de Ciencias Geológicas 3(2):238–247

Huízar-Álvarez R, Carrillo-Rivera JJ, Juárez Sánchez F (2016a) Fluoruro en el agua subterránea: niveles, origen y control natural en la región de Tenextepango, Morelos, México, UNAM. Inv. Geogr. Instituto de Geografía 90:40–58

Huizar-Alvarez R, Samira Ouysse, Espinoza-Jaramillo MA, Carrillo-Rivera JJ, Mendoza-Archundia E (2016b) The effects of water use on Tothian flow systems in the Mexico City conurbation determined from the geochemical and isotopic characteristics of groundwater. Environ Earth Sci 75:1060

Hurtado R, Gardea-Torresdey J (2004) Environmental evaluation of fluoride in drinking water at "Los Altos de Jalisco", in the central Mexico Region. J Toxicol Environ Health—Part A–current issues 67(20–22):1741–1753

Irigoyen-Camacho ME, Mejía-González A, Zepeda-Zepeda M, Sánchez-Pérez L (2013) Dental fluorosis prevalence in Mexican localities of 27 states and the D.F: six years after the publication of the salt fluoridation Mexican official regulation. Rev Invest Clin 65:237–247

Irigoyen-Camacho ME, García Pérez A, Mejía-González A, Huizar-Alvarez R (2016) Nutritional status and dental fluorosis among schoolchildren in communities with different drinking water fluoride concentrations in a central region in Mexico. Sci Total Environment 541:512–519

Jefferson F, Weinstein P, Chin-Hsiao T (2005) Environmental Medicine. In: Selinus O. (ed in Chief) Essential of medical geology impacts of the natural environmental on public healt. Chapter 22. Elsevier Acedmic Press. Burlintong, MA 01803

Jiménez-Cisneros BE (2007) Información y calidad del agua en México. Universidad Autónoma de Nuevo León Monterrey, Nuevo León, México. Trayectorias IX(24):45–56

Lesser Illades JM, Sánchez Díaz LF (1986) Hidrogeoquímica del acuífero de la Ciudad de México. Ingeniería Hidráulica en México 3:64–77

Liy H (2000) A Compendium of Geochemistry: from solar Nebula to the human brain, Princeton University Press, Princeton

Lindh U (2005) Biological functions of the elements. In: Selinus O (ed in chief), Essentials of medical geology impacts the natural environment on public health. Chapter 6. Elsevier Academic Press, Burlington, MA 01803

López Paraguay MZ (2013) Adsorción de arsénico y fluoruros en nanopartículas y su posterior separación del agua tratada. Unpublished PhD Dissertation. CIMAV. S.C. Chihuahua, México

Loyola-Rodríguez JP, Pozos-Guillén AJ, Hernández-Guerrero JC (1998) Bebidas embotelladas como fuentes adicionales de exposición a flúor. Salud Pública Mex 40:438–441

Martini M (1984) On the behaviour of fluorine in volcanic processes. Bull Volcanol 47(3):483–489

Mazari M, Mackay DM (1993) Potential for groundwater contamination in Mexico City. Environ Sci Technol 27(5):794–802

Macías-García C, Mazari-Hiriart M (2018) Estudio sobre el estado actual de mantos acuíferos, la Explotación de pozos, el abasto real y la demanda actual y Potencial del líquido en la ciudad de México. Consejo Económico Social de Ciudad de México (pp 356)

Mejía-González MA, González-Hita L, Briones-Gallardo R, Cardona BA, Soto-Navarro P (2014) Mecanismos que liberan arsénico al agua subterránea de la Comarca Lagunera, estados de Coahuila y Durango, México. Tecnología y Ciencias del Agua 5(1):71–82

Metz W (1998) Review of the scientific basis for establishing the essentiality of trace elements. Biol Trace Element Res 66:185–191

Michibata H, Uyama T, Ueki T, Kanamore K (2002) Vanadicytes cells hold the key to resolving the highly selective accumulation and reduction of vanadium in ascidian. Microsc Res Tech 56:421–434

Minning BA, Goldberg (1997) Arsenic (III) and arsenic (V) adsorption on three California Soils. Soil Sci 162:886–985

Navarro O, González J, Júnez-Ferreira H, Bautista CF, Cardona BA (2017) Correlation of arsenic and fluoride in the groundwater for human consumption in a semiarid region of Mexico. Procedia Eng 186:333–340

Nickson RT, McArthur JM, Ravenscroft P, Burgess WG, Ahmed KM (2000) Mechanism of arsenic poisoning of groundwater in Bangladesh and West Bengal. Appl Geochem 15:403–413

Niparajá AC (2005) Estudio del arsénico en Baja California Sur, México. https://issuu.com/niparajaac/docs/estudio_arsenico_niparaja

Nordberg M, Cherian G (2005) Biological Responses to Elements. In: Selinus O (ed in chief), Essentials of medical geology impacts the natural environment on public health. Chapter 8. Elsevier Academic Press, Burlington, MA 01803

Nordstrom DK, Jenne EA (1977) Fluoride solubility in selected geothermal waters. Geochemistry et Cosmochimistry Acta 41:175–188

Nriagu JO (1983) Lead exposure and lead poisoning. Lead and lead poisoning in antiquity. Wiley, N York, pp 309–424

Nriagu JO (1998) Tales told in lead. Sci 281:1622–1623

Ortega-Guerrero MA (2009) Presencia, distribución, hidrogeoquímica y origen de arsénico, fluoruro y otros elementos traza disueltos en agua subterránea, a escala de cuenca hidrológica tributaria de Lerma-Chapala, México: Revista Mexicana de Ciencias Geológicas 26(1):143–161

Ortega-Guerrero MA (2017) Evaporative concentration of arsenic in groundwater: health and environmental implications, La Laguna Region, Mexico. Environ Geochem Health 39:987–1003

Ozsvath DL (2009) Fluoride and environmental health: a review. Review Environ Science, Biotechnol 8:59–79

Pitre V (1994) Analysis of induced recharge from a waste water canal through fractured clays in Mexico City. M.Sc dissertation. Universidad of Waterloo, Canadá

Planer-Friederich B, Armienta MA, Merkel BJ (2001) Origin of arsenic in the groundwater of the Rio Verde Basin, Mexico. Environ Geol 40:1290–1298

Prashanth L, Kattapagari KK, Chitturi RT, Baddam VR, Prasad LK (2015) A review on role of essential trace elements in health and disease. JNTR Univ Health Sci 4:75–85

Ramírez-Ayala R, Azcona-Cruz I (2017) Efectos tóxicos del manganeso. Rev Especial Médica Quirúrgica 22:71–75

Rao NS (2003) High-fluoride groundwater. Environ Monit Assess 176:637–645

Ravenscroft P, McArthur JM, Hoque BA (2001) Geochemical and Palaeohydrological Controls on Pollution of Groundwater by Arsenic. In: Chappell CO, Abernathy, Calderon (eds), Arsenic Exposure and Health Effects, Chapter IV: 53–78 Elsevier Science Ltd. Oxford

Roberts C, Manchester K (1995) Metabolic and endocrine disease. In: The Archeology of Disease, London, Alan Sutton Publishing, pp 163–185

Schumann K (2001) Safety aspects of iron in food. Ann Nutr Metab 45:91–101

Salonen JT, Nyyssonen K, Korpela H, Tuomilehto J, Seppanen R, Salonen R (1992) High stored iron levels are associated with excess risk of myocardial infarction in Eastern Finnish Men. Circulation 86:803–811

Sánchez-Díaz LF (2007) Origen, transporte, distribución y concentraciones de los fluoruros en el sistema hidrogeológico volcánico Atemajac-Toluquilla, Jalisco, México. Unpublished PhD Dissertation, Universidad Nacional Autónoma de México

Saxena VK, Ahmed S (2001) Dissolution of fluoride in groundwater: a water—rock interaction study. Environ Geol 40:1084–1087

SEDESOL-INE (1994) Informe de la situación general en materia de equilibrio ecológico y protección del ambiente. Secretaría de Desarrollo Social, Instituto Nacional de Ecología, México

Selinus O, Alloway B, Centeno JA, Fonkelma RB, Fuge R, Lindh U, Smedley PL (2005) Preface. In: Selinus O (ed. in Chief) Essentials of medical geology impacts the natural environment on public health. Elsevier Academic Press, Burlington, MA 01803. USA

SEMARNAT (2002) Informe de la ecretaria del Medio Ambiente en México. Compendio de estadísticas ambientales, México: Secretaria del Medio Ambiente y Recursos Naturales
Silver CS, Rothman DS (1995) Toxic and health: the potential long-term effects of industrial activity. World Resources Institute, Washington
Smedley PL, Kinninburgh DG (2002) A review of the source, behaviour and distribution of arsenic in natural waters. Appl Geochem 17:517–568
Smedley PL, Kinninburgh DG (2005) Arsenic in groundwater and the environment. In: Selinus O (ed. in chief), Essentials of medical geology impacts the natural environment on public health. Chapter 11. Elsevier Academic Press. Burlington, MA 01803
Smith N (1990) The Production of Nature and The Production of Space In: Claudia Villegas Delgado (Transator), Uneven Development. Nature, Capital and the Production of Space. Basil Blackwell. Biblioteca básica de Geografía. Serie Trad- 2. SUA. Fac. Filos y Letras UNAM. México, 2006
Smyth TG (1992) El flúor en la prevención de la caries dental. Ediciones Díaz de Santos. SA
SSA (Secretaría de Salud) (1996) Norma Oficial Mexicana NOM-127-SSA1-1996. Salud ambiental. Agua para uso y consumo humano. Límites permisibles de calidad y tratamientos a que debe someterse el agua para su potabilización, México: Secretaria del Medio Ambiente y Recursos Naturales
SSA (Secretaría de Salud) (2004) Presencia de flúor en México. Centro Nacional de vigilancia y control de enfermedades epidemiológicas, estudio nacional, México
Stumm W, Morgan JJ (1981) Aquatic Chemistry. 2nd edn, John Wiley Sons, NY
Tóth J (1963) Theoretical analysis of groundwater flow in a small drainage basin. J Geophys Res 68:4795–4812
Tóth J (1978) Gravity-induced cross-formational flow of formation fluids, Red Earth Region, Alberta, Canada: analysis, patterns and evolution. Water Resources Res 14(5):805–843
Tóth J (1988) Groundwater and hydrocarbon mitigation. In: Back W, Rosensheim JS, Seaber PR (eds), Hydrogeology, Geology of North America, Vol. O-2:485–502
Tóth J, Corbet T (1986) Post-Paleocene evolution of regional groundwater flow-systems and their relation to petroleum accumulations, Taber area, southern Alberta. Canada. Canadian Petroleum Geology Bulletin 34(3):339–363
Tóth J (1999) Groundwater as a geologic agent: an overview of the causes, processes, and manifestations. Hydrogeol J 7:1–14
USEPA (2011) US environmental protection agency and the US department of Health And Human Services (HHS) http://water.epa.gov/action/advisories/drinking/fluoride_index.cfm
Varela-González GG, García-Pérez A, Huizar-Alvarez R, Irigoyen-Camacho ME, Espinoza-Jaramillo MM (2013) Fluorosis and dental caries in the hydrogeological environments of Southeastern communities in the State of Morelos, Mexico. J Environ Protection 4:994–1001
Varsányi I, Kovács OL (1997) Chemical evolution of groundwater in the River Danube deposits in the southern part of the Pannonian Basin (Hungary). Appl Geochem 12:625–637
Valenzuela-Vásquez L, Ramírez-Hernández J, Reyes-López J, Uribe A, Lázaro-Mancilla O (2006) The origin of fluoride in groundwater supply to Hermosillo City, Sonora, México. Environ Geol 51:17–27
Venegas SS, Ramírez S G, Romero GC, Reyes VP, Razo M, Gutiérrez NL, Arellano GF (1991) La Primavera geothermal field, Jalisco. The Geology of North America, Vol. P-3, Economic Geology, Mexico, The Geological Society of America
Wenzel WW, Blum WE (1992) Fluorine speciation and mobility in F-contaminated soils. Soil Sci 153(5):357–364
World Health Organization (2001) Arsenic and arsenic compounds, environmental health criteria. vol. 224, 2nd edn, World Health Organization, International Programme on Chemical Safety, Geneva
World Health Organization (2010) The WHO recommended classification of pesticides by hazard and guidelines to classification 2009. World Health Organization, Ginebra, Suiza
World Health Organization (2015) Water Sanitation and Health (WSH). Naturally occurring hazards: Fluoride. http://www.who.int/water_sanitation_health/naturalhazards/en/index2.html (Read 10/8/15)

Chapter 7
Groundwater Flow Systems and Their Importance in the Assessment of Transboundary Groundwater: The Mexico–U.S.A. Case

Gonzalo Hatch-Kuri and José Joel Carrillo-Rivera

Abstract The "Transboundary Aquifer" concept envisaged in the United Nations Resolution 63/124 "*The Law of Transboundary Aquifers*" has had a significant impact on the evaluation of transboundary aquifers around the world. In the Mexico–U.S.A. case, it has not been possible to officially determine the total number of shared aquifers, therefore, the evaluation of the systemic functioning of "Transboundary Groundwater" is to be settled; it represents an absent concept in the international transboundary water enactments. This work carries out an analysis based on scientific evidence and legal documents to determine the nature of the current conceptual discrepancies between the scientific definitions of "Transboundary Aquifer" and "Transboundary Groundwater." Results support the need to incorporate a systemic vision of the functioning of groundwater, as well as the scientific homologation of concepts and methodologies applied by those states interested in jointly assessing groundwater as to avoid water conflicts in the context of its incipient integrated management.

Keywords Transboundary aquifers · Transboundary groundwater · Water management · Political geography · U.S.A.–mexico border

G. Hatch-Kuri (✉)
Maestría En Gestión Integrada de Cuencas, Facultad de Ciencias Naturales, Universidad Autónoma de Querétaro, Av. Junípero Serra, Antiguo Aeropuerto, Campus Aeropuerto, S/N Santiago de Querétaro, Qro C.P. 76140, Mexico
e-mail: ghatch@comunidad.unam.mx

J. J. Carrillo-Rivera
Instituto de Geografía, Universidad Nacional Autónoma de México. Investigación Científica. Ciudad Universitaria S/N., Ciudad de México C.P. 04510, Mexico
e-mail: joeljcr@igg.unam.mx

M. M. Alconada-Magliano (ed.), *Intensified Land and Water Use*, Springer Earth System Sciences, https://doi.org/10.1007/978-3-030-65443-6_7

7.1 Introduction

Authors such as Wolf et al. (2005) indicated that 60% of the flow of all the world's rivers is located in 276 transboundary basins, which in turn are shared by 145 countries; besides, the national territory of 33 countries is entirely within a single transboundary basin. However, below the soil, groundwater circulates through geological strata (generically called aquifers), which constitutes 97% of the *physically accessible* continental fresh water; thus, to thoroughly comprehend a "transboundary" hydrological cycle, it is necessary to understand in which way (direction, quantity, and quality) water flows through the shared hydro-geographical units, that is, the surface basins and the so-called aquifers. In this regard, organizations such as the International Shared Aquifer Resource Management (ISARM), the International Groundwater Resources Assessment Centre (IGRAC), The International Hydrological Program (UNESCO-IHP), and The World Bank in conjunction with the International Association of Hydrogeologist (IAH), have an inventory of about 592 transboundary aquifers in the globe (IGRAC 2018). On a local analysis scale, Mexico and the United States officially recognized 11 Transboundary Aquifers; in 2016, the U.S.A. closed the *Transboundary Aquifer Assessment Program* (TAAP) which evaluated only four aquifers shared with Mexico in order to have updated hydrogeological databases and, recently, works have been published that point to the possible existence of more than 30 transboundary aquifers in the Texas–Mexico border section (Sánchez et al. 2018).

The Mexico–U.S.A. transboundary aquifers assessment works have been strongly influenced by the "Transboundary Aquifer" concept of the United Nations Resolution 63/124, *The Law of Transboundary Aquifers*. Its definition favors an abstract description of a particular transboundary geological reference in detriment of the object of concern—groundwater—, and especially its functioning, evident through what is called Groundwater Flow Systems (GWFS); that is, groundwater + geological unit + flow system pattern (Freeze 1969; Tóth 1999; Russell and Priebe 2017; Carrillo-Rivera and Ouysse 2013). From an analytical review of scientific evidence and the regulatory instruments of international law for transboundary waters, as well as the local water regulatory frameworks in the Mexico–U.S.A. border, a conceptual discrepancy was observed in the definition of two concepts used to suggest interest in groundwater, that is, "Transboundary Aquifer" and "Transboundary Groundwater."

The codification of groundwater as a transboundary watercourse requires conceptual and methodological clarity based on scientific evidence, in order to give legal certainty to the regulated objects, applied procedures, regulatory agreements, and the impact of their results in the context of joint management schemes of this water (Rivera and Candela 2018). For that matter, the application of GWFS helps by providing enough and necessary evidence to understand the operation of groundwater and its determination as a transboundary watercourse. In previous works (Hatch-Kuri 2017; Carmona et al. 2017; Hatch-Kuri 2018; Hatch-Kuri et al. 2019), the need to have a specific regulatory scheme for groundwater was examined and proposed in

Mexico, through the analytical review of some evaluation works of the Mexico–U.S.A. Transboundary Aquifers. Considering that at international and local level there are asymmetric ways to conceptualize scientifically and legally both terms, "aquifer" and "groundwater," this contribution, from an interdisciplinary perspective, combines the approaches of Political Geography and Hydrogeology as a methodological evaluation framework, which substantiates why is important to overcome the limitations imposed by the "Transboundary Aquifer" concept. It concludes with a proposal for a scientific and legal definition of "Transboundary Groundwater," as to discuss a future joint water management unit as a reference point, to continue the inventory of Mexico–U.S.A. transboundary groundwater.

7.2 Political Geography and Hydrogeology: A Required Dialogue

The groundwater is a constitutive part of the Hydrological Cycle, since it is estimated that some 97% of continental freshwater is in fact groundwater. Its technical withdrawal is connected, generally, to a complex local network of conducts for its natural distribution. This hidden nature represents a condition of zero visibility and social prestige, contrary to what happens with surface water. In addition to this, its study requires a high academic specialization where different methodologies from experimental disciplines such as Hydrogeology, Geology, Hydrology, Edaphology, Chemistry, Biology, Ecology, and Geomorphology converge, and which provide evidence to understand their movement and physicochemical behavior (i.e., isotopic and hydrogeochemical) in the soil and in the geological strata below the soil. However, groundwater has been a neglected element in the water knowledge dissemination, professional training, and decision making; clearly evident in the concept "Hydrological Cycle," which in the first instance, the phase of infiltration and dynamic residence of the water in the subsoil is vague in several iconographic schemes, while the interaction of the human being with the rest of the cycle, is a subject that requires complex and interdisciplinary approaches that integrate this dimension and, in turn, overcome the excessive epistemological fragmentation which distinguishes scientific water activity in general.

Political Geography as a social science has traditionally focused on the systematic study of the Theory of the State Function, the analysis of its fundamental components such as territory, population, forms of government, regional integration with other states, and the control of political-administrative boundaries, among others (Agnew 2002; Talledos 2014). Based on the premise that it is the responsibility of the state to articulate and command the processes and mechanisms that define the administration of the national territory, such as the management of transboundary natural properties, it poses the challenge of developing interdisciplinary research that provides new elements of analysis on the political implications of the different ways of exercising and applying sovereignty over waters shared by two or more countries—as revealed in the previous work of Ribeiro (2010, 2012) and Brooks and Linton (2011).

Similarly, it should be remembered that one of the contemporary epistemological problems that concern Geography is to think about the history of nature in terms of its social transformation, that is, the understanding of its social history (Moreira 2010). From this approach, authors such as Swyngedouw (2009) and Linton and Budds (2014) have developed an emerging theory that conceives water as a "hybrid" that synthesizes both its material composition (H_2O), but also expresses the different transformations it undergoes when human beings act and transform their spatial mobility. From this perspective, it is argued that water is "produced" and "reproduced" (acquiring new meanings and social interpretations), internalizing social conflicts, such as those related to the conceptual definition of water itself. It may be asked whether the Hydrological Cycle is truly a concept? Is it fully understood? Hence, it is a fundamental task to establish the necessary institutional arrangements that define its instrumentalization in the set of public policies that dominate the water sector.

The study of different conflicts, relationships, and processes that define the management of transboundary groundwater entails the specialists of Political Geography to the rigorous observation of several visions and methodological debates from modern Hydrogeology, as to identify and analyze various components of each process that leads to groundwater "hybridization" (Linton and Budds 2014). In this sense, the Theory of Tothian Groundwater Flow Systems, proposed by Tóth (1999), represents a constitutive concept of the scientific debate that deals with understanding the particular planetary functioning of one of the main components of the Hydrological Cycle, a theory that poses a systemic vision of groundwater and its interdependent relationship with other environmental components (soil, vegetation, climate, rocks, geomorphology, lakes, rivers, springs, oceans, etc.). It is of great interest to emphasize that this theory provides clarity in the conceptual distinction between "aquifer," as the geological reference whose properties determine the movement and volume of water that is present; and "groundwater" characterization, regarded as a flow from a systemic point of view. To this effect, and given the groundwater importance, it is significant to ask whether the conceptual definition of "Transboundary Aquifer," that is, water + rock, is the best concept to define the groundwater management unit, in transboundary conditions, where usually an understanding on how groundwater flow is not a priority.

Therefore, interdisciplinary collaboration approaches provide benefits for the framework of international and scientific efforts that have established an initial inventory of transboundary aquifers, as well as the scientific evaluation of shared groundwater flows. In this debate, interdisciplinarity approaches are essential, because they help by providing greater clarity in the conceptual definitions and, consequently, in the development of evaluation schemes for transboundary groundwater according to the governance models recently promoted as *Integrated Management of Transboundary Waters*. The development and application of the indicated scientific methodologies have a direct impact on the institutional forms that the states will adopt for the management and sovereign control of these international watercourses (Ribeiro 2012; Rivera and Candela 2018).

7.3 Transboundary Aquifer or Transboundary Groundwater?

Transboundary groundwater waits to be codified in international law as a shared watercourse, which prevents states from fully recognizing it as such and carrying out evaluation work in accordance with the nature of a groundwater systemic vision. Possibly, this situation owes its explanation to the lack of conceptual clarity that requires a full understanding of key concepts from the field of modern hydrogeology. This field considers groundwater as a dynamic entity, contrary to the aquifer which portrays a static view instead; but also to the mental requirement of the existence of diverse scientific methodologies and applied to the dynamic evaluation of this water, which avoids reaching to different results through a different margin of interpretation.

In terms of what international law establishes for transboundary waters, the existence of two regulatory instruments in force with binding effects is recognized, and their content describes situations of conditional nature for states to legally identify and recognize the existence of transboundary groundwater. On one hand, the *Convention on the Protection and Use of Transboundary Watercourses and International Lakes* (UNECE Water Convention 1992), indicates that groundwater can have transboundary status as long as it is capable of verifying that it is crossed by an international border. On the other hand, the United Nations Resolution 51/229 *Convention on the Law of the Non-navigational Uses of International Watercourses* (UN General Assembly Resolution 1997) considered that international watercourses are a single physical unit (surface water and groundwater); thus, such transboundary groundwater, understood as a transboundary aquifer for this Convention, are only those aquifers of unconfined nature. Note that in the distinction between both descriptions referring to groundwater arise two different concepts, on the one hand, "Transboundary Groundwater" and on the other, "Transboundary Aquifer."

Both definitions suggest different groundwater conceptions. While the UNECE Water Convention refers to groundwater from the condition that implies the definition and evaluation of its movement below the surface, in other words, its movement as a flow, requires the production of scientific evidence that verifies its transboundary dimension. This implies, in fact, that the states concerned should assign agreements for the assessment of groundwater functioning and thus be able to verify their transboundary condition. Therefore, the main concept used for the UNECE Water Convention is "Transboundary Groundwater." Conversely, the UN Watercourses Convention definition is different, and although part of the principle recognizes groundwater flowing through a particular geological unit (the aquifer), this situation implies that it is no longer groundwater the concept that fundamentally must be evaluated but, in contrast, the determination of the type of hydraulic characteristic (confined or unconfined) of shared aquifer is privileged, which also must meet an additional condition to be considered transboundary, that is the nature is to be unconfined or that the groundwater flow systems feed different surface water bodies existing on the aquifer, such as rivers, lakes, ponds. or lagoons (see Fig. 7.1, where an unconfined aquifer is shown). Certainly, this is a positive aspect because it recognizes that groundwater is a fundamental and constitutive part of the "Transboundary Hydrological Cycle," which is in permanent motion and, therefore, questions the feasibility of continuing the management of the transboundary waters from a fragmentary approach (surface water vs. groundwater). However, if the definition of "Transboundary Groundwater," is understood as a synonym for "unconfined transboundary aquifer." This leads to subsume an act, a posteriori, of the determination and evaluation of the operation of groundwater within the particular geological reference (against what UNECE Water Convention has suggested). Thus, the need to characterize is prevailing, first of all, the type of geological formations and verify their transboundary nature (not confined), which also excludes other types of geological units and their response to the water extraction as well as those in confined and semi-confined conditions. Thus, for UN Watercourses Convention, the groundwater systemic functioning determination has remained in the background; additionally, the groundwater flows presence in a so-called confined aquifer is dangerously out of any evaluation of transboundary watercourses.

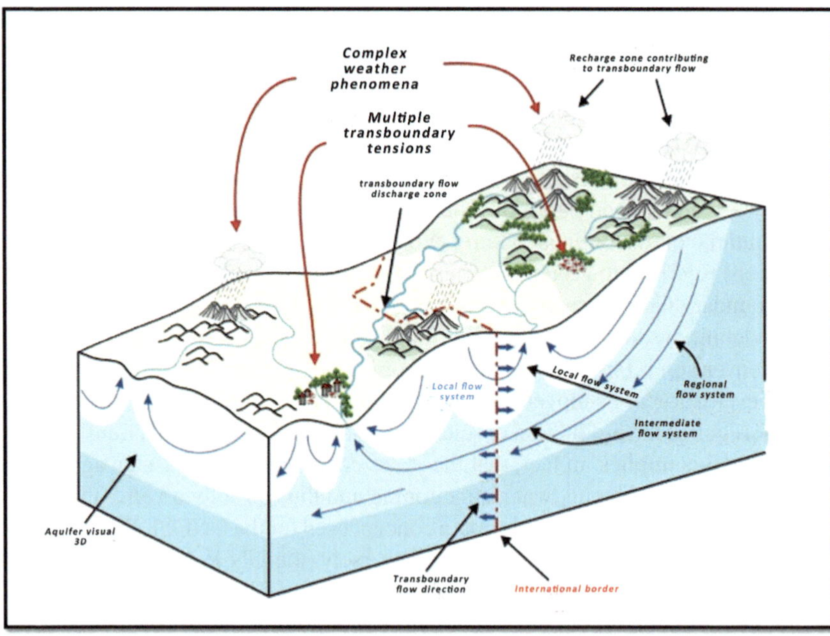

Fig. 7.1 Transboundary Regional Hydrological System (Adapted from UNESCO-PHI 2015)

However, looking to solve the conceptual differences described above, a plenary session of the United Nations Assembly witnessed in 2008 the presentation of the 63/14 resolution titled *The Law of Transboundary Aquifers*, as a result of valuable efforts to demonstrate groundwater importance and recognize it as an international watercourse and managing it under such precept. In this document there is no definition of "Transboundary Groundwater," if applicable in Article 2, four concepts related to "Transboundary Aquifer" are defined: (a) "aquifer" means a permeable water bearing geological formation underlain by a less permeable layer and the water contained in the saturated zone of the formation; (b) "aquifer system" means a series of two or more aquifers that are hydraulically connected; (c) "transboundary aquifer" or "transboundary aquifer system" means, respectively, an aquifer or aquifer system, parts of which are situated in different states; (d) "aquifer State" means a state in whose territory any part of a transboundary aquifer or aquifer system is situated.

These definitions highlight, essentially, the primary importance of defining, in the first instance, the main characteristics of the particular geological reference (lithology, stratigraphy, structure, porosity, hydraulic conductivity); to have a 3D vision (geometry) and information necessary to determine if it has a transboundary condition, that is, due to its extension, is verified to be shared by two or more states. Also, although it is considered that the aquifer may be hydraulically linked to other aquifers because groundwater flows among them, it is obvious that it leaves in the background the scientific evaluation of groundwater movement through flow systems. Additionally, this document overlooks the total thickness of permeable rocks that include those below the "less permeable layer" (condition [a] above). Note that in real conditions, rocks with aquifer possibilities may have several hundred meters in thickness and are often directly underutilized by extraction wells.

It is obvious that an adequate aquifer characterization, whose properties determine the water movement rate, is an imperative act, but it should be emphasized that one of the fundamental constraints that transmit previous definitions linked to the "aquifer" is the notion that groundwater remains "static" and is "stored" below the soil, in detriment of the scientific evidence that concludes that groundwater flow systems are located at different depths and travel with different spatial-temporal velocity that allow them to infiltrate and emerge. In other words, the question is what the motion of water is, instead of the geological reference of the aquifer (Freeze 1969; Tóth 1999; Russell and Priebe 2017; Carrillo-Rivera and Ouysse 2013). Certainly, those definitions had an influence of what is foreseen in the UN Watercourses Convention that, as indicated in the study of Burchi (2018), results from an academic and political consensus by specialists who integrate the ISARM and collaborated in the formulation the Resolution 63/124 through the International Law Commission.

For its part, UNECE, considering the mandatory content of UNECE Water Convention, published *Model Provisions on Transboundary Groundwaters* (2014) where it is specified what should be managed, assessed, and under what principles should be done. In his introduction, many authors acknowledge the existence of a conceptual discrepancy between "Transboundary Aquifer" and "Transboundary Groundwater," but note that UNECE Water Convention mandates that groundwater transboundary conditions should be evidenced, that is, the groundwater flow movement is shared by two or more states, affirmation that, needless to say, does not exclude the determination of the aquifer particular conditions; on the contrary, it is an intrinsic component to this convention (UNECE 2014: 3).

Another similar document entitled *Regional Strategies for the Evaluation and Management of Transboundary Aquifers in the Americas* (UNESCO-IHP 2015) aims to allow the American states to orientate themselves to execute strategies to evaluate their shared aquifers. Unlike UNECE *Model Provisions*, this guide recovers the "Transboundary Aquifer" concept of Resolution 63/124. This contribution lays the basis for a comprehensive scheme to manage aquifers through the application of conceptual models that include only the water-balance to define the groundwater availability in aquifers, and the creation of transboundary and geospatialized databases with geological, edaphological, topographic, geophysical, hydrology, chemical, isotopic, climatic, land-use indicators, etc. It should be noted that the water-balance methodology for the assessment and groundwater management has been questioned by the international scientific community for more than four decades (Bredehoeft et al. 1982; Freeze and Cherry 1979; Ward 1967). However, its application continues being promoted, creating severe inconsistencies in its application and negative effects for the environment and water users.

In summary, the discrepancy around the concepts of "Transboundary Groundwater" and "Transboundary Aquifer" may be seen in the set of documents analyzed, both with different connotations (see Table 7.1), as for UN Watercourses Convention and Resolution 63/124, which highlights the importance of the particular geological reference and its transboundary determination; for UNECE, the main concept to be defined and assessed is groundwater functioning, which implies identifying the spatial distribution of the components of flow as recharge-throughflow-discharge zones located in border areas, the place where water infiltrates, residence time of each flow component in the aquifer (age), chemistry, among others. Some of these aspects are intended in Resolution 63/124, but it should be noted that what is the subject to be evaluated is groundwater, not the aquifer per se (i.e., meaning the rock) therefore the paramount question remains: Is the concept of "Transboundary Aquifer" the most appropriate management unit to define the transboundary groundwater functioning?

In a recent publication the IGRAC (2018) warned about the global inventory of 366 "Transboundary Aquifers" and 226 "Transboundary Groundwater Bodies" and although it does not require both terms of conceptualization, surely this results from the cooperation of different scientific processes that have led to determine some aquifers as transboundary, whereas in other places the groundwater functioning is initially being assessed and, consequently, its determination as transboundary water flow systems (European case).

Resolution 63/124, which remains a draft status, is a benchmark for interested states to promote agreements for the transboundary aquifer assessment and management. In the Americas, the most relevant case is the Guaraní Aquifer Agreement, signed by Argentina, Brazil, Paraguay, and Uruguay, ratified in August 2010 and pending final approval. In Central America, the first Technical Report for the Ocotopeque-Citalá Transboundary Aquifer was published in 2016, under the

Table 7.1 Conceptual definitions related to "Transboundary Groundwater" in several international legal frameworks

Instrument	Type	Year	Conceptual definition
Convention on the Protection and Use of Transboundary Watercourses and International Lakes (UNECE Water Convention)	Binding	1992	Groundwater can be transboundary, if it can be verified
Convention on the Law of the Non-navigational Uses of International Watercourses (UN Watercourses Convention)	Binding	1997	Transboundary aquifers are only unconfined. Groundwater and surface water are a single physical unit
Resolution 63/124 *The Law of Transboundary Aquifers*	Pending approval	2008	Permeable water-bearing geological formation located on a less permeable layer, and the water contained in the saturated zone of the formation, which can be hydraulically related to other aquifers, forming a Transboundary Aquifer System which, due to its geological extension, can have different parts located in different states
Model Provisions on Transboundary Groundwaters (UNECE)	Binding	2014	Evaluate groundwater functioning and the determination of its transboundary condition

auspices of the IHP, institutional recognition was achieved within the "Trifinio Plan" between El Salvador, Honduras and Guatemala (UNESCO-IHP 2015). In the Mexico–U.S.A. case, there have been significant advances in the binational characterization of four transboundary aquifers, through fieldwork to report carried out, as well as those official and binational ones such as TAAP, although it has not yet been possible to formalize any general agreement for their joint management. However, the results of these scientific works reveal the marked influence of the "Transboundary Aquifer" concept present in Resolution 63/124.

7.4 The Resolution 63/124 the Law of Transboundary Aquifers, Its Influence on the Mexico–U.S.A. Transboundary Aquifers Assessment

To date, Mexico and the United States of America have not initiated procedures to adhere to the UNECE Water Convention or, where appropriate, to the UN Watercourses Convention, as reference frames to transboundary groundwater management. Nevertheless, both countries concluded in the last century two binding treaties for the political allocation of three transboundary basins (the Tijuana, Colorado, and Grande/Bravo rivers), *Convention between the United States and Mexico. Equitable Distribution of the Waters of the Rio Grande and Treaty Between the United States of America and Mexico for Utilization of Waters of the Colorado and Tijuana Rivers and of the Rio Grande* (1944 Water Treaty). These documents failed to cover the transboundary groundwater regulation. However, both treaties and institutions created for a shared rivers management, that is, the International Boundary and Water Commission, CILA in Mexico, and the International Boundary Water Commission or IBWC in United States, have worked, *ex profeso*, as diplomatic channels to deal with issues related to groundwater management on a common border of 3,169 km in length.

Noticeably, in binational treaties the concepts "Transboundary Groundwater" or "Transboundary Aquifer" are absent, but they are also replicated in water domestic regulatory frameworks from Mexico and U.S.A. southern border states (Arizona, California, New Mexico, and Texas). It should be noted that contrary to what happens in the first country, where groundwater is managed and controlled by the Federal Government, in the U.S.A. it is regulated by each state, who manage it according to various legal doctrines regarding the water rights policy each entity has adopted for this purpose (Hatch-Kuri 2017; Sánchez and Eckstein 2017). According to the above, it is initially observed that in both countries the domestic legal provisions are evidence of discrepancies in the way groundwater and aquifers are conceived and conceptualized, in the same way as in the international legal context reviewed in the previous section (see Table 7.2). This translates into the application of asymmetric groundwater management schemes throughout the Mexico–U.S.A. border.

In general terms, it can be seen that the "aquifer" concept is defined in all the legal frameworks examined, and it was even found in other similar concepts that allude to the aquifer, as is the case of "Groundwater Reservoir" (Texas Administrative Code 1977).

However, the most pronounced conceptual discrepancy lies in the way in which each country legally conceives an "aquifer" (initially, as receivers or containers of groundwater reserved for economic use). Whereas in Mexico, the concept "aquifer" highlights the uncertainty that exists to establish its real limits only considering the so-called conventional or administrative ones, an aspect that implies that this becomes a political convenience act for who is in charge of establishing them. In the United States of America, contrarily, their conceptual definition suggests that there should be scientific evidence to determine its main characteristics (thickness, definition of lateral and horizontal limits). This vision found, on the one hand, of Mexican administrative aquifers and, on the other, aquifers as geological formations, which have been mapped in Sánchez et al. (2016), to define the Mexico–U.S.A. transboundary aquifers total number. Their study reveals that, based on aquifers current legal definitions, there are 36 Mexican (administrative) and 36 (geological) aquifers in the United States of America, but that the set of information analyzed suggests that only in 16 aquifers there is sufficient and reasonable information to conclude its transboundary nature or that it extends through the territory of both countries.

Table 7.2 Conceptual definitions related to "Groundwater" and "Aquifer" in water domestic regulatory Mexico–U.S.A

Concept	State	Legal definition
Aquifer	Arizona (U.S.A.)	Means a geological formation that contains sufficient saturated materials to be capable of storing water and transmitting water in usable quantities to a well. Source: *Arizona Revised Statutes*
	California (U.S.A.)	Means a geological formation or structure that transmits water in sufficient quantities to supply pumping wells or springs. Source: *California Water Code*
	New Mexico (U.S.A.)	Means a geological formation that contains sufficient saturated material to be capable of storing and transmitting water in usable quantities to a well. Source: *New Mexico Statutes Chapter 72, Water Law*
	Texas (U.S.A.)	Geological formation, group of formations, or portion of a formation capable of yielding significant quantities of groundwater to wells or springs Source: *Texas Administrative Code*
	Mexico	Any geological formation or set of formations hydraulically connected to each other, through which water circulates or is stored water can be extracted for usage and exploitation and whose lateral and vertical limits are conventionally defined for purposes of evaluation, management, and administration of the national waters in the subsoil. Source: *Ley de Aguas Nacionales Mexicana*
Groundwater	Arizona (U.S.A.)	This concept does not exist in the law
	California (U.S.A.)	Water beneath the surface of the earth within the zone below the water table in which the soil is completely saturated with water. Source: *California Sustainable Groundwater Act*
	New Mexico (U.S.A.)	This concept does not exist in the law
	Texas (U.S.A.)	Water percolating below the surface of the earth Source: *Texas Administrative Code*
	Mexico	This concept is not defined in law

Another work by Sánchez et al. (2018) has identified and characterized the lithology, stratigraphy and some elements of the hydraulic conductivity in the Mexico–Texas border geological materials, to be able to specify their transboundary nature. The study concludes that 33 transboundary aquifers exist, located from the binational conurbation Ciudad Juárez/El Paso, to the Gulf of Mexico. This study sets a valuable precedent and is likely to be replicated in the rest of the U.S.A.–Mexico border. If applicable, the results would help to overcome the current official limitation that only recognizes 11 shared aquifers (see Fig. 7.2). However, both studies can

Fig. 7.2 Mexico–U.S.A. Transboundary Aquifers officially recognized (Adapted from CILA 2020)

be complemented with the transboundary groundwater flow systems determination and their functioning.

Together with these valuable academic contributions, TAAP was completed in 2016 at the United States of America, which assess four aquifers shared with Mexico: Río Santa Cruz, Río San Pedro (Sonora-Arizona) Bolsón del Hueco, and Bolsón de la Mesilla/Conejos-Médanos (Chihuahua-Texas and New Mexico). The TAAP, initiated in 2006, is developed under the Federal Law *United States–Mexico Transboundary Aquifer Assessment Act*, which considers these aquifers as strategic and central elements for regional border economic integration (Hatch-Kuri 2017). This program aimed to: (i) develop binational information and share databases on the quality and quantity of groundwater; (ii) evaluate the affordability and movement of transboundary groundwater, and its interaction with surface water; (iii) develop and improve the information of groundwater flow systems; (iv) analyze water quality trends; (v) apply models and information necessary to protect water quality and improve sources of supply, and (vi) provide useful information to decision makers. So far, these results have been published several documents as *Informe de Actividades Hidrogeológicas del Acuífero Conejos-Médanos,* estado de Chihuahua, and in *San Pedro River Aquifer Binational Report* (2016).

In a recent work by Hatch-Kuri et al. (2019) a critical analysis was made about the results of the binational assessment of the San Pedro River Transboundary Aquifer (Sonora-Arizona) published in the *San Pedro River Aquifer Binational Report* (Callegary et al. 2016). In this work of interdisciplinary approach (Political Geography and Hydrogeology) it was concluded that, although diplomatic and scientific cooperation is recognized, it resulted in a robust database of hydrogeological, geological, meteorological, climatological, and edaphic indicators of the aquifer studied. However, little has been achieved regarding the understanding of the groundwater characterization as a system and its functioning related to other environment fundamental components. In this way, aspects related to the lack of a specific definition for concepts raised in the Report, such as "secondary aquifers," "confined aquifer," "aquifers in fractured rocks," "fill aquifers," and "regional aquifer," avoid having clarity about the central concept evaluated the San Pedro River Transboundary Aquifer. Although the lack of reported data homologation and units standardization were evidenced, this responds to the fact that binational work is not possible unless there is a general agreement signed by both governments on the matter, for that reason, each technical subnational working group worked separately (Callegary et al. 2018).

Despite this, TAAP has set a precedent in terms of scientific and diplomatic cooperation in transboundary groundwater, in fact, this issue was addressed with special emphasis in April 2019 in the framework of the *Binational Summit on Groundwater at the U.S.A.–Mexico Border*, in El Paso, Texas, a forum convened by both sections of the IBWC, where it was evident that although there is an interest of both governments in continuing to cooperate on this issue, the weight of scientific, institutional, legal, and financial asymmetries strongly influence the progress made to date (Hatch-Kuri 2019a). Indeed, the conceptual discrepancies found are a tacit example of the scientific and legal asymmetries that prevail, not only at a local or regional scale, since they reflect what happens globally.

Authors such as Callegary et al. (2018) indicated that assessing transboundary groundwater is a problem to be solved, regarding the conceptual and methodological limitations existing in both countries. In this sense, in previous works (Carrillo-Rivera et al. 2016) it was stated that in Mexico, the National Water Law lacks a legal definition for the concept "Groundwater" as well as for "Transboundary Groundwater" and that the Norma Oficial Mexicana 011 of the Comisión Nacional del Agua (NOM-011-CONAGUA), is used to get an inaccurate definition of the called annual average availability of groundwater in an (administrative) aquifer. Originally reported by Ward (1967), then by Freeze and Cherry (1979) as well as by Bredehoeft et al. (1982) the content of the NOM-011 includes an extreme simplification of complex processes such as the determination of the water-balance. Such computation provides a volume of rainwater that infiltrates the soil reaching the aquifer (geological or administrative) universally known as recharge; thus, a coined and misleading concept "total average annual recharge" is meant to be calculated. It should be noted that among other inconsistencies, the same "recharge" water thus computed is used as "recharge" in several of the called (administrative) aquifers. Therefore, the alleged balance (equality in Table 7.3) that should be measured and

Table 7.3 Equation used to determine the so-called "annual average availability" in a Mexican administrative

Average annual availability of subsoil water in an aquifer =

Annual average total recharge – Committed natural discharge – Groundwater extraction attached to a basic hydrogeological reality based on field data are not reliably evaluated, since in Mexico there is a lack of the necessary infrastructure (physical and human) to determine (not estimate) with precision data such as evapotranspiration, infiltration, extraction, precipitation, recharge, ecological flow, etc. In this way, works such as Arciniega et al. (2017) reveal that the use of methodologies such as remote satellite perception to calculate the water-balance in Transboundary Aquifers could only partially improve the application of NOM-011-CONAGUA, since there is the need to overcome the methodological limitations imposed by the aquifer (administrative) concept and the misleading calculations proposed in the NOM, sending to the background, the definition and understanding of the transboundary groundwater flow systems. In contrast, as noted in their analysis by Callegary et al. (2018), the application of methodologies such as isotopic determination of groundwater residence time, elevation, and climatic conditions of recharge may add a different objective to that of the application of NOM-011-CONAGUA, these are scientific asymmetries that characterize groundwater assessment and that were explicit in *Binational Summit on Groundwater at the U.S.A.–Mexico Border.*

The set of works and legal framework reviewed here, suggests the unquestionable influence that "Transboundary Aquifer" is a concept taken from the Resolution 63/124, as both government and academic sectors have focused on the study of characteristics of transboundary aquifers Mexico–U.S.A. However, it is pending the determination of transboundary groundwater systemic functioning under the Tóthian Flow Systems method, a fundamental approach to be arranged in domestic or international laws as shared or international watercourse and, consequently, to establish joint management schemes for its protection and environmental conservation. According to specialists like Rivera (2019), who stated that "the notion of aquifers (geological boundaries) is slowly disappearing, and nested (multiscale) groundwater flow systems are becoming the units of study. Groundwater databases around the world will become interoperable and available online as open sources. In the social and legal areas, integrated surface water and groundwater models will merge with economic and social models creating unique water management models," which suggests that future transboundary aquifer assessment works might no longer circumvent the definition of flow system functioning. Certainly, references of these works are in Canada–United States (Pétré and Rivera 2015; Pétré et al. 2015) and other parts of the world, such as Asia (Han et al. 2010) or Europe (IGRAC 2018).

7.5 Toward a Common Conceptual, Scientific, and Legal Reference for Mexico–United States of America Transboundary Groundwater

Based on the seven pillars of transboundary cooperation outlined in the United Nations Resolution 64/692 *Water, Peace and Security: Transboundary Water Cooperation* (2008b), specifically the Item 29 states that "Effective transboundary water management starts at the national level, where coordination and cooperation between different ministries and water-related institutions are needed, as have sufficient financing and political commitment"; also, the set of recommendations derived from the results of the 2012 Consultation *Groundwater Governance.* A Global Framework for Action (The World Bank, UNESCO-IHP, FAO, and IAH) that calls on states to regulate groundwater in their water laws in order to protect and improve their management against the environmental and social challenges climate change poses, and its current irrational use and, finally, considering that TAAP was the product of a U.S.A. domestic law to evaluate four transboundary aquifers with Mexico (2006–2016).

The book "*Groundwater Law: A Proposal*" was written and published in Mexico (Carmona et al. 2017) where the general objective is the protection, preservation, and control of groundwater extraction, based on the flow's systemic operation of different hierarchies. Out of a total of 98 articles presented in 11 Chapters, it is of interest to highlight Chapter VII "On Transboundary Aquifers," for containing a regulatory framework for transboundary groundwater based on the principles of sovereignty, territorial integrity, and sustainable development. This is a unique proposal in Latin America. This Groundwater Law Proposal is the result of the integration, *ex profeso*, of an interdisciplinary team in groundwater studies (Hydrogeology, Geography, Law studies, and Political Sciences) belonging to the National Autonomous University of Mexico (UNAM, Ciudad de México), who at the current stage of the pending issue of the General Water Law, as mandated by Article 4 Constitutional (Human Right to Water and Sanitation), joined the efforts of the organized society to modernize the Mexican regulatory water framework. In that situation, three academic activities that accompanied the process for formulating academic proposals deserve mention. On the one hand, two Colloquiums "Groundwater in México" were held, the first in 2013, at the facilities of the Mexican Institute of Water Technology (IMTA), the second, in 2015 at the Senate of the Republic with support of the Special Commission for Climate Change (Garza et al. 2018), while in 2017, the International Forum "Asymmetries in the Management and Regulation of Groundwater in North America: Toward an integrated scheme in México" (Hatch-Kuri 2019b). In the context of the latter, the delivery of the International Workshop "Regional Strategies for Transboundary Aquifers Management" was coordinated by two specialists (Alfonso Rivera, Geological Survey of Canada, and Andrew Stone, American Groundwater Trust). In sum, these activities reinforced the content of the Mexican transboundary groundwater regulation proposal.

As stated in previous works (Hatch-Kuri 2018), the academic proposal of the Groundwater Law has been delivered to several sectors involved on the subject (public, private, academic, and non-governmental organizations), in exchange, it has requested the issuance of a professional opinion with the purpose of recovering necessary comments and suggestions to restate its content and eventually deliver a much more enriched version to work with the Federal Legislative Power. As a result, more than 10 opinions have been received. The conceptualization "Groundwater" and "Transboundary Groundwater" requires the following changes:

In Chapter 1, "General Provisions," the concept "Transboundary Aquifer" is defined as *the transboundary groundwater management unit whose dimensions correspond to the flow system and the corresponding regional flow system pattern. Its delimitation and administration will meet the contents of the applicable international instruments* (Carmona et al. 2017). As it is evident, this definition emphasizes the characterization of the groundwater movement in the particular geological reference, without forgetting that this is the means that determines the movement of water. Its formulation is based on the fact that Groundwater Flow Systems (GWFS) illustrate the interaction, in space and time, of all the components of the Hydrological Cycle by including: (i) groundwater, (ii) aquifer, (iii) flow patterns, and (iv) interaction between them (Freeze 1969; Tóth 1999; Russell and Priebe 2017; Carrillo-Rivera and Ouysse 2013). In this way, the scientific and legal concepts in the law should be explicitly referred to as follows:

- **Groundwater**: it refers to the water that occupies the porous, conduits, or fractured space of sediments and rocks in below the surface, product of the infiltration rainwater (which depends on the conditions of the climate, soil, and vegetation); also the water that rises to the surface feeding streams, rivers, ecosystems, springs, sea, and wetlands.
- **Aquifer**: the geological referent in which groundwater moves and whose properties (porosity, storage coefficient, and hydraulic conductivity—permeability—) favor the one-time stored volume and groundwater movement.
- **Groundwater Flow Patterns:** groundwater flow is generated by gravitational forces through elevation differences on the water table, whereas the flow patterns develop in local, intermediate, and regional systems, modified by the permeability heterogeneities in subsoil rocks. The topography has a ubiquitous effect on these patterns causing its movement at greater depths. *Local systems* have lower temperature, pH, and salinity, and the residence time is from months to a few years. They are more vulnerable to contamination and changes in climatic conditions; *regional systems* have high pH, temperature, and salinity, their time of residence is thousands of years, are less vulnerable to be contaminated as well as to changes in climatic conditions. The *intermediate flows* have an extension limited by the convergence of at least two regional systems. The nature of the flows is set by comparative determinations.
- **Groundwater interactions with aquifer and flow patterns** are of the types: (i) physically controlled by Darcy's Law; (ii) chemical, water quality due to the dissolution of the minerals in subsoil rocks; (iii) isotopic, due to the elevation recharge zones and their age or residence (time) in the subsoil, and (iv) biological, due to its interaction with ecosystems, which are manifestly contrasted in the discharge, throughflow and recharge zones of each flow pattern.

With the above concepts, the possession of a conceptual and methodological clarity transferred to the legal framework is pursued, which deepens the definition of *Transboundary Groundwater Flow Systems* and which are precisely illustrated in the scheme of Fig. 7.1 (arrows that simulate water flow within the aquifer) and Fig. 7.3 (the Tóthianos flow systems), or where appropriate, have also been analyzed in previous work (Rivera 2015). Considering a systemic vision of groundwater in

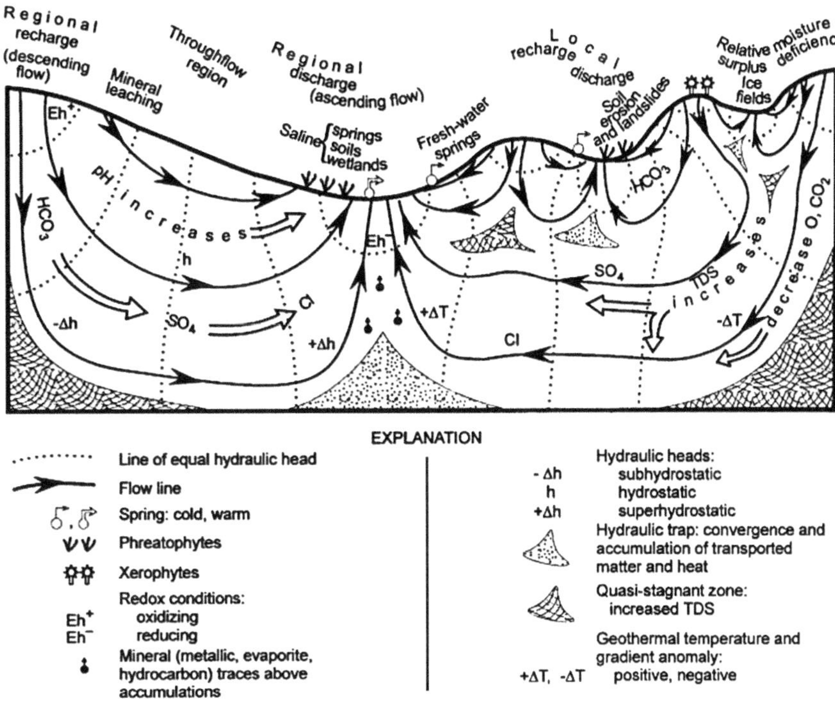

Fig. 7.3 An Aquifer and Groundwater flow systems (adapted from Tóth 1999)

the subsoil strengthens the content of the definitions of "aquifer" or "transboundary aquifer," analyzed in the previous sections, overcoming its current limitations.

Under the GWFS conceptual paradigm, its benefit is expressed in the clarity of the spatial location of key components, such as groundwater recharge, throughflow, and discharge areas in transboundary conditions, a situation that requires rethinking the desirability of continuing using the "Transboundary Aquifer" concept, such as the management unit for these shared groundwaters (Rivera 2019). With this rethinking approach, the Mexican State would be expected to supplant the current notion of administrative aquifer, to the extent that the results of the scientific evaluations determined by GWFS throughout the country, would be the unit of domestic and international groundwater management and not the transboundary aquifers.

Finally, although the findings of Rivera and Candela (2018), and Sánchez and Eckstein (2020) who conclude that the best management schemes for transboundary groundwater is when these are managed locally, certainly, in Mexico, this is a major impediment, until the content of the Article 27 Constitutional conferring the groundwater patrimonial domain to the Head of the Federal Executive Power in turn is modified along with an updating of the methodology of evaluation (GWFS instead of Water-Balance), the groundwater management will be little democratized, or where appropriate, may be subject to political sovereignty and territorial of the

Mexican subnational governments, thinking about the governance schemes in United States and Canada. Meanwhile, a first fundamental advance is to incorporate updated concepts in the current Mexican domestic legislation, as to be certain in the exercise of sovereignty and groundwater management in transboundary conditions.

7.6 Final Considerations

The action of naming concepts, objects, and people is a political action, at the same time it implies to exercise power. Thus, from Political Geography a conceptualization of water as "hybrid" implies to understand that processes as an action of power and, in this sense, concepts as "transboundary aquifer" and "transboundary groundwater," are loaded with different connotations and are the direct result of scientific consensus that is transmitted to decision makers. Contributions made from Hydrogeology are helping to define key concepts that characterize aquifers and the systemic functioning of groundwater from scientific evidence but, these contributions have not yet been fully translated into acts of power or authority embodied in international transboundary water laws due to a lack of political consensus and understanding between specialists and decision makers.

However, every day there is an increasing recognition of the fact that groundwater is the fundamental component of the "Transboundary Hydrological Cycle"; from a system view approach, this implies water movement to, from, and within aquifers, as well as the relationship with other environmental components. The codification of groundwater as an international or shared watercourse is an effort that undoubtedly requires strong science and sound political decisions to be translated into the formulation of local and integrative management schemes for the effective protection and conservation of groundwater. Thus, the transboundary aquifers assessment in the future must provide scientific elements or evidence that help solve the question referring to which country has affected groundwater more, in order to clarify which of the two countries should assume major responsibility in groundwater deterioration, not only in quantity but of the quality of the transboundary groundwater, beyond official data.

Groundwater Flow Systems and their components (groundwater + aquifer + flow system pattern + their interaction) point to the strengthening of the "Transboundary Groundwater" concept. This work shows evidence that the concept "Transboundary Aquifer" and its effect on Mexico–United States of America Transboundary Aquifers scientific evaluation works have provided evidence of the definition of shared hydrogeological units, but not in groundwater flow systems functioning. For that reason, standardizing and harmonizing scientific concepts and legal definitions in domestic water regulations in both countries is an initial step to make clear where the states should exercise their sovereignty and what they should negotiate in order to avoid conflicts within the context of integrated transboundary water management.

References

Agnew J (2002) Making political geography. Routledge

Arciniega S, Breña A, Hernández J A, Pedrozo A (2017) Balance hídrico en acuíferos transfronterizos basado en percepción remota. In: IAP, F.G.R.A. Libro de Resúmenes del I Congreso Latinoamericano and V Congreso Nacional de Manejo de Cuencas Hidrográficas

Arizona Revised Status (2020) Title 45 "Waters", Chapter 2 "Groundwater", Article 9: §45–561

Bredehoeft JD, Papadopulus SS, Cooper HH (1982) The water budget myth. In: Scientific basis of water resource management, Studies in Geophysics. National Academy Press, Washington, DC, pp 51–57

Brooks DB, Linton J (2011) Governance of transboundary aquifers: balancing efficiency, equity and sustainability. Int J Water Resour Dev 27(3):431–462

Burchi S (2018) Legal frameworks for the governance of international transboundary aquifers: pre-and post-ISARM experience. J Hydrol Reg Stud 20:15–20

California Water Code (2020) Division 13 "California water districts", Chapter 1, general provisions, § 37900 (1)

Callegary JB, Megdal SB, Villaseñor ET, Petersen-Perlman JD, Sosa IM, Monreal R, Noriega FG (2018) Findings and lessons learned from the assessment of the Mexico-United States transboundary San Pedro and Santa Cruz aquifers: the utility of social science in applied hydrologic research. J Hydrol Reg Stud 20:60–73

Callegary JB, Minjárez Sosa I, Tapia Villaseñor EM, dos Santos P, Monreal Saavedra, R, Grijalva Noriega FJ, Huth AK, Gray F, Scott CA, Megdal SB, Oroz Ramos LA, Rangel MM, Leenhouts JM (2016) San Pedro River aquifer binational Report: international boundary and water commission, p 170

Carmona LMC, Carrillo-Rivera JJ, Hatch-Kuri Gonzalo, Huizar-Álvarez R, Ortega-Guerrero M (2017) Ley del agua subterránea: una propuesta. Universidad Nacional Autónoma de México, Ciudad de México, México

Carrillo-Rivera JJ, Peñuela L, Huizar-Álvarez AR, Ortega-Guerrero A, Vallejo J, Hatch-Kuri G (2016) Capítulo 10. Conflictos por el agua subterránea. Geografía de México. Una reflexión espacial contemporánea 1, pp 151–166

Carrillo-Rivera JJ, Ouysse S (2013) Evaluation of groundwater flow system functioning in Mexico to reduce drought impacts. Drought in Arid and Semi-Arid Regions. Springer, Dordrecht, pp 269–280

Freeze RA (1969) The mechanism of natural ground-water recharge and discharge: one-dimensional, vertical, unsteady, unsaturated flow above a recharging or discharging ground-water flow system. Water Resour Res 5(1):153–171

Freeze RA, Cherry JA (1979) Hydrogeology. Prentice-Hall Publications, p 604

Garza G, Carrillo-Rivera JJ, Huizar-Álvarez R (2018) Coloquios sobre el agua subterránea en México, Senado de la República, LXIII Legislatura, p 155

Han Z, He J, Niu L (2010) Transboundary aquifers in the Great Mekong River Basin. International conference "Transboundary aquifers: challenges and new directions" (ISARM2010). UNESCO-IAH-UNEP Conference, Paris

Hatch-Kuri G (2017) Paso del Norte: La competencia por las aguas transfronterizas. El Colegio de Chihuahua-Universidad Autónoma de Ciudad Juárez, Ciudad Juárez, México

Hatch-Kuri G (2018) A joint management of transboundary aquifers: from asymmetries to environmental protection. Frontera norte 30(59):129–154

Hatch-Kuri G (2019a) Cumbre Binacional Sobre Aguas Subterráneas Transfronterizas México-Estados Unidos. TecH20 Center, El Paso, Texas, 10 y 11 Abril Investigaciones Geográficas, No. 99, jul. 2019

Hatch-Kuri G (2019b) Toward comprehensive groundwater management in Mexico. Dossier Water, Voices of Mexico, 108, Summer: 57–58

Hatch-Kuri G, Carrillo-Rivera JJ, Huizar-Álvarez R (2019) Evaluación crítica del acuífero transfronterizo río San Pedro. Reg Cohes 9(1):61–85

IGRAC (2018) Population and aerial statistics for 199 Transboundary Aquifers. IGRAC, Delft
Linton J, Budds J (2014) The hydrosocial cycle: defining and mobilizing a relational-dialectical approach to water. Geoforum 57:170–180
Moreira R (2010) Para onde vai o pensamento geográfico? por uma epistemologia crítica. Edit, Contexto
New Mexico Statutes Water Law (2016) Chapter 72 "Water Law", Section 72, Article 5A Groundwater Storage and Recovery, § 72-5A-3
Norma Oficial Mexicana 011-CONAGUA (2015) Conservación del recurso agua-Que establece las especificaciones para determinar la disponibilidad media anual de las aguas nacionales, DOF: 27/03/2015, SEGOB México
Pétré MA, Rivera A (2015) Synthesis of knowledge of the Milk River Transboundary Aquifer (Alberta. Natural Resources Canada, Canada-Montana, USA)
Pétré MA, Rivera A, Lefebvre R (2015) 3D unified geological model of the Milk River Transboundary Aquifer. Alberta, Canada-Montana, USA
Ribeiro WC (2010) Geografia política e gestão internacional dos recursos naturais. Estudos avançados 24(68):69–80
Ribeiro WC (2012) Soberania: conceito e aplicação para a gestão da água. Scripta Nova. Rev Electrón Geogr Cienc Soc 16(418):28
Rivera A (2015) Transboundary aquifers along the Canada–USA border: Science, policy and social issues. J Hydrol Reg Stud 4:623–643
Rivera A (2019). What is the future of the Groundwater? Groundwater 57(5):661–662. https://doi.org/10.1111/gwat.12902
Rivera A, Candela L (2018) Fifteen-year experiences of the internationally shared aquifer resources management initiative (ISARM) of UNESCO at the global scale. J Hydrol Reg Stud 20:5–14
Russell HF, Priebe EH Comp (2017) Regional-scale Groundwater Geoscience in Southern Ontario: Ontario Geological Survey, Geological Survey of Canada, and Conservation Ontario open house. http://doi.org/10.4095/299750. Report number: OF 8212
Sánchez R, Eckstein G (2017) Aquifers shared between Mexico and the United States: management perspectives and their transboundary nature. Groundwater 55(4):495–505
Sánchez R, Eckstein G (2020) The path towards groundwater management in the borderlands of Mexico and Texas. Wiley Interdisc Rev Water 7(1):e1399
Sánchez R, López V, Eckstein G (2016) Identifying and characterizing transboundary aquifers along the Mexico–US border: an initial assessment. J Hydrol 535:101–119
Sánchez R, Rodríguez L, Tortajada C (2018) The transboundariness approach and prioritization of transboundary aquifers between Mexico and Texas. Ambio 47(7):760–770
Swyngedouw E (2009) The political economy and political ecology of the hydro-social cycle. J Contemp Water Res Educ 142(1):56–60
Talledos E (2014) La geografía: un saber político. Espiral 21(61):15–49
Texas Administrative Code (1977) Water Code, §330.3(8) and §335.1(8)
Tóth J (1999) Groundwater as a geologic agent: an overview of the causes, processes, and manifestations. Hydrogeol J 7(1):1–14
United Nations General Assembly Resolution 51/229 (1997) Convention on the Law of the Non-navigational Uses of International Watercourses (08 July 1997). Retrieved from https://treaties.un.org/doc/source/docs/A_RES_51_229-Eng.pdf
United Nations General Assembly Resolution 63/124 (2008a) The law of transboundary aquifers (11 December 2008). Retrieved from http://www.un.org/en/ga/search/view_doc.asp?symbol=A/RES/63/124&Lang=S
United Nations General Assembly Resolution 63/124 (2008b) Water, peace and security: transboundary water cooperation (04 March 2010). Retrieved from https://undocs.org/pdf?symbol=en/A/64/692
United Nations Economic Commission for Europe (UNECE) (1992) Convention on the Protection and use of transboundary watercourses and international lakes (17 March) https://www.unece.org/fileadmin/DAM/env/water/pdf/watercon.pdf

United Nations Economic Commission for Europe (UNECE) (2014) Model provisions on transboundary groundwaters. United Nations Economic Commission for Europe, Geneva

United Nations Educational, Scientific and Cultural Organization (UNESCO-IHP) (2015) Estrategia regional para la evaluación y gestión de los Sistemas Acuíferos Transfronterizos en las Américas. UNESCO, Montevideo, Uruguay

Ward RC (1967) Principles of Hydrology. McGraw Hill. ISBN 10: 0070940479; p 403

Wolf A, Kramer A, Carius A, Dabelko G (2005) Gestionando conflictos por el agua y cooperación. In: Renner M, French H, Assadourian E (Eds.), La situación del mundo 2005: Redefiniendo la seguridad mundial. Informe Anual del Worldwatch Institute sobre el progreso hacia una sociedad sostenible, Barcelona, Icaria, Centro de Investigación para la Paz, 155–178

Chapter 8
Landscape Functioning as a Basis for Establishing Sustainable Intervention: Soils and Groundwater Flows

Margarita María Alconada-Magliano

Abstract The ecosystems and the introduced production systems of a region are distributed in a dynamic mosaic of operations. Ecosystems regulate and are a consequence of the presence of groundwater. However, in general, interventions in the landscape are carried out without considering how the elements that integrate it (soil, water, vegetation, geomorphology) are linked, and without establishing the cause–effect relationships of the natural and anthropic processes that generate land degradation and prevent the sustainability of production systems. Regarding the soil, its study is carried out without establishing the origin of its properties and without defining the quality and nature of groundwater in environments where groundwater is shallow. The theory of groundwater flow systems and its regional hydraulic continuity allows an understanding of how the elements of the landscape are linked and provides a basis to be able to decide sustainable interventions. In this chapter, the consequences and limitations of soil studies according to current procedures are analyzed. In addition, the need to incorporate study and interpretation criteria that considers the inescapable relation between local soil in a regional context is exposed, thus having an integral vision of landscape functioning. This is demonstrated by a case study, in northwest Buenos Aires province, Argentina, where soil, groundwater, and vegetation relationships have been studied.

Keywords Northwest Buenos Aires · Argentina · Groundwater flow · Sustainable management · Water quality · Soils

8.1 Introduction

The description of the environment and its functioning must be studied from a landscape perspective taking into consideration all the elements that make it up (soil, water, vegetation, lithology, and geomorphology). By understanding the cause–effect

M. M. Alconada-Magliano (✉)
Edafología, Facultad de Ciencias Agrarias y Forestales, Universidad Nacional de La Plata, Calle 60 y 119 s/N (1900), La Plata, Argentina
e-mail: margaalconada@yahoo.com.ar

M. M. Alconada-Magliano (ed.), *Intensified Land and Water Use*, Springer Earth System Sciences, https://doi.org/10.1007/978-3-030-65443-6_8

relationships of the natural and anthropic processes produced in the environment, it is possible to decide sustainable landscape intervention and to select those management practices that effectively contribute to reverse or control an environmental problem that restricts production and life quality.

The ecosystems and productive systems of a region are distributed in a functioning mosaic. The ecosystems regulate and are a consequence of groundwater, being it possible for different ecosystems to be linked or complement one another in their hydrologic functioning. This is due to the existence of regional hydrologic continuity of groundwater systems flows (Tóth 1962, 1995, 2000, 2016; Engelen and Jones 1986; Carrillo-Rivera 2000). These authors proposed a dynamic groundwater understanding, instead of the static analysis of the aquifer unit and of the hydric balance, showing that there is groundwater interdependence between zones in a basin or between neighboring basins, between groundwater and superficial water and also showing that these are linked to other landscape element (relief, lithological unit, soil, and vegetation). Moreover, these authors stated that different flow systems may be found in the same aquiferous unit or that the same flow system may circulate through two or more aquiferous units. Carrillo-Rivera (2000) explained that in numerous places around the world the existence of underground hydraulic interdependence between separate superficial basins that may be superficially disguised or not easily detected has been proven which may lead to interpretation mistakes when analyzing, for example, the origin of waterlogging.

It is evident that groundwater flow systems are the linking natural element of the other landscape elements being this especially noticed in sedimentary plains. It is thus important to consider that the dynamics of groundwater, when studying soil and vegetation, exerts influence on them. However, this dynamic understanding in the study of soil, when defining its use capability and its relationship with natural or artificial vegetation, is not the approach that prevails. On the other hand, the soil and vegetation are frequently studied in a static and local way; the prevailing procedure for studying groundwater is hydric balance, and the superficial basin is used as the environment management unit. These criteria to study the environment are the ones generally considered when deciding how the landscape is organized, which explains, at least in part, the increasing degradation of soil and water, worldwide known (FAO 2011; Pla-Sentis 2017), and in Argentina in particular (Casas and Albarracín 2015).

In this chapter, the consequences and limitations of soil studies according to current procedures are analyzed. In addition, the need to incorporate study and interpretation criteria that consider the inescapable relation between local soil in regional context is exposed, thus having an integral vision of landscape functioning. This is demonstrated by a case study, in northwestern of Buenos Aires province, Argentina, where soil, water, and vegetation relationships have been studied.

8.2 Scope of Soil Studies and Consequences

8.2.1 Definition of Soil and Its Study

The soils are the basis of ecosystems, its origin and properties are the result of the combination of biotic and abiotic factors defined by Jenny (1941) as formation factors (original material, time, biota, relief, and climate). Porta et al. (1994) defined the soil as a natural body with distinctive, repetitive, and previsible properties. This is due to the fact that observable and measurable edaphic properties are the consequence of how the forming factors are linked. Therefore, it is possible to know ahead the occurrence of certain types of soil and properties that will repeat if the factors combine in an analogous way. If some factors change in intensity or direction, the soil is likely to change. However, the regional and local relationships are not always clearly set when studying the soil up to 1.5–2.0 m depth or up to the presence of mechanical impedance (limestone, water table, rock) at a given time. Then regional and local relationships are not clearly set when soils are classified by any taxonomic criteria (Soil Survey Staff 2014; WRB 2014) that are based on the occurrence or absence of certain edaphic properties, regardless of groundwater as a dynamic element in soil evolution. Therefore, the previsibility is not always correctly determined when only local factors are considered and the origin of soil properties is not established. Thus, when defining the relief factor of the soil being studied there is frequently only indication of topographic location and whether a water table is present at the time of study or if there are hydromorphic characteristics (such as Fe-Mn motting and concretions). This will not suffice in environments where: (i) water table fluctuates near the surface, (ii) water table is under the frequent soil study depth, (iii) there are discharge zones of intermediate and regional flow. Those flows may not become evident for a long period. As explained before, it is possible for two different flow systems to coexist in the same place, and to appear more superficially depending on the changes of recharge and discharge flows (Tóth 2000, 2016; Carrillo-Rivera 2000). This fact should be taken into consideration when the landscape is intervened in an area hydrologically functioning as complementary to another one (recharge and discharge zones of the same flow of groundwater).

8.2.2 Current Limitations in Soil Studies

When regional and local hydrological changes are not considered together with how these affect soil and vegetation, partial views or diagnosis errors could occur in situation such as: (i) defining origin of floods (Alconada-Magliano et al. 2011b); (ii) adequacy of the vegetation as "bio-drainage" technique in hydric excess regions (Heuperman 2003; Heuperman et al. 2002; Tomar 2007; Gupta2007; Alconada-Magliano et al. 2009; Galetti 2014; Nosetto and Jobbágy 2014; Bonnesoeur et al. 2019); (iii) need to take advantage of water table for production in zones with water scarcity and manage the farmland by environments (Martini and Baigorri 2004; Saks et al. 2012; Álvarez et al. 2015; Lardone 2016); (iv) consequences of raising levels of water table due to deforestation (Burkart 2009; Morello et al. 2009); (v) implementation of forestry, agroforestry and silvopastoral systems, mainly in salinity environments (Jobbágy et al. 2006; Taleisnik and López-Lauenstein 2011; Ewens et al. 2012; Sosa 2012; Domínguez-Daguer and Laclau 2014; Martinez-Calsina et al. 2015; Laclau et al. 2015; Chará et al. 2019); (vi) accurately determining the origin of natural or by anthropic interventions edaphic salinity (Nosetto et al. 2007, 2009; Jobbágy and Nosetto 2015; Lavado and Taboada 2017; Taboada et al. 2017; Rubio et al. 2019); (vii) consequences of drainage systems (PMI 1999); (viii) selection of irrigation systems; (ix) define vulnerability to agrochemicals and other contaminant of groundwater (Gaona et al. 2019), (x) define the role of wetlands in different regions (Custodio-Gimena 2001; López-Geta and Fornés-Azcoiti 2009; Peñuela-Arévalo et al. 2015). The absence of a dynamic functioning view, consequently, prevents us from defining adequacy, effectiveness, and sustainability of any type of landscape intervention that changes the hydrology of a place of land, environment, or region. It also leads to the fact that changes that could occur when faced with other local and regional climatic scenarios cannot be anticipated.

8.2.3 Origin and Consequences of Environmental Intervention

Social, economic, productive, and environmental consequences have been reported worldwide. Pla-Sentis (2006) pointed out that development and agricultural expansion policies during the last decades in Latin America have been conducive to water and soil degradation, thus affecting hydrological basin, causing decrease in productivity, increase in production costs, flooding, landslide, sedimentation in dams, and desertification. According to this author, this happens because the cause–effect relations in the above mentioned degradation processes are not adequately identified or evaluated thus recommending a hydrological study approach. Pla-Sentis (2014, 2017) also pointed out that the misuse of water and soil derives from a growing and developing population with scarce research and empirical practices of water and soil conservation. Moreover, he considered this to be the principal source of catastrophes and argues that the consequences of soil degradation are of similar impact as those of global warming and biodiversity loss; these three processes are closely related.

This coincides with degradation and contamination of the environment mainly related to agronomic activities as described in Argentina by Casas and Albarracín (2015). It is common in Argentina, the management of soils is decided according to economic context and/or soil use aptitude which arise from available soils mapping being the most accurate ones those at the scale of 1:50,000 (GeoINTA 2013a). In line with what has been previously mentioned, the mapping was carried out at a certain time, regardless of either edaphic properties, origin, or groundwater quality, even when close to the surface, or without establishing hydrological relationships between soils close to each other, or without considering the local and regional hydraulic continuity (Tóth 1962, 2000, 2016; Engelen and Jones 1986; Carrillo-Rivera 2000).

Even when research and technology developments, mainly in other fields (genetics, agrochemicals, machinery, and others), have enabled increasing yield in main crops, environmental problems derived from misuse of soil and water become more serious in many regions. FAO (2011) points out that the need for food for the estimated population in 2050 should increase by 70%, which makes it relevant to improve procedures for sustainable production. In view of previous comments, it becomes evident that it is necessary to take on hydrological approaches to study the environment in general and soils in particular. Regarding the soil, it is necessary to incorporate physico-chemical properties of groundwater in those environments subjected to their local and regional influence, and/or to infer their origin and behavior considering edaphic properties associated to water table conditions (hydromorphic features, salinity, alkalinity, pH, cationic, and anionic composition). According to all the above said it is necessary to: (i) connect the different elements which compose the local and regional landscape; (ii) analyze soil distribution patterns and ecosystems; (iii) observe how soils, according to their origin, are modified by natural and anthropic factors. Those would enable us to reverse the current increase in the degradation of soil and water and to propose sustainable interventions.

8.3 Tóth's Theory of Flow Systems in the Study of the Soil

8.3.1 Definition and Relevance of Groundwater Flow Systems as Unit of Study

The required hydrological approach to understand natural and anthropic processes which occur as a consequence of soil and water management (Pla-Sentis 2006, 2014) can be addressed applying Toth´s theory of groundwater flow systems (Tóth 2000, 2016). An understanding of these systems contributes to the planning of a sustainable management of linked ecosystems. This author described three basic systems of groundwater flow types: local, intermediate, and regional, and three zones within these: recharge, throughflow, and discharge. These are defined by the distance and depth of the water that makes up each flow. Thus, as regards a given place: how deep the water penetrated and how long it flowed between the recharge zone (descending flow) and discharge zone (ascending flow). The throughflow zone (lateral and horizontal flow) is between these two. A flow system is local if it is determined by the local rainfall and it has a short flow. It is regional when it is formed by rainfall from other places and travels hundreds to thousands of kilometers until it discharges. The flows travel separately in the underground media like oceanic currents. The zones that integrate the flow are complemented in their hydrological behavior and link different ecosystems (Carrillo-Rivera 2000).

Each zone has completely different and contrasting water conditions which can be identified by environmental indicators: soil, geomorphology, vegetation, quality of water, hydraulic recharge and others (Carrillo-Rivera 2000; Tóth 2000, 2016). It is necessary to define the physico-chemical quality of the water to establish how the systems flows work. This is possible because the water has "*traveling memory in its molecules*": total dissolved solids, anions, and cations concentration, whether it is fresh or saltwater, hot or cold, young or old, if it fell at sea level or on mountains, and if it has contaminants (Fagundo-Castillo 1990; Fagundo-Castillo and González-Hernández 2016). The temperature, pH, redox potential (Eh), alkalinity, and tracing elements are incorporated in each flow system path as shown in Fig. 8.1a. The physico-chemical quality of the water results from the intemperation of the geological material across which it travels, through reactions of hydrolysis, ionic exchange, precipitation, solubilization, etc. (Fagundo-Castillo 1998; Fagundo-Castillo and González-Hernández 2016).

Figure 8.1 shows (a) a general scheme of groundwater systems (types, zones); (b) the water physico-chemical characterization which prevails in each of these. It is observed that pH, salinity (EC electrical conductivity), alkalinity (SAR sodium adsorption ratio, Na concentration), temperature (T), redox potential (Eh) are increased from recharges zones to discharge zones. The magnitude of the change depends on the flow type. Also, the anionic composition modified with prevailing HCO_3^- in recharge zones and Cl^- in discharge zones. However, this general water scheme may show variation due to coexisting water flows from different origins in the same site and due to different interactions between the soil, rainfall, groundwater, vegetation, climate, management, and other natural or anthropic factors. This is analyzed in item 8.4.

Tóth (2000, 2016) and Carrillo-Rivera (2000) thus made a difference between the superficial basin that responds to a topographic pattern linked to a main river and the "*Unitary basin*" that is the basic unit of flow systems. This basin results from regional hydraulic continuity defined by hydrogeological contrast of the underground that generates flows not necessarily belonging to the superficial basin under study. Thus, Tóth (2000) proposed the basic diagnostic unit of groundwater be "*groundwater flow systems*" that he defined as "a natural and coherent unit, in space and time, consisting of the groundwater of specific physico-chemical quality, that flows through geologic material related with a certain geomorphology, vegetation and soil."

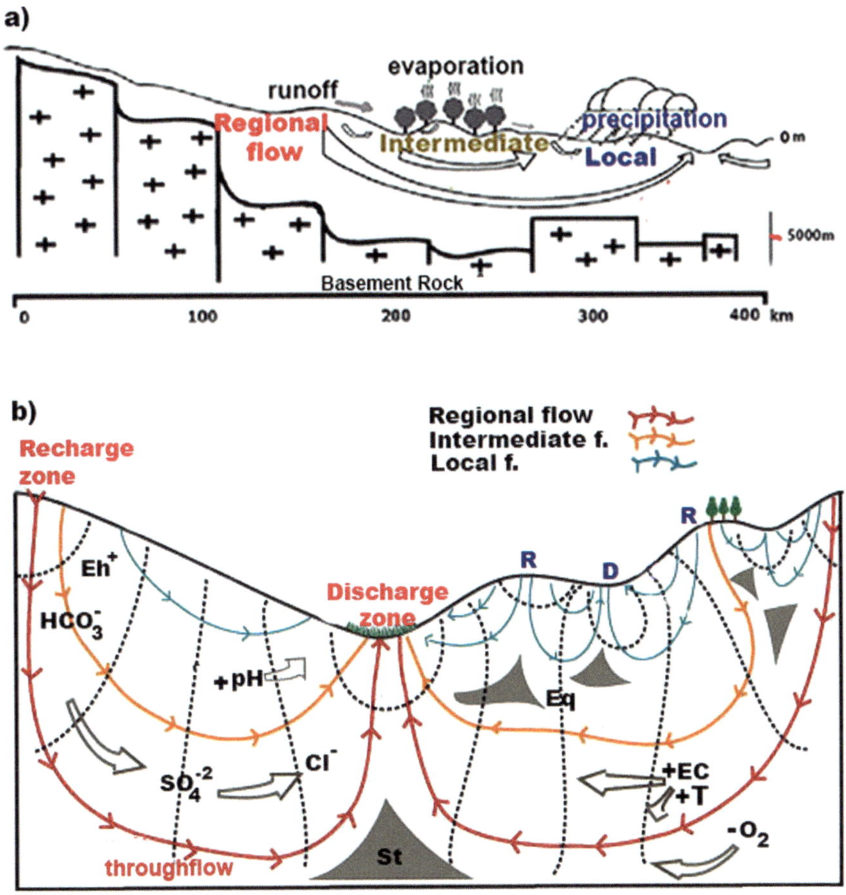

Fig. 8.1 **a** Example of groundwater flow systems in sediment over basement rock. **b** Theoretical profile showing fundamental groundwater flow systems; the vertical scale depends on the geological framework (i.e., basement rock position) (Adapted from Tóth 2000). EC, electrical conductivity (salinity increases with depth and travel); T, temperature (increases with depth and travel); Eq, equipotential line; St: stagnation zone (O_2 decrease with increase of salinity or total soluble solids)

8.3.2 Soil and Groundwater Flow Systems Relations

Figure 8.2 shows an example. The soils shown in Fig. 8.2 have been developed from the Late Pleistocene/Early Holocene loess and loessoid (equivalent to reworked loess or loess-like deposits) (Zárate 2003), in a temperate humid climate, being the relief a factor what modifies the soil type and vegetation (Fig. 8.2a). If the soil is studied at a point, it is impossible to know the functioning soil. It is necessary to know this point under study is linked to the different flow systems and the zone within these (Fig. 8.2b). For example, if the soil at the point above mentioned is *Typic*

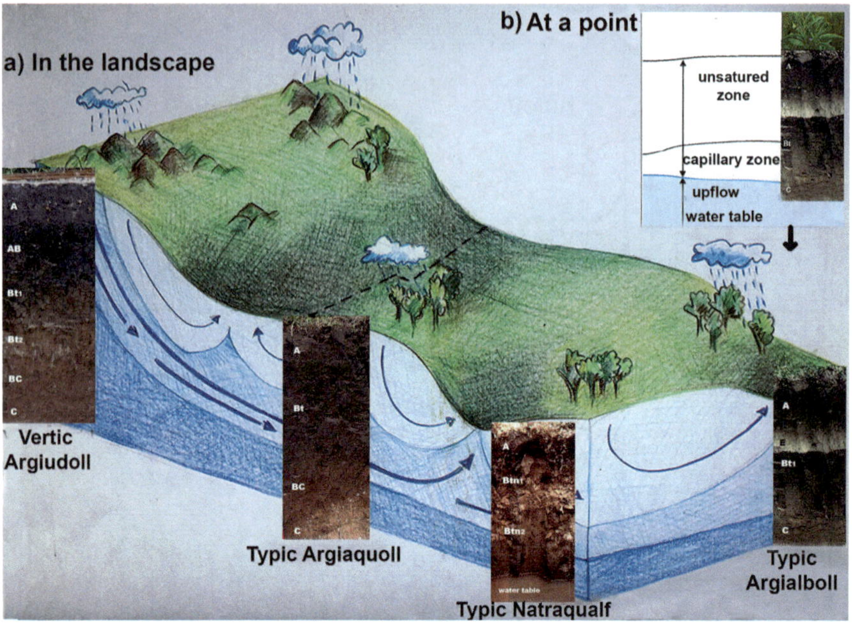

Fig. 8.2 **a** Landscape observation: Sequence of soils in a loess landscape, with humid temperate climate, in different zones of groundwater flows (study of soil in the landscape) **b** Observation at a point (partial soil study)

Argialboll by traditional soil classification studies (Soil Survey Staff 2006). This soil is commonly called "*fresh low*" due to its low salinity up to depth of study, this soil belongs to the local flow discharge zone. However, deeper down Na^+ content of the soil increases (horizon A has 0.4 me.l^{-1} Na^+, Btss at 53 cm deep has 2.2 me.l^{-1} Na^+) (Hurtado et al. 2006). It is possible then to estímate that more deeply there is a more alkaline flow like an intermediate flow. Therefore, it is possible that two water flows of different origin coexist in this place. This is consistent with the existence of very alkaline soils in the region, as seen in Fig. 8.2a. Considering how the soils are distributed in the landscape enables us to establish regional links that are logical in geological sedimentary material (Engelen and Jones 1986). In Fig. 8.2a, the soil in the lowest areas of the regional relief are *Typic Natraqualf* which are commonly called "*alkaline low*," belonging to intermediate or regional flows discharge zones where the Na^+ content is high since the water travels long distances.

In higher topographic positions of the region the *Vertic Argiudoll* soil is found linked to a local flow recharge zone. In somewhat lower positions, *Typic Argiacuoll* soil is found probably linked to a local flow throughflow zone. In this example, the local and regional topographic positions and regional geological uniformity enables us to easily establish the relationships between soil type, local and regional topographic position and groundwater characterization. In other regions, however, mainly where there are lithological discontinuities in the original material of soils and/or groundwater flows of different origin coexist, the relationship between the landscape perspective in the study of the soils becomes especially relevant.

8.4 The Landscape as a Unit of Study and Soil Intervention. Study Case: Longitudinal Dunes (Médanos Longitudinales) in Northwest Buenos Aires Province (the Sandy Pampas)

8.4.1 Geological Framework

The Sandy Pampas (SP) show the occurrence of an extensive aeolian sand cover (Fig. 8.3a). This region consists of a large and complex sand dune system that corresponds to the "Sand Sea" as it was named by Iriondo (1990 in Zárate 2003). Within the Pampean Plain (Llanura Pampeana) or Humid Pampa (1,000,000 km^2) in Argentina the sub-region called Sandy Pampa (SP) is identified. It is located in Northwestern Buenos Aires province and is found below 100 ma.s.l. Its surface is 5,500 km^2 with relatively homogeneous geomorphology, regional slope 0.25%, without defined drainage pattern, surface water flows slowly among a microrelief of crests and stabilized dunes, within which larger geomorphological units are found: the *Longitudinal Dunes* (Médanos Longitudinales) (3,800 km^2) to the North and the *Parabolic Dunes* (Médanos Parabólicos) (1,700 km^2) to the South (Fig. 8.3b). The distribution of Longitudinal Dunes, parallel to sub-parallel oriented NNE-SSW, transversal to the regional topographic slope (W-E) exert important control over hydrological behavior. There are 20,000 ha of superficial water bodies of extremely varied extent, from 0.01 ha to 15,000 ha, being the smaller ones (between 0.05 and 10 ha) the most common (Dangavs 2005; Gabellone et al. 2003). Among the shallow lakes the most important systems are the Hinojo-Tunas (lake complexes) (H-T in Fig. 8.3b). The drainage type is arreic and with shallow water table (<5.0 m in depth) which is linked to the occurrence of hydric excesses alternating with hydric deficit periods. However, the rising of water table is not in correspondence with local precipitation but with the contribution of long path groundwater flows (Alconada-Magliano et al. 2011b).

The Sandy Pampas limit to the N-NE with the Salado River which was artificially linked by means of channels, to the SE with the Vallimanca Creek and to the South with the Encadenadas Lagoons (chained lagoons). At present these three hydric sub-regions are linked and make up the so-called Salado River Basin (170,000 Km^2) (PMI 1999). Geologically, it presents a Precambrian crystalline rock basement that emerges toward the South in the Tandilia and Ventania mountain systems (500 and 1,100 m a.s.l.) and is then progressively buried toward the North by younger sediments, as thick as 2–6 km (Zárate and Rabassa 2005). González (2005) pointed out that SP is a region that consists of a large and complex sand dune system that on the surface

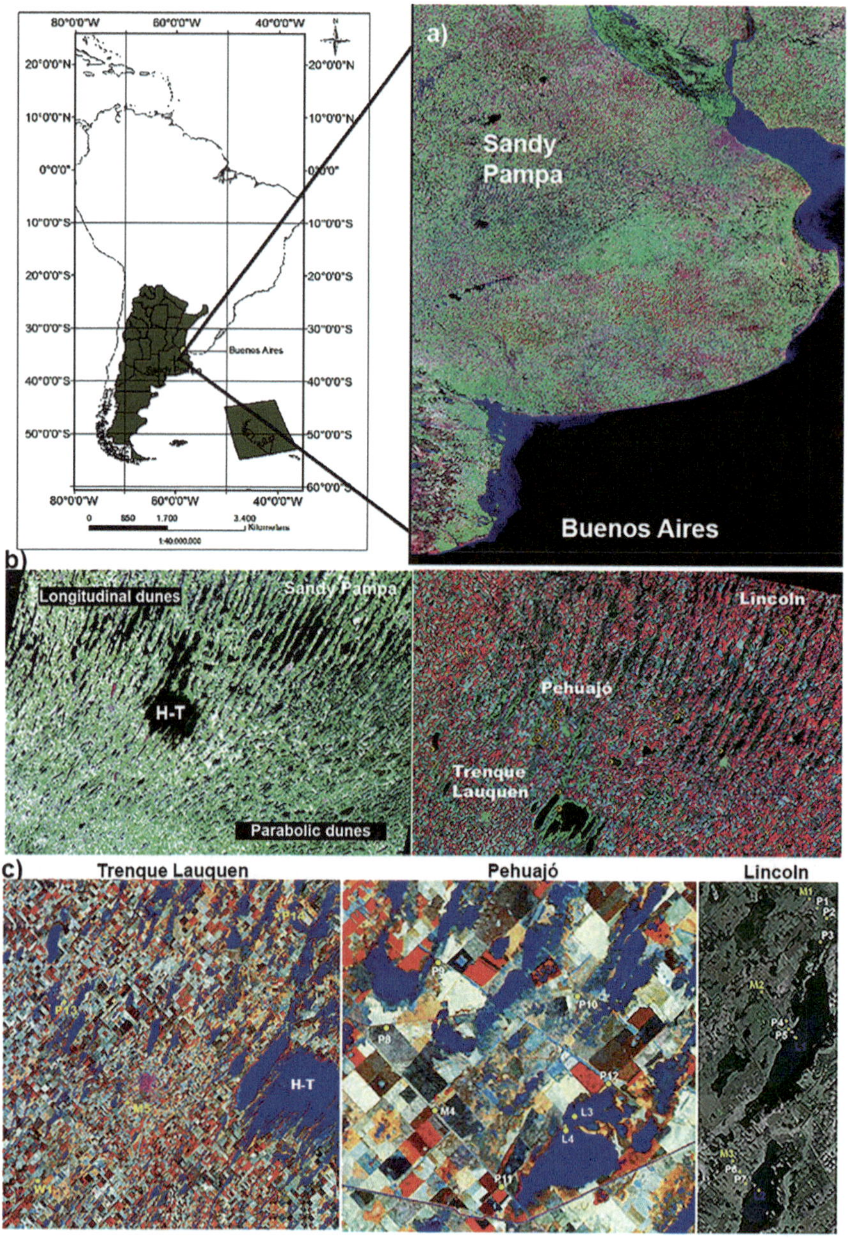

Fig. 8.3 **a** Location of Sandy Pampa in the province of Buenos Aires, Argentine. **b** Longitudinal Dunes and Parabolic Dunes, in Sandy Pampa. H-T lacustrine system Hinojo-Las Tunas in Trenque Lauquen. **c** Study sites in Trenque Lauquen, Pehuajó and Lincoln

presents the E1 formation (aeolian sand cover) called Invading Dunes, of Holocene Age. It is located over the Pampanea, Araucana, and Puelche formations (from the Pliocene to the Pleistocene) which altogether make up the aquiferous system with regional hydraulic continuity not confined. The deeper geological formations are not integrated with the currently active groundwater flow systems (González 2005).

8.4.2 Characteristics of the Area

The region that is called "Longitudinal Dunes" (Fig. 8.3b) is composed of dunes probably belonging to a Late Pleistocene age and it is an eolian system of lineal accumulation of sand and silt, fixed by vegetation in correspondence with semi-arid climates. They are approximately 100 km in length, 2–5 km wide, and 6 m high, which cut the regional slope 0.25‰ NW-SE. Among the dunes there are inter-dune zones that are flat to slightly concave from 0.5 to 5 km in width. The height and original shape of the dunes have been softened by the action of the wind and at present are flattened geoforms with slopes varying between 0.08 and 0.3% (Dillon et al. 1985, 1987). These authors described within this system of larger geoforms (Longitudinal Dunes) smaller landforms called: *dunes, mantles, shallow mantles, inter-dune depressions, and buckets*, from higher to lower topographic positions. This determines a microrelief with dunes, flat extended hills with little depressions and shallow lakes and ponds whose drainage is slow and anarchic. Figure 8.4 shows the distribution of some of these geoforms and associated soil and vegetation in the county of Trenque Lauquen during a period with and without hydric excess.

The region is characterized by alternating dry cycle and hydric excess which have been recorded since 1576; the last humid cycle started in 1970, with great

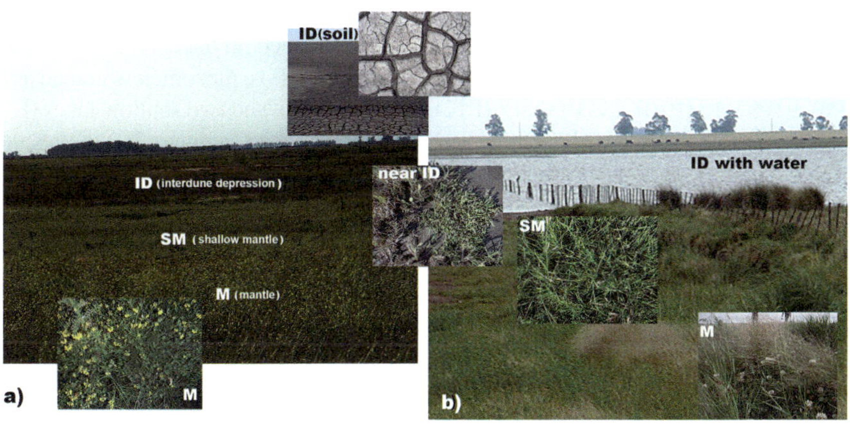

Fig. 8.4 Geomorphological units: Inter-dune depression, shallow mantle and mantle. Sequence of vegetatión: **a** without water and **b** with water

floods in the 1980s (Moncaut 2003). The climate in this region went from semi-arid to subhumid-humid (Etcheverry 2003), water condition which currently persists. This has led studies to focus on the occurrence of water in the surface and only partially on groundwater (Gabellone et al. 2003). In more recent studies the influence of groundwater is confirmed in the occurrence of hydric excess, flooding and the appearance of shallow lakes and ponds in some Sandy Pampa places and other regions hydrogeologically related (Alconada-Magliano et al. 2011b; Alconada-Magliano and Damiano 2017).

8.4.3 Studied Locations in the Region of Longitudinal Dunes

In this chapter we present some sites chosen in three places: Lincoln (Li), Pehuajó (Pe), and Trenque Lauquen (TL) whose complete results are presented in Alconada-Magliano (2008), Alconada-Magliano et al. (2011a, b), Alconada-Magliano and Damiano (2017), and Alconada-Magliano et al. (2017). The chosen sites allowed us to exemplify the holistic and dynamic analysis criteria previously discussed.

Figure 8.3c shows the study sites in Trenque Lauquen, Pehuajó, and Lincoln. In Lincoln (Li) work was done at the "El Labrador" farmland; it is a landscape of flat, extended hills which alternate with lower parts and shallow lakes, varying in height between 82.4 and 90.0 m a.s.l. Twelve sites were subject to soil and water sampling and included in three sections (S) distributed in approximately 12 km of total length: S1: Upper section, Mill 1 (M1), piezometers 1 (P1), 2 (P2), and 3 (P3) (distance from M1 to P3 1,605 m, slope 0.16%). S2: Medium section, Mill 2 (M2), piezometers 4 (P4), and 5 (P5), and Shallow Lake 1 (L1) (distance from M2 to L1 1,815 m, slope 0.14%). S3: Lower section, Mill 3 (M3), piezometers 6 (P6) and 7 (P7), and shallow lake 2 (L2) (distance from M3 to L1, 1,055 m, slope 0.12%) (Alconada-Magliano and Damiano 2017). In Pehuajo (PE), results from "El Tostado" farmland are presented. Eight sites were subject to soil and water sampling located on flat extended hills and shallow lakes, at heights between 86.0 and 87.0 m a.s.l. Five piezometers named as 8 (P8), 9 (P9), 10 (P10), 11 (P11), 12 (P12), one mill (M4), and two shallow lakes (L3, L4), are analyzed (Alconada-Magliano 2008). In Trenque Lauquen (TL) the results of two piezometers (P13, P14), one mill (M5), and deeper water wells are analyzed (Alconada-Magliano 2008).

8.4.4 Procedure to Study Soil and Groundwater

Soil. The analytical and morphological properties of the soil were analyzed by means of traditional field studies (Etchevehere 1976; Schoeneberger et al. 2012) and laboratory procedures (Black 1965; Page et al. 1982). Then, they were taxonomically classified by Soil Taxonomy (Soil Survey Staff 2006). In the Northwest of Buenos Aires province, given the presence of shallow water table and large extension of saline and alkaline soils, focus was on the field study of hydrohalomorphic features (Fe–Mn mottling and concretions, compacting, colors, distribution of humidity in the profile, roots growth, presence of sodium humates, accumulation of salts, or silt crusting), and on physico-chemical properties such as: paste pH, electrical conductivity in the soil saturation extract (EC $dS.m^{-1}$), and soluble cations to define the sodium adsorption ratio (SAR). Additionally, in some places exchangeable cations and soluble anions were measured (Alconada-Magliano et al. 2016).

Also, the vegetation composition and coverage were studied. It was considered how these vary according to weather conditions, water table depth, hydric excesses, waterlogging, or hydric deficit. The location of the soil on the local landscape (topographic position or physiographic unit) and in the regional landscape, became of transcendental importance to understand the hydrological link between environments. With satellite images on numerous dates, these relationships between the local soil and the regional landscape were analyzed (23 satellite images in Alconada-Magliano 2008). In these satellite images the evolution of the hydric and vegetation coverage, land use and soil distribution pattern were analyzed and compared to weather conditions (rainfall in each period studied in the satellite image). By means of these images, a first approach of the zones that could be complementary in their hydrological functioning was obtained following Tóth´s theory (Alconada-Magliano 2008).

This study procedure in which the local conditions of an environment are linked to its regional environment acquires relevance in regions such as Northwestern Buenos Aires province (NWBA), as indicated in points 8.4.1 and in 8.4.2, which are characterized by the presence of different geological materials originating from the soils and determining the geomorphological units described by Dillon et al. (1987). In addition, the presence of intermediate water flows (long distance) has been recognized (Alconada-Magliano et al. 2011b). This is set in evidence in the elevation of the water table and the permanence of large water covers on the surface in inter-dune depressions (Fig. 8.4), in permanent and semi-permanent shallow lakes and ponds that can change their covered surface independently of local rainfall.

Physico-chemical analysis of water and definition of groundwater flow systems. Natural water acquires its chemical composition through a complex process of physico-chemical interactions involving geological, hydrogeological, geomorphologic, climatic, pedological, microbiological, anthropic, and other factors (Fagundo-Castillo 1990; Fagundo-Castillo and González-Hernández 2016). Consequently, more information of an environment allows a better understanding of the functioning of the landscape. In relation to water, the following properties must be measured, mainly pH, temperature (T^o), electrical conductivity (EC), and soluble cations to define the sodium adsorption ratio (RAS). Also, in order to establish more precisely the origin of water, it is necessary to also know the anionic composition. Then, the waters under study are characterized, classified, and grouped by means of hydrogeochemical models such as those proposed by Fagundo-Castillo 1998; Fagundo-Sierra et al. 2001; Fagundo-Castillo et al. 2005a, b; Fagundo-Castillo and González-Hernández 2016. These authors defined the hydrogeochemical patterns of water according to Kurlov (PHM, Na^+:K^+:Ca^{+2}:Mg^{+2}:Cl^-: HCO_3^-:SO_4^{-2} ratio), classify the waters by Piper-Hill and Stiff diagram, and use the MODELAGUA model to know the origin of the water. Another frequently used model for hydrogeochemical processing is that presented by Parkhurst et al. (1980) and Parkhurst (1995).

In the examples presented in this chapter, water was studied as follows. In Lincoln, average results of the analyses of the water extracted on 9 dates for 2 years (108 samples) of piezometers, mills, and shallow lakes, are presented (Fig. 8.3c). In Pehuajó and Trenque Lauquen only one date of the selected sites is presented. The cationic and anionic composition (Na^+, K^+, Ca^{2+}, Mg^{2+}, Cl^-, $SO_4{}^{2-}$, HCO^{3-}), were measured by ICP-MS (Mass Spectrometry and Atomic Emission Spectrometry by Plasma) (APHA-AWWA-WPCF 1989), and hydrogeochemical modeling was carried out using models that complement each other, to define what has been previously commented (Fagundo-Castillo 1998; Fagundo-Sierra et al. 2001; Fagundo-Castillo et al. 2005a).

In order to define the groundwater flow systems according to Tóth's criteria (Tóth 2000, 2016) establishing the types of flows (local, intermediate, or regional) and zones within them (recharge, throughflow, and discharge) (Fig. 8.1) it is necessary to know the physico-chemical characterization of the water, indicated in the previous paragraph, as well as the relationships with other landscape elements such as soil, vegetation, topography, and geomorphology.

It should be noted that in the definition of flow systems (types and zones), the values or criteria are not absolute but must be established for the region under study, as presented in Edmunds et al. (2002), Fagundo-Castillo et al. (2005a), Cardona and Carrillo-Rivera (2006), Carrillo-Rivera et al. (2007), Fagundo-Castillo et al. (2008), Peñuela-Arévalo and Carrillo-Rivera (2010, 2013), among others.

Objective of defining flow systems. The hydrological functioning of an environment conditions the results of interventions in the landscape. The knowledge of how this functioning occurs allows us to define production systems and management practices which adapt to that hydrological functioning. The intensity of the studies that are carried out as well as their orientation will depend on the characteristics of the environment and the purpose of the intervention in the landscape. Specifically in the study area, it can contribute to improving the alternating situations of frequent excesses and water deficits, by providing necessary information for the implementation of management practices proposed for water control in the bibliography. For example, for the control of water excesses, it is essential to define the flow systems of an area (flows and zones) in order to define the sites where the bio-drainage technique can be performed, as proposed by Heuperman et al. (2002), Heuperman (2003), Ambast et al. (2007), Tomar (2007), Gupta (2007), Bonnesoeur et al. (2019) for different parts of the world and in Argentina by Alconada-Magliano et al. (2009), Galetti (2014), Nosetto and Jobbágy (2014). Also, by knowing the water flows that circulate in a site, the relationships between topographic position, type of soil, management, trees, and crops that can take advantage of groundwater could be adjusted more precisely, as indicated for the crops of soy, corn, and sunflower in several papers (Baigorri et al. 2003; Barraco 2009; Barraco et al. 2010; Saks et al. 2012; Álvarez et al. 2015; Lardone 2016) and silvopastoral systems (Domínguez-Daguer and Laclau 2014; Martinez-Calsina et al. 2015; Laclau et al. 2015; Chará et al. 2019). More recently, land management "*by environments*" has been proposed in order to obtain sustainable productions (Calviño 2017; López and Finello 2017; Uranga 2017). Although these criteria of study and management constitute a real advance with respect to traditional agricultural management, they do not include the functioning of groundwater. However, as described in this chapter through examples, flow systems are an inseparable element of the environment, and need to be defined, so as to establish the plots or sites of differential management within a farmland.

8.4.5 Soil and Environment Relations

Table 8.1 shows characteristics of the sites studied in the localities of Lincoln, Pehuajó and Trenque Lauquen, in the region of Longitudinal Dunes in the Sandy Pampas. The general landscape of the region is a plain with extended flat hills alternating with small depressions, lows, and ponds. Specifically in the studied sites, the following topographic positions are described within a general physiognomy of the plain environment: high hill (HH), extended flat hill (EFH), low hill medium (LHM), low (L), bucket (B) and alternation of high hill, low and bucket (HH/L/B) which correspond to the geomorphological units of *Dunes, Mantles, Shallow mantles, Inter-dune depressions, and Deflation buckets,* described in the region by Dillon et al. (1987) (Fig. 8.4). As indicated above, these geomorphological units are due to the way in which the two original materials of the soil are presented: the Invading Dunes Formation (thickness 0–30 m, E1) in surface and the Pampeana Formation (loess, thickness 5–150 m, E3) underneath that (Dillon et al. 1987; SAGyP-INTA 1989; González 2005).

In the high hill positions (HH), the *Typic Hapludoll* and *Argic tapto Hapludoll* soils prevail in Lincoln; and *Entic Hapludoll* in Pehuajó and Trenque Lauquen (Table 8.1). These soils are located in areas where the E1 deposit (Invading Dunes) is thicker and corresponds to the Mantle geomorphological unit, in flatter positions, defined in the edaphic cartography of the region (SAGyP-INTA 1989; GeoINTA 2013b) as extended flat hills (EFH); *Entic Hapludolls* soils were also described at the sites studied in Pehuajó. However, in the Lincoln and Trenque Lauquen studied sites, soils with hydromorphic features, classified as *Acuic Hapludoll,* also appear. In the sectors where they also have halomorphism and a lithological discontinuity in the edaphic profile is present, the described soils were classified as *Natric tapto Hapludoll,* as identified in Lincoln. These soils are developed in the Shallow mantles geomorphological unit, where the thickness of the E1 deposit is thin, and the soil was also formed from the E3 sediment, the Pampeana Formation (loess) (Dillon et al. 1987).

The topographic height where these soils are described do not differ significantly and are even similar or equal (Table 8.1) and correspond with the environments defined in the regional edaphic cartography as extended flat hills. In Lincoln the soil classified as *Typic Hapludoll* on the M3 mill hill at 88.0 m a.s.l. is observed and close to it at the same topographic height, *Natric tapto Hapludoll* soil in the P6 piezometer is described. In Pehuajó, all soils are classified as *Entic Hapludoll,* and they are located in an extended flat hill position and topographic heights which vary only between 86.0 and 87.5 m a.s.l. In Trenque Lauquen the soils classified as

Table 8.1 Sampling sites: In cross-sections (S) S1 Upper (M1, P1, P2, P3), S2 Middle (M2, P4, P5, L1), and S3 Lower (M3, P6, P7, L2), of Lincoln (Li), and selected sites of Pehuajó and *Trenque Lauquen*. *Landscape position*, Po (HH, high hill; EFH, extended flat hill; LHM, low hill medium; L, low; B, bucket); HH/L/B (alternation of HH, L and B); *Topographic height* (m a.s.l.); Ground cover and/or dominant vegetation. Soil; *Typic Udipsamment* (TU); *Entic Hapludoll* (EH), *Typic Hapludolls* (TH), *Acuic Hapludoll* (AH); *Argic-Natric tapto*Hapludoll* (ATH-NTH); *Typic Natraqualf* (TNf); *Typic Natraquoll* (TN), (**Tapto mean buried*). Water table depth in m (m WT) (average measurement on 9 dates for two years). Type of water manifestation (WM): P piezometer, M mill, W well, and L shallow lake

	S	WM	Po	m amsl	Soil	m WT	Ground cover/Dominant Vegetation
Li	S1	M1	HH	90.6	HT	6.00	Soybean, corn,weat
		P1	EFH	89.4	AH	1.29	*Melilotus* sp., *Bromus unioloides, Lotus* sp.
		P2	LHM	88.4	NTH	0.60	*Spergularia* sp., *Lotus* sp., *Distichlis* sp, salts
		P3	L	88.0	TNf	0.47	*Distichlis* sp., *Spergularia* sp., *Sarcocornia* sp.
	S2	M2	HH	89.5	ATH	6.0'	Wheat, corn, free surface water
		P4	EFH	88.4	NTH	1.47	corn
		P5	L	87.2	TNf	0.49	*Distichlis* sp. *Chloris berroi*
		L1	B	85.9			lagoon shore without vegetation
	S3	M3	HH	88.0	TH	6.00	Soybean, corn
		P6	EFH	88.0	NTH	1.27	Soybean, corn
		P7	LHM	87.0	TN	0.82	*Distichlis* sp. *Cynodon* sp. *Salts*
		L2	B	85.9			Shallow lake shore without vegetation
Pe	M4		HH	86.9	EH	3.30	*Ulmus* sp., *Salix* sp.
	P8		EFH	86.8	EH	2.30	Pasture, soybean
	P9		EFH	87.5	EH	2.50	Pasture, soybean
	P10		EFH	86.0	EH	1.50	Pasture, soybean
	P11		EFH	86.3	EH	3.50	Pasture, soybean
	P12		EFH	85.9	EH	2.50	Pasture, soybean

(continued)

Table 8.1 (continued)

	S	WM	Po	m amsl	Soil	m WT	Ground cover/Dominant Vegetation
	L3		B	82.0	TN		*Distichlissp, Sarcocornia perennis, Bolboschoenus paludosus*
	L4		B	78.0	TN		Same as above
TL	M5		LHM	96.0	AH	6.00	Natural grassland: *Cynodon* sp., *Stipa* sp., *Cardus* sp.
	W1		HH	110.0	EH/TU	40.00	Soybean, pasture of *Eragrostis curvula*
	P13		HH	97.2	EH	3.00	Soybean, corn
	P14		HH/L/B	87.0	EH	2.90	Soybean, natural grassland

Entic Hapludoll and are located in different positions and heights that vary between 87.0 and 110.0 m a.s.l. In addition, in the highest positions the *Typic Udispsament* soil was also recognized, which corresponds to the geomorphological unit of Dunes. This soil has been developed from the original material called E1, Invading Dunes (Dillon et al. 1987).

In environments defined as low hill medium, in Lincoln, soils classified as *Typic Natraqualf* are described, whereas in Trenque Lauquen *Acuic Hapludolls* soils are described, although these are in a higher topographic position (96.0 m a.s.l.) (Table 8.1). Alconada-Magliano (2008) described in Trenque Lauquen *Typic Natraqualf* soils in the geomorphological unit of Inter-dune depressions. These soils were developed from the material corresponding to the Pampeana Formation (E3, loess).

Therefore, it can be seen that there is no direct correspondence between topographic height, topographic position, and soil type for all studied localities and even within the same locality. This becomes evident if the physico-chemical properties of soils and the quality of groundwater are analyzed in greater detail, as presented in point 8.4.6.

Soils studied in Lincoln

In Table 8.1 characteristics of the Lincoln environments were presented and in Fig. 8.3c the location of the three cross-sections studied in a satellite image, as explained in point 8.4.3, were shown. In Table 8.2, some properties of the soils in Lincoln, located in piezometers P1, P6, and P7 are presented. The environment, vegetation, and water table quality are also indicated on the date the soil was studied (they were studied on different dates). These soils are presented in Fig. 8.5, with the vegetation, water table depth for the studied date and average measurement for two years in piezometers (on 9 dates, in m), type and zone of groundwater flow to which each site belongs, and which is analyzed in more detail in point 8.4.6.

It is observed in Table 8.2a that the soils have high alkalinity and/or salinity whose magnitude is associated with the position they occupy in the general relief. In all cases it increases at the bottom of the profiles. The alkalinization observed in soil was related to the groundwater quality that affects (Table 8.2b), vegetation, and mineralogy recognized on the site (Alconada-Magliano et al. 2016, 2017), as well as, to the mineralogy identified in other sites in the region by Etchichurry et al. (1988) and González (2005). Thus, it is emphasized that in the *Acuic Hapludoll* soil of the piezometer located in the highest position of the general relief (P1), the sodicity is elevated from 44 cm (Table 8.2a). The pH 10 in the soil is associated with the high content of HCO_3^- (14.7 me.l^{-1}) and Na^+ (Na^+ 7.1 cmol.kg^{-1}; ESP 47) at the bottom of the profile, and corresponds to the contents of both ions in the water table (HCO_3^- 22.2 me.l^{-1}, Na^+ 29.5 me.l^{-1}) (Table 8.2b). From 44 cm deep, on the Bwn1 horizon, evaporitic minerals, gypsum, calcite and mainly halite are identified, which evidence the rise of water by evaporation even in elevated positions (P1, extended flat hill). Halite (NaCl) is found in all profiles at different depths associated with water table rise (Alconada-Magliano et al. 2016).

Although the *Natric tapto Hapludoll* soil of piezometer P6, also in an extended flat hill position, does not differ markedly in chemical composition from the soil in

Table 8.2 Sequence of horizons in soil profiles of P1, P6, and P7 in Lincoln. **a** In the soils *Acuic Hapludoll* (AH), *Natric tapto Hapludoll* (NTH), and *Typic Natraquoll* (TN): Depth of horizons (cm), paste pH, electrical conductivity (EC, $dS.m^{-1}$), Organic material (0 M %), exchangeable cations ($cmol.kg^{-1}$),cationic exchange capacity (CEC, $cmol.kg^{-1}$), exchangeable sodium percent (ESP, $cmol.kg^{-1}$), granulometry % (Sa, Sand; Si, silt; Cl, clay), textural class (L, loamy; SL, Sandy loam; SCL, Sandy clay loam); and in the water table: pH and CE ($dS.m^{-1}$). **b** In the soil and water from (a): soluble cations ($me.l^{-1}$), sodium adsorption ratio (SAR) and soluble anions ($me.l^{-1}$) and factor of chloride concentration in water (FCCl)

(a)						Exchangeable cations ($cmol.kg^{-1}$)						Granulometry %			
Site	horizon	Depth cm	pH	EC	OM	Ca^{+2}	Mg^{+2}	K^{+}	Na^{+}	ESP	CEC	Sa	Si	Cl	texture
P1 AH	A	0–26	6.5	0.3	1.94	7.3	2.2	1.8	0.4	2.8	14.5	46.5	33,5	20.0	L
	AB	26–44	8,4	0.4	0.71	3.7	1.9	1.0	1.1	13	8.3	58.0	29.5	12.5	SL
	Bwn1	44–74	10	2.5	0.22	5.2	4.2	2.7	6.8	48	14.1	49.0	33.5	17.5	L
	Bwn2	74–93	10	1.9	0.16	3.6	4.4	3.1	7.1	47	15.1	47.5	30.0	22.5	L
WT 93+			7.9	3.0											
P6 NTH	A	0–30	6.3	0.3	3.17	7.6	2.9	1.6	0.3	2.3	13.2	46.0	35.0	19.0	L
	AC	30–45	8.5	0.6	0.40	2.8	2.6	1.3	1.2	18	6.8	58.0	30.5	11.5	SL
	2Bt	45–61	9.6	1.9	0.36	–	–	4.2	7.6	36	21.1	43.0	38.0	19.0	L
WT 151+			8.0	4.5											
P7 TN	A	0–29	8.8	47.4	2.14	–	–	1.7	29	100	7.8	53.5	31.5	15.0	SL
	2Bt1	29–49	8.6	10.8	0.55	–	–	3.2	17	66	25.6	37.5	38.5	24.0	L
	2Bt2	49–59	8.8	11.1	2.05	–	–	2.9	16	74	21.1	45.5	28.0	26.5	SCL-L

(continued)

Table 8.2 (continued)

(a)						Exchangeable cations (cmol.kg^{-1})						Granulometry %			
Site	horizon	Depth cm	pH	EC	OM	Ca^{+2}	Mg^{+2}	K^{+}	Na^{+}	ESP	CEC	Sa	Si	Cl	texture
WT 59+			7.7	11.8											

(b)			Soluble cations					Soluble anions me.l^{-1}				
	horizon	Prof	Ca^{+2}	Mg^{+2}	K^{+}	Na^{+}	SAR	CO_3^{-}	HCO_3^{-}	Cl^{-}	SO_4^{-2}	FCCl
P1 AH	A	0–26	0.6	0.4	1.0	1.5	2.2	1.9	1.2	0.5	1.9	
	AB	26–44	0.5	0.5	0.4	3.8	5.6	3.2	1.2	2.4	3.2	
	Bwn1	44–74	0.5	2.3	1.2	24.3	20.6	16.4	5.9	7.8	16.4	
	Bwn2	74–93	0.4	1.9	1.1	19.1	17.9	14.7	2.9	3.5	14.7	
WT 93+			0.7	2.4	0.8	29.5	23.4	29.5	0.7	4.9	23.9	25.8
P6 HTN	A	0–30	0.9	0.3	0.7	1.1	1.1	0.8	1.8	1.9	0.8	
	AC	30–45	0.5	0.5	0.5	6.6	7.7	3.1	2.4	1.9	3.1	
	2Bt	45–61	0.7	0.5	0.9	20.2	20.6	13.9	2.4	2.5	13.9	
WT 151+			0.5	3.6	0.9	42.7	30	0.4	16.24	18.32	9.07	96.4
P7 NT	A	0–29	19.5	57.5	7.7	414	59.6	41.0	285.1	172.1	41.0	
	2Bt1	29–49	3.3	5.9	2.2	102	40.9	4.1	48.7	38.5	4.1	
	2Bt2	49–59	3.8	6.6	2.2	112	42.1	4.1	51.5	39.2	4.1	
WT 59+			0.6	15.8	1.9	109	38	0.0	11.88	76.86	42.87	388.7

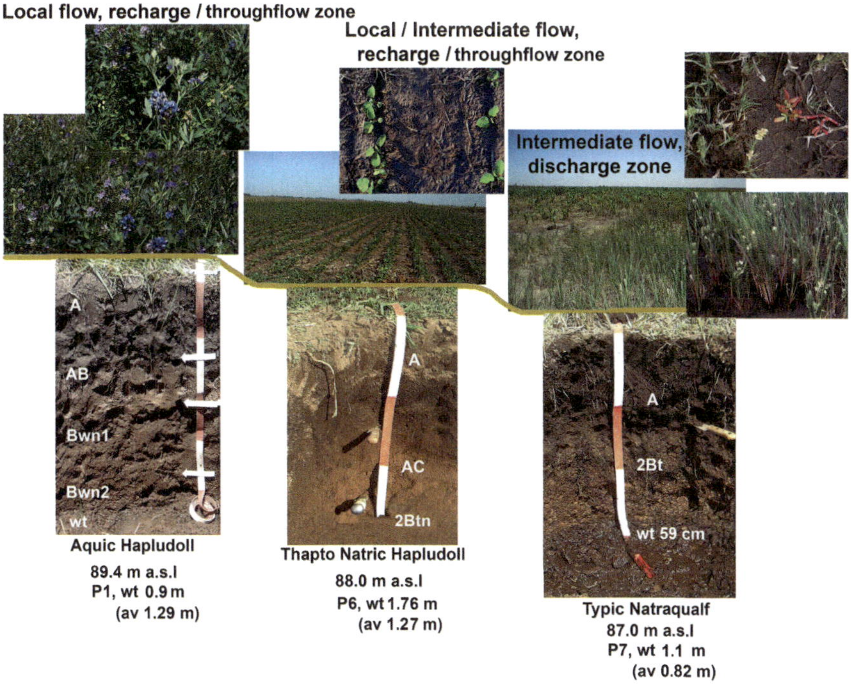

Fig. 8.5 P1, P6, and P7 Piezometers: sequence of soils, vegetation and type and zone of groundwater flow. Topographic height (m a.s.l.) and the water table depth (wt in m) are indicated for the studied date and also for the average value of 9 dates over two years in piezometers (av)

P1, it differs in the sequence of horizons and textures (Table 8.2a) because it was formed from the two original materials (lithological discontinuity) as commented at the beginning of this point. Also, the soil in P6 differs from organic matter content as measured in P1, which can not be explained by the type of soil but by the implemented management and variations in the depth of the water table (Table 8.1).

The soil *Typic Natraquoll* of piezometer P7, located in a lower sector (S3, low hill medium), formed from sediment E3, Pampeana Formation (loess), is characterized by a very high salinity and alkalinity from the surface (Table 8.2). In these soils, as mentioned for the soil of P1, there was a clear correspondence between the levels of salinity and alkalinity of the soil and those of the inciding water table.

As detailed in section 8.4.3, piezometers P6 and P7 are located in the same cross-section (S3, Lower), between the Mill, M3 (in the highest part) and the shallow lake, L2 (lower part). It should be noted that the distance between the M3 and L2 sites is 1,055 m with a slope of only 0.12%.

The relation between groundwater incidence and soil in Lincoln can be summarized as follows: In places where the material of Formation E1 (Invading Dunes, aeolian sand cover) is thicker as presented in the Mantles geomorphological unit

(GU), *Typic Hapludolls* or *Acuic Hapludoll* (P1) soils are recognized. As this material decreases in thickness, the Pampean Formation (E3, loess) appears, and in the soil a lithological discontinuity is recognized. These are classified as *Argic tapto Hapludoll* or *Natric tapto Hapludoll* (GU: Shallow mantles; piezometer P6). When E1 disappears, as observed in low environments such as in the piezometer P7, as well as in the geomorphological unit called Inter-dune depressions, the soil is classified as *Typic Natraqualf* (Tables 8.1 and 8.2). In proximity to these soils it is common that the horizon A increases in thickness and in organic matter content, classified as *Typic Natraquoll.*

Consequently, the study area in Lincoln has an irregular distribution of soils that affects the type and development of vegetation even in places very close to one another. In Fig. 8.6 wheat crop (*Triticum aestivum*) with irregular growth and sectors without vegetation is present in an extended flat hill landscape of the S1 cross-section in Lincoln. This occurs by decreasing the thickness of the superficial edaphic material developed in Formation E1 and consequently having a greater proximity to the water table.

Regarding the depth of the water table, it was observed that in thc 9 datcs analyzed in the Lincoln sites over 2 years, it varied significantly in all the sites (P1 a P7). Thus, the coefficient of variability (VC) between dates was greater than 37% in all sites, reaching CV from 43.5 to 73.2% in the piezometers of the lowest positions. For example, in the 2-year-period studied, in the 3 piezometers of Fig. 8.5 and Table 8.2,

Fig. 8.6 **a** Effect on wheat cultivation of the thickness of material from Formation E1 (Invading Dunes) in the environment of the P1 piezometer. **b** The absence of vegetation is observed due to the proximity of the water table

the water table varied in P1 between 0.66 and 1.87 m, in P6 between 0.44 and 2.02 m, and in P7 between 0.27 m and 1.30 m (Alconada-Magliano et al. 2017).

Another aspect to highlight in the soils studied in Lincoln is the differences found in the type of clay formed. In all profiles, there was an increase in their deeper layers in the proportion of expansive type clay (montmorillonite), between 45 and 60% of the total clay measured (Alconada-Magliano et al. 2016) which is related to the increase in hydrohalomorphism (Table 8.2). In this regard, it is related to what is indicated by Fagundo-Castillo (1990) and Fagundo-Castillo and González-Hernández (2016) that the alteration of primary minerals and formation of secondary minerals is influenced by residence time and water quality (soil–water interaction). In all the studied soils, the existence of quartz, plagioclase (anorthite and albite), and calcite is recognized in all depths. The gypsum was recognized only in the hills of the P1 and P6 piezometers, and the halite, although it was observed in all profiles, only from 30 to 45 cm deep (Alconada-Magliano et al. 2016, 2017). Etchichurry et al. (1988) for the region in general report on soils and surface sediments the following minerals: quartz (between 19 and 26%); plagioclase (between 15 and 40%); K-feldspars (between 6 and 8%); weathered rock (between 14 and 47%); augite, volcanic glass, biotite (between 0.6 and 1.71%); hypersthene, hornblende, lamprobolita (<0.4%); and rutile, zircon, and garnet (<0.10%). The mineralogical composition, and especially the albite, with water table with high Na^+ content (Table 8.2b), could explain the formation of expansive type clay discussed above. Zapata-Hernández (2004) indicated that these are formed from the hydrolysis of silicates with the CO_2 dissolved in the rainwater or by the H_2CO_3 of the groundwater.

From the above, regarding Lincoln soils, it can be stated that there is not always a direct relationship between the degree of hydrohalomorphism of the soils with the position they occupy in the landscape and topographic height. These hydrohalomorphism depend especially on the presence of lithological discontinuities, and on the type of groundwater flow that affects the soil. It also depends on the area that the soil occupies within the identified groundwater flow (recharge, throughflow, or discharge). Thus, the interaction between the original material, soil, groundwater, and vegetation is modified. Soils as those described in Lincoln are representative of what happens at regional level. These evolve according to local and regional conditions, especially linked to intermediate type water flow (long travel) as described for the northwest of Buenos Aires province (Alconada-Magliano et al. 2011b).

Consequently, this hydraulic linkage between the local and regional landscape must be considered when interpreting the suitability of land use and to establish sustainable management practices. However, it is not always fully considered. Specifically in the regional soil cartography available at 1:50,000 scale (GeoINTA 2013b), the "soil Series" and "Subgroup class of the soil" according to Soil Taxonomy (Soil Survey Staff 1975, 1999, 2006, 2010 or/and 2014), define the aptitude and limitations of land use up to the depth of study for the date observed. Although it is undoubtedly very valuable information, when used to carry out local interventions, it should be considered that the work scale is insufficient. In addition, in order to define the best land use or its management contribution to improving a general situation, such as water excesses and deficits, local and regional links between soils and environments must also be considered. In the cartographic units included in the available soil maps, soils associated with the soil under study are also descripted. However, this is not always taken into account when planning the general management of a region or a particular farmland. For example, what the consequences of a forest plantation on cultivated plots may be, or the effects that drainage systems can have on hydrogeologically linked environments. Inferring from the soil the origin of the pedological processes that are observed, as well as which may be the evolution of the soil compared to other climatic scenarios, allows for sustainable interventions in the landscape.

The following paragraphs explain through examples what was discussed above: In the map of soils (1:50,000) (GeoINTA 2013b) of the cartographic sheet called Smith (3563-30-2), which includes the Lincoln´s sector shown here, in a general relief of high hill, in a pronounced hill position, an *Entic Hapludoll Norumbenga Serie* soil is described, and low water retention is indicated as use limiting. Associated with this soil, in the same cartographic unit, the *Argic tapto Hapludoll Ortiz de Rosa Serie* (OR) is identified in a hill medium position. Regarding this soil, it is indicated that it is not saline or alkaline, and as its limitation, drainage problems are highlighted due to an increase in humidity in depth (hydromorphic features). However, the fact that at one meter depth the exchangeable sodium percent (ESP) increases to 7.8 and the pH to 8.5 (in water, relation 1: 2.5) is not mentioned. Likewise, in areas close to these, but in a position of high planes, the OR Series is also described, but next to the *Natric tapto Hapludolls Carlos Salazar Series* soil, in the low plain sectors. The latter soil has high sodium alkalinity and poor drainage. It is evident that these soils are hydrogeologically linked and consequently, the management carried out at one site can affect those other sites that are linked in their hydrological functioning.

Soils studied in Pehuajó

The soils studied in Pehuajó located next to one another (Fig. 8.3c), in an environment of extended flat hills and topographic heights between 86 and 87.5 m a.s.l., are all soils classified as *Entic Hapludoll* (Table 8.1). The original material of these soils is the Invading Dunes Formation (E1) and their present low development (edaphic profile sequence A-AC-C). They have a generally deep A horizon of about 0.35 m, with low salinity and alkalinity (<0.9 $dS.m^{-1}$, pH < 6 and SAR < 1), except in the site soil piezometer P9 having moderate surface alkalinity (SAR 6.7). However, on this site, for the date the soil was studied, the water table was located more deeply (2.5 m, Table 8.1). Only in the soil of the P10 piezometer there was an increase in humidity as from horizon C (0.60 m deep), which appeared compacted and with moderate hydromorphism, associated with a water table closer to the surface than in the rest of the piezometers (Table 8.1). Notwithstanding, the water table depth varied significantly between dates measured over a period of 2 years. The minimum and maximum depths measured were: P8 between 1.42 and 2.44 m; P9 between 2.43 and 3.17 m; P10 between 1.18 and 2.39 m; P11 between 2.89 and 3.59 m; P12 between 2.01 and 2.81 m. These variations were only partially associated with rainfall occurring on the site (data shown in Alconada-Magliano 2008; Alconada-Magliano et al. 2016). This can be explained by the presence of flows of different origin, as discussed in

Fig. 8.7 P8, P11, and P10 Piezometers: sequence of soils, vegetation and type and zone of groundwater flow. Topographic height (m a.s.l.) and the water table depth (wt in m) are indicated for the studied date and also for the average value of 5 dates over two years in piezometers (av)

point 8.4.6. Figure 8.7 shows an example of the sequence of soils at the site studied in Pehuajó, piezometers P8, P11, and P10. The vegetation present, water table depth for the studied date and average measurement for two years in piezometers (5 dates, in m), and the type of groundwater flow and area within this flow are shown (analyzed in point 8.4.6).

Regarding the correspondence of the soils studied here (Table 8.1) with those described in the edaphic maps available at scale I: 50,000 (cartographic sheet 3563-35-Juan José Paso, GeoINTA 2013b), it can be indicated that they resemble *Entic Hapludoll Piedritas Series soil.* This series of soils is frequently associated with other series of soils that do not present problems of hydromorphism or alkalinity. However, in areas close to these, cartographic units are described where the Piedritas Series is associated with soils with halomorphic properties such as soils classified as *Natric tapto Hapludoll Carlos Salas Series*, the *Natric Duracuoll Serie Salazar*, and the *Natric tapto Pichincha Series.* The latter is located in the lowest positions of an environment where lows alternate with hills. As commented for Lincoln, these associations of soils in short distances allow the affirmation that there is an underground hydraulic connection that affects the evolution of the soils. The comments can be confirmed by means of studies of water quality as analyzed in the following point (8.4.6).

Soils studied in Trenque Lauquen

Trenque Lauquen soils develop in a general environment of wide plains with alternating micro-hills and micro-lows, in which permanent or temporary shallow lakes are located. The soils presented in Table 8.1, located in higher positions, are classified as *Typic Udipsamment, Entic Hapludoll,* and *Acuic Hapludoll.* The *Typic Udipsamment* soil that was developed in the geomorphological unit of Dunes where the aeolian sand cover is thicker (E1, Invading Dune) (Dillon et al. 1987), presented high surface salinity (greater than 6 $dS.m^{-1}$). Likewise, the salinity was very high on the *Entic Hapludoll* soil surface of the site of P13 and P14 (12 and 43 $dS.m^{-1}$), developed in a local topographic position of hill and half hill, respectively, although at regionally different sites (P13, in La Vidaña at 97.2 m a.s.l. and P14, in Planas Hugo at 87.0 m a.s.l.) (Table 8.1). On the contrary, the *Acuic Hapludoll* soil developed in a low hill medium (LHM) position showed neither salinity nor alkalinity. In the study area, local environments of hills, half hills and low with *Entic Hapludoll, Typic Natraquoll,* and *Typic Natraqualf* soils were also described at short distances. The latter soil has a very high alkalinity from the surface (Alconada-Magliano 2008).

This complex distribution of soils where the salinity and alkalinity of the soil under study does not directly correspond to the position that the soils occupy locally, is confirmed in the soil mapping available for the region (GeoINTA 2013b). In Trenque Lauquen (cartographic sheets 3763-4 and 5, 3563-34 and 35), cartographic units (CU) with soils that occupy positions close to one another in intricate patterns of distribution and suitability for use are described. For example, the CU called Pas16 Complex, corresponds to a general environment of extended plains with lows, and is composed of the *Entic Hapludoll Piedritas Series soils* in hills, *Argic tapto Hapludoll Cañada Seca Series* in the middle hills, and *Natric Duracuoll Salazar Series* in low planes. Another representative example of cartographic units of soils described for the region (GeoINTA 2013b), is the CU called Complex He3, corresponding to gently undulating sandy plains with *Entic Hapludoll Piedritas Series* soils in the middle hills, *Entic Hapludoll* Norumbenga Series in the hills, and *Acuic Hapludoll* Henderson Series in the lows. In some other CU, soils similar to those indicated for CU Complex H3 are described; however, in the low ones, the *Typic Natraquoll Drabble Series* soil is also described.

As indicated for the sites studied in Lincoln and Pehuajó, and the comments regarding the distribution of soils described in the soil mapping of the region scale 1:50,000 (GeoINTA 2013b), there is an obvious hydraulic linkage between soils and local and regional environments. The way in which the groundwater affects (downward, lateral, or upward flow), the depth of the water table and the thickness of the original soil materials determine how they evolve and in turn, how this affects the development of vegetation. This will be analyzed in greater detail in the following point (8.4.6).

8.4.6 Groundwater Quality

Table 8.3 shows the chemical composition and temperature of groundwater extracted at Lincoln (Li), Pehuajó (Pe), and Trenque Lauquen (TL) sites (Alconada-Magliano 2008; Fagundo-Castillo et al. 2014; Alconada-Magliano et al. 2016).

Table 8.4 shows waters type by standard Piper-Hill and Stiff diagrams (Stiff 1951), hydrogeochemical patterns (PHM) by Kurlov (Fagundo-Castillo 1998), water groups to which they belong (Parkhurst et al. 1980; Parkhurst 1995; Fagundo-Sierra et al. 2001), types of flow and zones within these according to criteria of the theory of flow systems of Tóth (2000, 2016) (Fig. 8.1). Thus, in the present study the types of flow (local, intermediate, or regional) and zones (recharge, throughflow, and discharge) (Table 8.4) were defined from the physical and chemical characterization of the water and landscape elements (Tables 8.1, 8.2, and 8.3).

Three major water groups were defined in Lincoln from bicarbonate-sodium to chloride-sulphate-sodium. The identified water groups and flow systems to which they belong (Table 8.4) are as follows: *Group 1*, HCO_3–Na, HCO_3 > Cl–Na and HCO_3 > Cl > SO_4–Na with average EC 4.16 dS.m^{-1} (M1, M2, P1, P4 and P6, all located in the highest position in cross-sections). Also, water samples of P12 from Pehuajó and W1, P13, and P14 from Trenque Lauquen belong to this group. *Group 2*, Cl > HCO_3 > SO_4–Na, average EC of 5.0 dS.m^{-1} (M3, highest elevated sector of T3, sampling to 6 m depth), water samples of P8 and P11 from Pehuajó belong to this group. *Group 3*, Cl > SO_4–Na and Cl–Na average EC 13.33 of dS.m^{-1} (P2, P3, P5, P7, L1, and L2), water samples of P10, M4, L3, and L4 from Pehuajó and M5 from Trenque Lauquen belong to this group. Consequently, in the studied sites in Lincoln, water evolved from *bicarbonate sodium* in P1, the highest elevated sector, with external environmental characteristics of recharge, to *chloride sulphate sodium* in lower sectors with discharge characteristics in P2, P3, P5, P7, L1, and L2. In all the cases the prevailing and commonly found cation is sodium (AHP, 901, Table 8.4).

It should be noted that although the Pehuajó and Trenque Lauquen water samples by type and AHP belong to the assigned water groups, the salinity is different. In general, it is significantly less than the measurement in Lincoln. The exception was obtained in P8 of Pehuajó, where the salinity and alkalinity were also very high (7.7 dS.m^{-1} and SAR 22.6) (Table 8.3). In Lincoln, even in the piezometers located in the highest position in cross-sections (P1, P4, y P6), water salinity is greater than 3.0 dS.m^{-1} and SAR is greater than 20. In lower sites the salinity increases until 20 ds.m^{-1}and the SAR until 45.5 (Table 8.3). It was also observed that while in Lincoln different water qualities are associated with different topographic positions and soils, in Pehuajó the piezometers are located in the same topographic position defined as an extended flat hill (EFH, Table 8.1). In Pehuajó it is noteworthy that extreme values of salinity and alkalinity were obtained in piezometers close to each other (P12: 0.1 dS.m^{-1} and SAR 0.4; P8: 7.7 dS.m^{-1} and SAR 22.5) (Tabla 8.3). In this case, the coexistence of two groundwater flows, local and intermediate type, becomes evident.

In Lincoln, the presence of intermediate (long travel) flows that coexist with local flows in the highest positions is most evident. In Lincoln the low variability

Table 8.3 Average values for nine sampling dates for water in cross-sections (S) S1 Upper (M1, P1, P2, P3), S2 Middle (M2, P4, P5, L1), and S3 Lower (M3, P6, P7, L2), of Lincoln (Li). F8 to F12, M4, L3 y L4 of Pehuajó (Pe). P13 y P14, M5, and B1 of Trenque Lauquen (TL). P: piezometer, M: mill, L: lagoon, W: water well. Physical and chemical characteristics of sampled water: pH, electrical conductivity (EC dSm^{-1}), temperature (T °C), relation of sodium absorption (SAR), cations and anions ($me.l^{-1}$)

S	N^0	pH	T	CE	Cl^-	SO_4^{2-}	HCO_3^-	Na^+	K^+	Mg^{+2}	Ca^{+2}	SAR
Lincoln												
S1	M1	7.9	17.4	3.92	10.7	9.9	18.2	32.7	0.9	3	2.7	20.8
	P1	8.1	17	3.01	4.6	5.3	20.3	27.2	0.7	2,1	0.7	22.9
	P2	7.9	17.2	10.21	45.1	42.3	16.7	92.6	1.3	8,3	3.3	38.7
	P3	7.7	17.4	19.5	105.5	81.2	16.2	171.8	2.2	21.1	7	45.5
S2	M2	8	17.4	5.92	22.1	14.8	22.1	53.1	1.2	4.4	1.6	31.2
	P4	8.2	17.5	3.86	10.3	7.4	19.7	36.1	0.7	1.5	0.4	37.6
	P5	7.7	17.2	15.97	84	53.6	22.5	138.3	2.2	16.2	4.6	43
	L1	8.4	18.7	8.64	49.8	30.3	6.8	76.5	1.8	7	7	35.2
S3	M3	7.9	19.2	5.01	22.9	12.3	14.7	41	1.4	6	3.3	20
	P6	8	16.9	4.14	15.2	8.6	16	36.3	0.9	3.1	1	25.9
	P7	7.8	17.1	10.56	61.4	33.6	12.6	88.1	1.7	13.3	4.6	29.8
	L2	8.3	17	15.06	95.4	51.8	6.2	127.7	2.4	17	8.3	34.6
Pehuajó												
	M4	7	19.1	5.15	35.2	9.7	9	32.8	1	11.2	6	9.6
	P8	7.6	27	7.69	20.3	5.5	14.1	49.1	1	6.3	1.6	22.6
	P9	7.4	27	1.51	12.4	3.7	2.1	6.1	0.5	0.7	0.7	5.9
	P10	7.2	19.0	1.54	23.7	4.0	3.0	21.1	1.2	2.0	1.6	13.1
	P11	7.2	19.5	0.57	4.6	1.6	2.1	4.7	0.9	0.2	0.8	4.9
	P12	7.0	17.7	0.18	0.4	0.2	1.4	0.4	0.2	0.2	0.8	0.4
	L3	9.0	25.2	4.19	28.7	11.0	3.0	35.2	0.9	4.2	2.1	17.2
	L4	9.9	21.0	10.42	74.6	34.8	8.1	97.8	2.0	10.1	1.5	38.1
Trenque Lauquen												
	M5	7.2	19.10	31.60	295.8	74.0	9.8	292.6	3.9	79.1	18.5	38.4
	W1	8.1	18.20	3.01	17.0	7.2	4.7	25.8	0.4	3.5	2.0	13.2
	P13	7.4	17.80	1.62	0.8	0.2	14.1	1.8	3.0	1.4	1.7	1.2
	P14	7.9	18.60	0.87	1.9	0.9	6.4	6.8	0.7	2.6	2.1	3.7

Table 8.4 Type of water, average hydrogeochemical pattern (AHP: relation Na^{+}:K^{+}:Ca^{+2}:Mg^{+2}–Cl^{-}:HCO_3^{-}:SO_4^{+2}), water group (G); flow hierarchy (I) intermediate, local (L); zones (Z): recharge- throughflow (R-T); discharge (D) for the same samples in Table 4.3. In cross-sections (S) S1 Upper (M1, P1, P2, P3), S2 Middle (M2, P4, P5, L1), and S3 Lower (M3, P6, P7, L2), of Lincoln (Li). F8 to F12, M4, L3 y L4 of Pehuajó (Pe). P13 y P14, M5 and B1 of Trenque Lauquen (TL). P: piezometer, M: mill, L: lagoon. W: water well. G1: M1, M2, P1, P4, and P6; G2: M3; G3: P2, P3, P5, P7, L1, and L2

S	N°	Type	AHP	F	Z
Li, S1	M1	$HCO_3 > Cl > SO^4{-}Na$	901-532	L	R-T
	P1	$HCO_3{-}Na$	901-172	L	R-T
	P2	$Cl > SO_4{-}Na$	901-514	I	T
	P3	$Cl > SO_4{-}Na$	901-514	I	D
Li, S2	M2	$HCO3 > Cl > SO_4{-}Na$	901-442	I	T
	P4	$HCO_3 > Cl{-}Na$	901-352	L	R-T
	P5	$Cl > SO_4{-}Na$	901-514	I	D
	L1	$Cl > SO_4{-}Na$	901-613	I	D
Li, S3	M3	$Cl > HCO_3 > SO_4{-}Na$	901-532	L	T-R
	P6	$HCO_3 > Cl > SO_4{-}Na$	901-442	L	R-T
	P7	$Cl > SO_4{-}Na$	901-613	I	D/T
	L2	$Cl > SO_4{-}Na$	901-613	I	D
Pe	M4	$Cl{-}Na > Mg$	712-811	I	T
	P8	$Cl > HCO_3{-}Na$	811-541	L	T/D
	P9	$Cl > SO_4{-}Na$	901-712	L	T
	P10	$Cl{-}Na$	901-811	L	D/T
	P11	$Cl > HCO3{-}Na$	910-631	L	T
	P12	$HCO_3 > Cl{-}Ca > Na$	451-27	L	R
	L3	$Cl > SO4{-}Na$	901-712	I	D
	L4	$Cl{-}Na$	901-811	I	D
TL	M5	$Cl > SO_4{-}Na$	802-802	I	D
	W1	$HCO_3 > Cl{-}Na$	901-361	L	T
	P13	$HCO_3{-}Na > Ca$	721-190	L	R
	P14	$HCO_3 > Cl{-}Na > Mg$	513-271	L	T

coefficient (VC %) among the measurements made on 9 dates (2 year period) of several water properties from piezometers should be noted. For example, in the P1 to P5 the salinity measured in EC had a VC lower than 9%, being only very high in the shallow lakes (VC > 65%) and in the mill 3, M3 (VC 42%) (Alconada-Magliano et al. 2016). Likewise, the low VC of the SAR (6.3%) is worth noting if all the values measured in the 7 piezometers (P1 a P7) during the 2 years period on 9 dates are considered. In the remaining variables measured there was a similar behavior.

In contrast, there were significant variations in the depth of the water table (VC between 30 and 73% according to considered piezometers) (Alconada-Magliano et al. 2016). In any of the cases there was correspondence with rainfall occurred in the study site. Moreover, even in piezometers of hills (P1, P4 and P7), a marked enrichment in salts was observed regarding the quality of the rainwater collected at the site (Table 8.5), The comments confirm the coexistence of water flows of different origin in the region as indicated in Alconada-Magliano et al. (2011b) and Alconada-Magliano and Damiano (2017).

In Table 8.5 the average values obtained in Lincoln, Pehuajó and Trenque Lauquen, in piezometers, mills, wells, and shallow lakes, which arise from the samples showed in Table 8.3 and those from other studies are compared (Alconada-Magliano 2008; Fagundo-Castillo et al. 2014). Also included in this table is the chemical composition of rainwater collected in Lincoln (Alconada-Magliano et al. 2017), and the frequent quality of freshwater according to Custodio-Gimena and Llamas (1983). The general enrichment in salts in all the waters of the region, being particularly greater in Lincoln is clearly noted. It is worth highlighting the very high content of Cl^- even in those waters where bicarbonates prevail, as well as the prevalence of the Na^+ ion with respect to the other cations, in all sites. Also, the factor of Cl^- concentration with respect to rainwater (FCCl), obtained by procedures described in Fagundo-Castillo (1998) and Fagundo-Sierra et al. (2001), is high in all positions reaching a maximum in piezometers P3 of 555, similar to that obtained in shallow lakes (Alconada-Magliano et al. 2017).

Waters sampled in the region studied have caused, during travel, high weathering of geological materials they go through, as described by Fagundo-Castillo (1990) and Fagundo-Castillo and González-Hernández (2016). Based on the results obtained in the region studied belonging to the Sandy Pampa, it can be indicated that the environment studied at Lincoln at the regional level functions as a discharge zone of intermediate flows which are linked to local flow (point 8.4.7). In this regard, Macchiavello and Sueiro (2012), specifically for Lincoln and Villegas, find significant variability in the water table quality and regional behavior. Alconada-Magliano et al. (2011b) recognize the existence, in the Sandy Pampa, of local and intermediate flows, estimating that the regional ones (of greater depth and distance of travel) would lead to the Atlantic Ocean. Thus, this has been considered to be due to lack of evidence, in this latest study, of waters with physico-chemical quality and temperature as high as those reported in the literature for deeper wells in the province of Buenos Aires (Pesce and Miranda 2000).

Table 8.5 Average physicochemical properties of water: pH, conductivity (CE dS.m^{-1}), cations and anions (me.l^{-1}), $CaCO_3$ (mg l^{-1}), electrical conductivity (EC), sodium adsorption ratio (SAR), in piezometers (P), wells-mills (W-M), and lagoons (L) in all locations: Trenque Lauquen (TL), Pehuajó (Pe), y Lincoln (Li) Adapted from Alconada-Magliano et al. (2017)

	Location	pH	CE	HCO_3^-	Cl^-	SO_4^{-2}	Ca^{+2}	Mg^{+2}	Na^+	K^+	SAR
P	TL	7.8	3.3	10.9	16.0	4.7	1.7	2.9	27.6	1.6	15
	Pe	7.3	3.8	5.7	27.0	4.5	4.2	5.9	25.0	1.0	11
	Li	7.9	9.6	17.7	46.6	33.1	3.2	9.3	84.4	1.4	35
W-M	TL	7.8	7.3	6.2	63.5	17.7	5.1	16.7	68.7	1.1	17
	Pe	7.8	4.0	15.2	22.5	6.6	1.7	4.9	37.2	0.8	20
	Li	7.8	4.9	18.6	18.2	12.3	2.4	4.4	42.3	1.1	24
L	TL	7.5	17.4	4.92	139.5	52.3	8.1	21.9	165.2	2.8	38
	Pe	8.5	14.9	5.34	123.2	26.1	4.3	16.1	141.2	2.7	37
	Li	8.3	11.8	6.5	72.6	41.0	5.5	12.0	102.1	2.1	35
Fresh wáter		6.5–8.5	0.1–0.2	0.8–6	0.3–7	0.02–5.2	0.5–19	0.08–8.2	0.04–6.5	0.002–0.25	
Rainwater Lincoln		7.2	0.063	0.3	0.2	0.31	0.2	0.05	0.10	0.03	1.4

8.4.7 Association Between Landscape Elements and Groundwater

According to the theory of Tóth (2000) (Fig. 8.1), depending on the chemical composition of the water, by their types and facies (PHM, Table 8.4), water from discharge zones prevails in the studied region, since the Na^+ cation and Cl^- anion are dominant. However, if the electrical conductivity (EC), temperature (T°), and associations between the local and regional landscape are also considered, recharge and throughflow areas of local and intermediate flows can also be identified (Table 8.4). Nevertheless, it should be noted that with respect to the general scheme of relations between landscape elements and water characteristics of Toth's theory (Fig. 8.1), in the studied region, there are not always simple correspondences to detect. Thus, the differences observed in topographic heights, soil types, as well as in the depth of groundwater (Table 8.1), did not show a predictable relationship with the characteristics of water at the sites studied (Tables 8.3, 8.4, and 8.5). This is due to the fact that flows of different origin and soils that have developed from two geological materials coexist in the region (Formations: E1, Invading Dunes and E3, Pampeana, loess).

In this way, the greater general salinity and alkalinity of Lincoln's waters compared to those of Pehuajó and Trenque Lauquen, can be explained by intermediate underground water flows that change the way they affect the environments depending on regional climate events, as well as by the zone (recharge, throughflow, or discharge) that these environments occupy in the flows. Also, the salinity and alkalinity measured at a site can be modified if mixtures of intermediate water flows with local flows occur. The latter is at least partly regulated by the thickness of the Invading Dune Formation (E1) material. If this is deep, it allows local recharges to occur and makes it difficult for shallow groundwater to rise by capillarity and/or hydraulic pressure, depending on whether the site is also a discharge zone for some flow (upward water flow) (Fig. 8.6).

In Pehuajó it was observed that soils that do not differ from the point of view of their taxonomic classification, have underground waters with different quality and direction. For example, in the soils identified as *Entic Hapludolls* in positions of extended flat hills and high hills, the incidence of waters belonging to the three groups identified in the studied region was recognized. Likewise, the zone (recharge, throughflow, and discharge) to which these soils belong (Tables 8.1 and 8.4) was different. Other soils, also located in environments called extended flat hill (EFH) in Lincoln, such as piezometers P4 and P6, were classified as *Natric tapto Hapludolls* and groundwater belongs to group 1. Only in the lowest positions, where the Invading dunes are absent and the soils are classified as *Tipic Natracualf*, there is a correspondence with the water table quality (G3).

Additionally, the significant variations that were measured in the water table depth can be highlighted, without significantly changing the quality of the water for the period studied for 2 years at Lincoln (point 8.4.6). Consequently, its effect on the evolution of soil and vegetation is different. However, it does not change the way in which the soil is classified by Soil Taxonomy, although its use capability changes.

For example, the depth of the water table studied (9 dates), in the piezometer P2 varied between 0.15 and 1.16 m, in the piezometer P4 between 0.94 and 2.10 m, and in the piezometer P6 between 0.44 and 2.02 m deep. In all three places the soil is classified as *Natric tapto Hapludolls* and the water quality was different. In P2 it belongs to G3 and in P4 belongs to G1.

Based on all said above, it can be indicated that in environments such as those presented in Pehuajó, of extended flat hills with *Entic Hapludolls* soil, the differences in the behavior and quality of the water tables are not easy to predict. The contrasting quality of the waters of the piezometer P8 (G2, Cl > HCO_3–Na, EC 7.69 dS.m^{-1}) with respect to that of the piezometer P10 (G3, Cl–Na, EC 1.54 dS.m^{-1}) and P12 (G1, HCO_3 > Cl–Ca > Na, EC 0.15 dS.m^{-1}) (Tables 8.3 and 8.4) stands out. The types of flows in the three piezometers were identified as local, but with different zones within them, between recharge and discharge (Table 8.4). It is clear that in these cases, fluctuations in the water table will significantly affect the growth of the crop. At the time of the study, a soybean crop was developed in all piezometers (Table 8.1).

Consequently, in some places there was a poor predictability of the behavior and quality of the water table from the analysis of surface elements of the landscape (soil type, geomorphology, topographic position, and vegetation). In cases where better correspondence was observed, it was due to the absence of the E1 Formation, Invading Dunes. When this material was not present, the intermediate flows that characterize the region become evident. In the places where flows of different origin coexist, changes in water, soil, and vegetation can be observed, depending on local and regional rainfall events, as well as on the zone to which each site corresponds when one or more flows coexist.

Figure 8.8 summarizes the relationships between landscape elements described in a farmland of Trenque Lauquen (Alconada-Magliano 2008; Alconada-Magliano et al. 2011a). The geomorphological units are indicated according to the criteria of Dillon et al. (1987), vegetation, and groundwater flow systems to which these environments are linked, according to Tóth's theory (2000, 2016). Figure 8.9 also shows associations of the landscape but in a larger area between Mari Lauquen and Berutti, both localities belonging to Trenque Lauquen county (Alconada-Magliano et al. 2011a). In general, this sequence of soils and associations are representative of the Sandy Pampas. The characteristics of the aquiferous system of the region under study (Gónzalez 2005), types of soils and water qualities presented here, allow us to affirm that there is regional hydraulic continuity, and that there are several groundwater flow systems that coexist, as described by various authors (Tóth 1962, 2000, 2016; Engelen and Jones 1986; Carrillo-Rivera 2000).

Frequent relations of soil and groundwater

Although, as indicated, it is not always possible to foresee through elements of the surface landscape the types of flows that coexist in an area, below are the main types of soil and vegetation that could be established in advance as linked to a general scheme of operation with groundwater.

Soils in water recharge zones: As characteristic of recharge areas, soils classified as *Typic Udipsamment, Entic Hapludoll,* and *typical Hapludol* can be indicated,

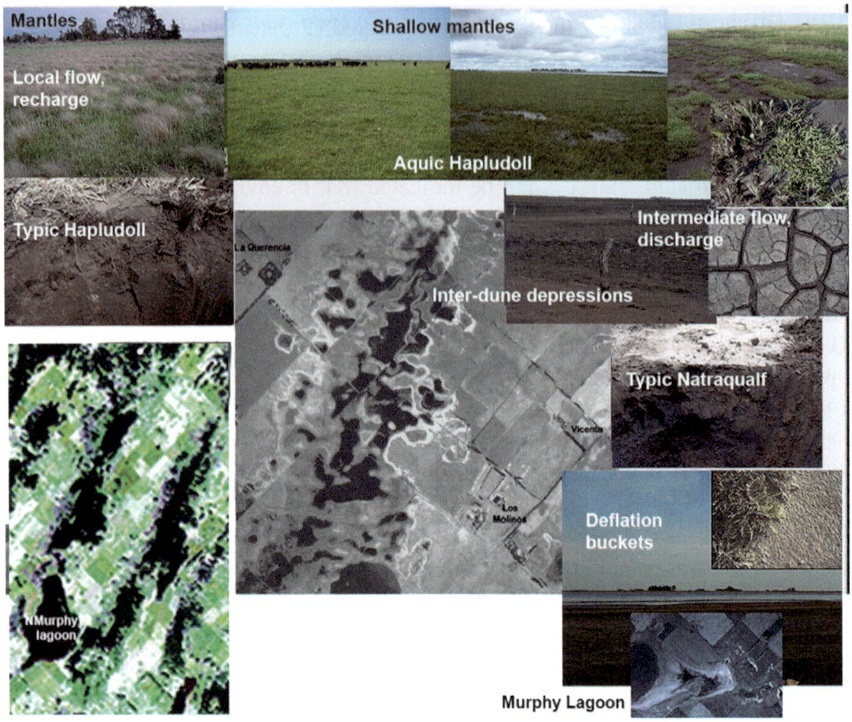

Fig. 8.8 Relations between landscape elements in a Trenque Lauquen farmland: geomorphological units, type of soil and vegetation

Fig. 8.9 Relation between soils, environments and recognized flows between Mari Lauquen and Berutti (both in Trenque Lauquen county)

without hydromorphism features, and without salinity or alkalinity. However, these soils can also be located in throughflow zones of local and intermediate flows. The thickness of the surface deposit of the Invading Dunes Formation (E1) modifies the depth at which the water table is located. Nevertheless, the waters of the wells and mills measured in areas with this type of soil reveal that below the frequent depth of soil study, other water flows circulate, mainly of intermediate type with greater salinity and alkalinity.

Soils in water discharge zones: In this area the soils receive water with upward flow (hydraulic pressure) and consequently manifest hydromorphism processes. The degree of halomorphism depends on the type of flow. Soils with hydromorphism and without or little halomorphism, are characteristic of discharge soils of local flows. In the northwest of Buenos Aires province (the Sandy Pampas), soils classified as *Acuic Hapludoll* and *Entic Hapludoll* with deep hydromorphism are recognized in these zones. Regarding the characteristic soils of intermediate flow discharge zones, in the Sandy Pampas region, soils classified as *Typic Natraquoll* and *Typic Natraqualf* were identified, both developed from the Pampeana Formation, loess (E3). This formation is the one that prevails throughout the province of Buenos Aires, but in this case, alkaline and saline groundwater near the surface are present. The *Argic tapto Hapludoll* and *Natric tapto Hapludoll soils*, described in the study area were not linked to a particular area; these soils were identified in recharge, throughflow, and discharge zones (Tables 8.1 and 8.4).

Vegetation and water flows

The vegetation that is sensitive to salinity, alkalinity, and hydromorphism, such as the *mesophyte prairie* of *Stipa trichotoma, Briza subaristata, Stipa neesiana, and Botriochloa laguroide* (Gabellone et al. 2003), can be identified as a water recharge environment. The natural vegetation of discharge zones of local flows presents resistance to hydromorphism, such as the *hydrophilic steppe* of *Ludwigia peploides, Solanum malacoxylon, Polygonum punctatum, Glyceria multiflora, Echinochloa helodes* (Gabellone et al. 2003). In intermediate flow discharge zones, the characteristic natural vegetation is the *halophyte steppe* resistant to hydrohalomorphism, with dominance of *Distichlis* sp. and in the most saline-sodium conditions, *Spartina* sp. and *Sarcocornia perennis*, as well as the presence of sodium humates in the soil surface (dark thin layer of dispersed organic matter). Figure 8.10 shows an environment of the geomorphological unit of Mantle with *mesophic prairie* (dominance of *Stipa* sp. and *Briza* sp.), and an environment close to the geomorphological unit of Inter-dune depression with *halophic steppe* (dominance of *Distichlis* sp., *Sarcocornia perennis*, and sodium humates).

However, vegetation does not always provide enough information on the types of groundwater flows that circulate at a site. This is particularly true in areas such as those in the present study, where local and regional rainfall modifies the water table depth, it also generates changes in surface water quality (local rainfall) and therefore, in the composition of the vegetation. For example, not only can changes take place in the plant species that dominate a particular community but also, characteristic species of other communities may appear when the degree of salinity, alkalinity, and/or hydromorphism is modified due to changes in the water regime of the soils.

Fig. 8.10 **a** Environment of the Mantle geomorphological unit with *mesophic prairie* (dominance of *Stipa* sp. and *Briza* sp.). **b** Environment close to the geomorphological unit of Inter-dune Depression with *halophic steppe* (dominance of *Distichlis* sp., *Sarcocornia perennis*, and sodium humates)

Based on what has been said, it can be indicated that although the characteristics of the soil and vegetation allow in general the identification of zones of recharge, throughflow, and discharge, they are not sufficient in all cases to define what type of flow they belong to (local, intermediate, or regional). It is important that in regions where the water table is shallow, the physical-chemical quality of the water that circulates at depths that could influence the evolution of the soil, the vegetation, and the sustainability of agricultural-forest management practices are studied. In cases where water cannot be measured, it can be inferred by the cationic, anionic, and physico-chemical properties of the edaphic profile due to the close correspondence recognized in Lincoln studies (Table 8.2) (Alconada-Magliano et al. 2016).

In all cases, it is necessary that the interpretations of water quality and relations with the elements of the landscape are made within the reference of the regional geomorphology and geology.This makes it possible to explain, for example, the presence of throughflow or discharge zones in a local hillside environment, as is often found in the Sandy Pampa. Thus, as presented in this chapter, soils with agricultural aptitude are recognized but whose productive sustainability depends on the general local and regional behavior of groundwater flows.

8.5 Final Considerations

The consequences in the intervention of a landscape can only be foreseen if all the elements that integrate it and the way in which these elements are linked are known. In the soil study, an understanding of the origin of the observed and measured properties must be incorporated, taking into account the regional context from a holistic and operational perspective. This is possible if the study of groundwater is included as an element that links the rest of the landscape elements.

The groundwater quality incident on a soil can be measured by collecting the water in piezometers. However, it is also possible to infer its quality from the study of the content of soluble cations and anions in the edaphic profile, mainly in those sectors of the profile subject to changes due to variations in the depth of the water table. It is also appropriate in order to know the functioning of the environment, to study the water quality of other manifestations of groundwater, such as wells, mills, and shallow lakes. This provides information on the coexistence of underground water flows of different origin that can be mixed with the water table due to climatic events of local or regional water excesses. Likewise, the study of the evolution of visible water covers in satellite images in different climatic conditions helps to understand how the local landscape works and it is linked to the regional one.

The sites selected in the Dunes Longitudinal sector of the Sandy Pampas in the Northwest of Buenos Aires province, Argentina, the differences observed in topographic heights, soil types, as well as variations in groundwater depth, did not always have a direct relationship with the characteristics and variations observed in the water quality of the studied sites. This is due to the fact that flows of different origin coexist in the region and that the soils have been developed from one and/or two original materials. Thus, for example, in the sector studied at Lincoln, small differences in topographic height or in the thickness of the original materials of the soil, generated marked differences in the way in which the flow of underground water of intermediate type affects the profile and is linked to local flows. In Pehuajó, very similar topographic heights and soil types were not associated in all the sites studied with equal water quality. The hydraulic link between sites is evident at local and regional level. This is consistent with the recognized aquiferous systems for the region.

The suitability of land use must be defined by the soil under study, as well as by the associated soils in the study sector and by other hydrogeologically linked regions. This allows a dynamic understanding of local functioning with the regional context that is necessary to carry out sustainable agricultural and forestry planning, as well as any intervention that affects water dynamics. Likewise, this holistic and dynamic understanding of the landscape is the starting point for defining management that contributes to the control of the excesses and water deficits characteristic of the region.

A new paradigm in the study of soil that includes the study of the functioning of groundwater, defining local and regional relationships, becomes evident and necessary to define sustainable production schemes that adapt to environments, contribute to their limitations, or take advantage of them.

References

Alconada-Magliano MM (2008) Procesos de inundación en el sector de Médanos Longitudinales del Noroeste de la provincia de Buenos Aires, Argentina, su relación con la vegetación, suelo, agua y clima. Opciones de desarrollo, unpublished Doctoral thesis in Geography, UNAM, México, 580 p, Anexo 103 p

Alconada-Magliano MM, Bussoni A, Rosa R, Carrillo Rivera JJ (2009) El bio-drenaje para el control del exceso hídrico en Pampa Arenosa, Buenos Aires, Argentina. Investigaciones Geográficas, Boletín del Instituto de Geografía, UNAM 68:50–72. http://www.redalyc.org/articulo.oa?id=56912236005

Alconada-Magliano MM, Damiano F (2017) Caracterización de suelos y del agua subterránea para definir intervenciones sustentables. In: Actas XX Congreso Geológico Argentino, Tucumán. Simposio Pedogénesis y cartografía de suelos como herramientas para la planificación, pp 1–7

Alconada-Magliano MM, Damiano F, Carrillo-Rivera JJ, Fagundo-Castillo JR (2017) Arsenic and fluoride in water in Northwestern Buenos Aires: their association with natural landscape elements. J Geogr Reg Plan 10(2):8–27

Alconada-Magliano MM, Damiano F, Fagundo-Castillo JR (2016) Estudio del suelo en el paisaje regional como base para definir su manejo agropecuario–forestal. Actas XXV Congreso Argentino Ciencia. Río Cuarto, Editorial UNIRIO, pp 1626–1645. https://www.unrc.edu.ar/unrc/comunicacion/editorial/repositorio/978-987-688-173-9.pdf

Alconada-Magliano MM, Fagundo-Castillo JR, Carrillo-Rivera JJ (2011a) La definicion de flujos de agua subterránea en la elección de prácticas de manejo del suelo. Actas IX Congreso Cubano de Geología, Taller sobre Aguas subterráneas y Contaminación, 17 p. http://www.redciencia.cu/geobiblio/paper/2011_Alconada_GEO12-O11.pdf

Alconada-Magliano MM, Fagundo Castillo JR, Carrillo Rivera JJ, Hernández PG (2011b) Origin of flooding water through hydrogeochemical identification, the Buenos Aires plain. Argentina. Environmental Earth Sciences 64(1):57–71

Álvarez C, Saks M, Lienhard CP (2015) Productividad del cultivo de maíz en la región semiárida pampeana. efecto del ambiente y nutrición. https://inta.gob.ar/sites/default/files/inta-productividad_del_cultivo_de_maiz_en_la_region_semiarida_pampeana__0.pdf

Ambast SK, Gupta SK, Gurbachan S (2007) Agricultural land drainage: reclamation of waterlogged saline lands. Central Soil Salinity Research Institute, Karnal, India, 231 p. http://www.cssri.org/Books.htm

APHA-AWWA-WPCF (1989) Standard methods for the examination of water and wastewater. Washington, V17

Baigorri HEJ, Cisneros JM, Estenssoro M, Martini E, Rang S, Sardoy JC, Begnis S (2003) Aprendiendo a utilizar la napa freática en la producción de soja. http://www.elsitioagricola.com/

Barraco M (2009) Productividad de soja (Glycine max L. Merrill) según propiedades edáficas de Hapludoles de la región de la Pampa Arenosa, unpublished Magister in Soil Science thesis, UBA, Facultad de Agronomía, 113 p

Barraco M, Álvarez C, Scianca C, Díaz-Zorita M (2010) Productividad de soja en suelos con capacidad agrícola limitada en la Pampa Arenosa. EEA INTA GENERAL Villegas. Memoria técnica 2009–2010:49–51. https://www.researchgate.net/publication/268013930

Black CA (1965) Methods of soil analysis. Agronomy No9 Part, 1770 p

Bonnesoeur V, Locatelli B, Ochoa-Tocachi BF (2019) Impactos de la forestación en el agua y los suelos de los andes: ¿qué sabemos? https://agritrop.cirad.fr/591482/1/Bonnesoeur%202019%20Impacto%20de%20la%20Forestacion%20en%20el%20Agua%20y%20Suelos.pdf

Burkart R (2009) El papel de las áreas naturales protegidas en un territorio forestal. In: Morello JH, Rodríguez AF (eds) El Chaco sin bosques: la pampa o el desierto del futuro, Orientación Gráfica Editora, pp 347–355

Calviño P (2017) Manejo de ambientes: Sudeste bonaerense. Resumenes Simposio Fertilidad 2017, Rosario, Buenos Aires, pp 41–43. https://www.fertilizar.org.ar/subida/evento/Simposio2017/ActasSimposio2017.pdf

Cardona A, Carrillo-Rivera JJ (2006) Hidrogeoquímica de sistemas de flujo intermedio que circulan por sedimentos continentales derivados de rocas volcánicas riolíticas. Ingeniería hidráulica en México V XXI(3):69–86
Carrillo-Rivera JJ (2000) Application of the groundwater-balance equation to indicate interbasin and vertical flow in two semi-arid drainage basins. Hydrogeol J V8(5):503–520
Carrillo-Rivera JJ, Cardona A, Edmunds WM (2007) Groundwater flow functioning in arid zones with thick volcanic aquifer units: North-Central Mexico. In: International symposium on advances in isotope hydrology and its role in sustainable water resources management (HIS-OIEA),Vienna V1:199–211
Casas R, Albarracín G (2015) El deterioro del suelo y del ambiente en la Argentina, Buenos Aires, Editorial FECIC, Tomo I, 604 p y Tomo II, 452 p
Chará J, Reyes E, Peri P, Otte J, Arce E, Schneider F (2019) Silvopastoral systems and their contribution to improved resource use and sustainable development goals: evidence from Latin America. FAO, CIPAV and Agri Benchmark, Cali: 60 p
Custodio-Gimena E (2001) Aguas subterráneas y humedales, 30 p. http://aguas.igme.es/igme/publica/sim_hidro_Murcia/tomo%20XXIV/01.pdf
Custodio-Gimena E, Llamas MR (1983) Hidrología Subterránea. (Segunda Ed) Ediciones Omega Barcelona, Tomo I, 1192 p
Dangavs N (2005) Los ambientes acuáticos de la provincia de Buenos Aires. XVI Congreso Geológico Argentino, La Plata, Capítulo XIII:219–236
Dillon AA, Hurtado M, Giménez J (1987) Influencia de la Geomorfología y Estratigrafía en los Anegamientos de la Pampa Arenosa. Actas X Congreso Geológico Argentino, Buenos Aires, Argentina, pp 301–304
Dillon AA, Hurtado M, Giménez J, Castillo RJ (1985) Consideraciones Geomorfológicas y Estratigráficas como Base del Carteo de Suelos de un Sector de la Pampa Arenosa (Pcia. Buenos Aires). Actas Primeras Jornadas Geológicas Bonaerense, Tandil: 737–749
Domínguez-Daguer D, Laclau P (2014) Silvicultura Sitio-específica en Suelos Hidromórficos de la Depresión del Salado. Jornadas de Salicáceas 2014 y Cuarto Congreso Internacional de Salicáceas en Argentina, La Plata, Buenos Aires, Argentina Domínguez-Daguer D, Laclau P (2014) Silvicultura Sitio-específica en Suelos Hidromórficos de la Depresión del Salado. Jornadas de Salicáceas 2014 y Cuarto Congreso Internacional de Salicáceas en Argentina, La Plata, Buenos Aires, Argentina. https://docs.google.com/viewer?a=v&pid=sites&srcid=ZGVmYXVsdGRvbWFpbnxkb2N1bWVudG9zc2FsaWNhY2Vhc3xneDozOWZhN2Q2YjgzNDJmMWQw
Edmunds WM, Carrillo Rivera JJ, Cardona A (2002) Geochemical evolution of groundwater beneath Mexico city. J Hydrol 258:1–24
Engelen GB, Jones GP (1986) Developments in the analysis of groundwater flow systems. IAHS 163:67–106
Etchevehere, PH (1976) Normas de Reconocimiento de Suelos. Publicación Nro. 152. INTA Castelar, 222 p
Etcheverry R (2003) Introducción. En: Maiola O, Gabellone N, Hernández M (eds) Inundaciones en la región pampeana. Ed UNLP, Argentina, pp 16–26
Etchichurry MC, Tófalo OR, Forzinetti ME (1988) Composición de la fracción psamítica de sedimentos actuales de la provincia de Buenos Aires y su significado tectónico. Actas Segundas Jornadas Geológicas Bonaerenses, Bahía Blanca, pp 419–428
Ewens N, Gezan S, Felker P (2012) Five year field evaluation of Prosopis alba clones on pH 9–10 soils in Argentina selected for growth in the greenhouse at seawater salinities (45 dS m − 1). Forests 3:95–113. https://www.mdpi.com/1999-4907/3/1/95
Fagundo-Castillo JR (1990) Evolución química y relaciones empíricas en aguas naturales. Efecto de los factores geológicos, hidrogeológicos y ambientales. Hidrogeología, Granada 5:33–46
Fagundo-Castillo JR (1998) Patrones hidrogeoquímicos y relaciones matemáticas en aguas naturales. Ingenieria Hidraúlica 19(2):62–78

Fagundo-Castillo JR, Alconada-Magliano MM, Carrillo-Rivera JJ, González-Hernández P (2014) Caracterización de los flujos de agua subterránea a partir de su salinidad. Tecnología y Ciencias del Agua V(3):63–80

Fagundo-Castillo RJ, Carrillo-Rivera JJ, Antigüedad I, González P, Peláez R, Melián C, Suárez M, Hernández R, Cáceres D (2005a) Caracterización hidrogeoquímica del sistema de flujo. Local-regional de la sierra del Rosario (Cuba). Rev Latinoamericana de Hidrogeologia 5:75–90

Fagundo-Castillo JR, González P, Alvarez-Varela E, Tillán-Ochoa G, Vinardell-Grandal E, Fagudo-Sierra J, Suárez-Muñoz M, Melián-Rodríguez C (2005b) Solución de problemas ambientales mediante la Modelación Hidrogeoquímica. No Especial Revista CENIC Ciencias Químicas 36. https://revista.cnic.edu.cu/revistaCQ/articulos/soluci%C3%B3n-de-problemas-ambientales-mediante-la-modelaci%C3%B3n-hidrogeoqu%C3%ADmica

Fagundo-Castillo JR, Carrillo-Rivera JJ, Antigüedad-Auzmendi I, González-Hernández P, Leláes-Diaz R, Hernández-Diaz R, Cáceres-Govea D, Hernández-Santana JR, Suárez-Muñoz M, Melán-Rodriguez C, Rodríguez-Piña M (2008) Chemical and geological control of spring water in eastern Guaniguanico mountain range, Pinar del Río. Cuba Environ Geol 55(2):247–267

Fagundo-Castillo JR, González-Hernández P (2016) Procesos Biogeoquímicos. Editorial Universitaria, 224 p. http://bida.uclv.edu.cu/bitstream/handle/123456789/7751/Procesos%20biogeoquimicos%20-20Fagundo%20Castillo%2C%20Juan%20Reynerio.pdf?sequence=1&isAllowed=y

Fagundo-Sierra JR, Fagundo JR, González P, Suárez M (2001) Modelación de las aguas naturales. Contribución a la Educación y la Protección Ambiental, 2, VII Taller de la Cátedra de Medio Ambiente, ISCTN, La Habana, ISBN 959-7136-13-9

FAO (2011) The state of the world's land and water resources for food and agriculture (SOLAW)—managing systems at risk. Food and Agriculture Organization of the United Nations, Rome and Earthscan, London, 308 p. http://www.fao.org/3/a-i1688e.pdf

Gabellone N, Sarandón R, Claps C (2003) Caracterización y zonificación ecológica de la Cuenca del Río Salado. In: Maiola O, Gabellone N, Hernández M (eds) Inundaciones en la región pampeana. Ed UNLP, Argentina, pp 123–136

Galetti MA (2014) Las inundaciones: mitigar o remediar. https://inta.gob.ar/sites/default/files/script-tmpinta_inundaciones_y_forestacin_mgaletti_oct2014.pdf

Gaona L, Bedmar F, Gianelli V, Faberi AJ, Angelini H (2019) Estimating the risk of groundwater contamination and environmental impact of pesticides in an agricultural basin in Argentina. Int J Environ Sci Technol:1–14. https://doi.org/10.1007/s13762-019-02267-w

GeoINTA (2013a) Suelos de la República Argentina. http://www.geointa.inta.gob.ar/2013/05/26/suelos-de-la-republica-argentina

GeoINTA (2013b) Suelos de Buenos Aires 1:50.000. http://www.geointa.inta.gob.ar/?s=buenos+aires and http://anterior.inta.gov.ar/suelos/cartas/

González N (2005) Los ambientes hidrogeológicos de la provincia de Buenos Aires. XVI Congreso Geológico Argentino, Capitulo XXII:359–374

Gupta SK (2007) Strategies for prevention and management of waterlogging and soil salinization and their limitations. In: Ambast SK, Gupta SK, Singh G (eds), Agricultural land drainage: reclamation of waterlogged saline lands. Central Soil Salinity Research Institute, Karnal, India, pp 43–56

Heuperman AF (2003) Bio-drainage; a situation analysis. Institute of Sustainable Irrigated Agriculture (ISIA) at Tatura, Australia, for the International Programme for Technology and Research in Irrigation and Drainage (IPTRID). www.wca-infonet.org/cds_upload/1058154421850_BIO_DRAINAGE.pdf

Heuperman, AF, Kapoor AS, Denecke HW (2002) Biodrainage. Principles, experiences and applications. International Programme for Technology and Research in Irrigation and Drainage, FAO. Knowledge synthesis report 6. http://www.iptrid.org

Hurtado M, Giménez J, Cabral M (2006) Análisis Ambiental del partido de La Plata. Aportes al Ordenamiento territorial Ed. 1ra, Buenos Aires: CFI, 134 p

Jenny H (1941) Factors of soil formation. A system of quantitative pedology. Ed. Mc. Graw-Hill, USA, 281 p

Jobbágy E, Nosetto MD (2015) Las napas freáticas superficiales como agente de degradación de suelos. En: Casas R, Albarracín G (Editores) El deterioro del suelo y del ambiente en la Argentina, Buenos Aires, Editorial FECIC, Tomo I, pp 287–300

Jobbágy EG, Vasallo M, Farley KA, Piñeiro G, Garbulsky MF, Nosetto MD, Jackson RB, Paruelo JM (2006) Forestación en pastizales: Hacia una visión integral de sus oportunidades y costos ecológicos. Agrociencia X(2):109–124

Laclau P, Valdés Smukas G, Villaverde R (2015) Sistemas silvopastoriles de la Depresión del Salado. Estructura y biomasa de montes de sauce Actas 3° Congreso Nacional de Sistemas Silvopastoriles, VIII Congreso Internacional de Sistemas Agroforestales, pp 67–72

Lardone AV (2016) Efecto de propiedades edáficas sobre la productividad de los cultivos de maíz y soja en la Pampa Arenosa. Semiárida Revista de la Facultad de Agronomía UNLPam 26(2):55–64 https://cerac.unlpam.edu.ar/index.php/semiarida/article/view/2145

Lavado RS, Taboada M (2017) Génesis y propiedades de los suelos halomórficos. En: Taleisnik E, Lavado RS (eds) Ambientes Salinos y Alcalinos de la Argentina, Buenos Aires, Argentina, Editorial Orientación Gráfica, pp 9–27

López G, Finello M (2017) Ambientes del centro de Santa Fe, ¿vale la pena el manejo diferenciado? Resúmenes Simposio Fertilidad 2017, Rosario, Buenos Aires, pp 44–50. https://www.fertilizar.org.ar/subida/evento/Simposio2017/ActasSimposio2017.pdf

López-Geta JA, Fornés-Azcoiti JM (2009) La Geología e Hidrogeología en la Investigación de Humedales. Serie Hidrogeológica y Agua Subterránea N° 28, Instituto Geológico Minero de España, 310 p

Macchiavello A, Sueiro N (2012) Comportamiento de napas y calidad de agua en los partidos de General Villegas y Lincoln. INTA EEA General Villegas, Publicación N°108. https://inta.gob.ar/sites/default/files/script-tmp-inta_mt2012_macchiavello_comportamiento_napas.pdf

Martinez-Calsina L, Lara JE, Suárez FA, Ballón M, Pérez PG, Vega H, Torres JC, Corbella R, Plasencia A, Caldez L, Banegas N, Luchina J, Nasca JA, Perez HE, Bottegal D, Zimerman M (2015) Producción de carne en un Sistema Silvopastoril de Algarrobos y Grama Rhodes de la Llanura Deprimida de Tucumán, Argentina. Actas 3° Congreso Nacional de Sistemas Silvopastoriles, VIII Congreso Internacional de Sistemas Agroforestales, pp 48–52

Martini E, Baigorri HEJ (2004) Claves para la caracterización de un ambiente con influencia de napa freática. In: Soja Actualización 2004. Información para la Extensión INTA Marcos Juárez 89:41–46

Moncaut CA (2003) Inundaciones y sequías tienen raíces añejas en la pampa bonaerense (1576–2001). In: Maiola O, Gabellone N, Hernández M (eds) Inundaciones en la región pampeana. Ed UNLP, Argentina, pp 27–48

Morello J, Rodríguez AF, Pengue W (2009) Análisis descriptivo del proceso de desmonte y habilitación de tierras en el Chaco Argentino. In: Morello J, Rodríguez AF (eds) El Chaco sin bosques: la pampa o el desierto del futuro. Orientación Grafica Editora, Buenos Aires, pp 291–311

Nosetto MD, Jobbágy EG (2014) Plantaciones forestales: sus servicios e impactos hidrológicos Cómo compatibilizar la producción con la conservación. https://ri.conicet.gov.ar/bitstream/handle/11336/14643/CONICET_Digital_Nro.18032.pdf?sequence=1&isAllowed=y

Nosetto MD, Jobbágy EG, Jackson RB, Sznaider GA (2009) Reciprocal influence of crops and shallow ground water in sandy landscapes of the Inland Pampas. Field Crops Res 113(2):138–148

Nosetto MD, Jobbágy EG, Tóth T, Di Bella CM (2007) The effects of tree establishment on water and salt dynamics in naturally salt-affected grasslands. Oecologia 152:695–705

Page A, Miller A, Keeney D (1982) Methods of soils analysis. Part 2, 2oEd. Madison, WI, USA, Soil Sci. Soc. Amer, 1159 p

Parkhurst DL (1995) Users guide to PHREEQC-a computer program for speciation, reaction-path, advective transport and inverse geochemical calculations. Water Resout Invest Rep 99–4529. https://pubs.usgs.gov/wri/1995/4227/report.pdf

Parkhurst DL, Thorstenson DC, Plummer LN (1980) PHREEQC. A computer program for geochemical calculations. US Geol Surv Water Resour Inv 90–92:210 p

Peñuela-Arévalo LA, Carrillo-Rivera JJ (2010) Importancia de los sistemas de flujo de agua subterránea para una adecuada gestión. Caso de estudio: Sur de la Cuenca de México. Revista Latino Americana de Hidrogeologia 7:5–10
Peñuela-Arévalo LA, Carrillo Rivera JJ (2013) Definición de zonas de recarga y descarga de agua subterránea a partir de indicadores superficiales: centro-sur de la Mesa Central, México. Investigaciones Geográficas, Boletín del Instituto de Geografía. UNAM Núm 81:18–32
Peñuela-Arévalo LA, García-Rubio G, Carrillo-Rivera JJ (2015) Importancia del agua subterránea en la conservación de los ecosistemas. Biodiversidad en el Distrito Federal: Estudio de Estado. Sección Medio Físico. GDF, DGZVS y Conabio
Pesce H, Miranda F (2000) Catálogo de manifestaciones termales de la República de Argentina. Volumen II - Región Noreste, Centro y Sur. SEGEMAR, Buenos Aires, 159 p. https://repositorio.segemar.gov.ar/handle/308849217/2517
Pla-Sentis I (2006) Problemas de degradación de suelos en América Latina: Evaluación de causa y efectos. X Congreso Ecuatoriano de la Ciencia del Suelo. http://www.secsuelo.org/wp-content/uploads/2015/06/1.-Problemas-de-Degradacion1.pdf
Pla-Sentis I (2014) Advances in the prognosis of soil sodicity under dryland and irrigated conditions. Int Soil Water Conserv Res, WASWAC, China 2(4):50–63
Pla-Sentis I (2017) Anthropogenic and climate change factors in present and future soil and water conservation problems. (Zlatic and Kostadinov, ed.) Soil and Water Resources Protection in the Changing Environment. Advances in GeoEcology Catena Verlag GMBH
PMI, Plan Maestro Integral Cuenca del Río Salado (1999) Ministerio de Economía de la Provincia de Buenos Aires-Halcrow-Banco Mundial, 1300 p
Porta J, López-Acevedo M, Roquero C (1994) Edafología para la Agricultura y el Medio Ambiente. Editorial Mundi Prensa, Madrid, España, 804 p
Rubio G, Pereyra FX, Taboada MA (2019) Soils of the Pampean region. In: Rubio G et al (eds) The soils of Argentina, Chapter 6, Ed Springer, pp 81–100
SAGyP-INTA (1989) Mapa de suelos de la provincia de Buenos Aires (Escala 1:500.000). Proyecto PNUD ARG/85/019, Buenos Aires, 544 p
Saks MG, Quiroga A, Fernández R, Zalba P (2012) Contribución de las napas al rendimiento de maíz en Molisoles de la planicie medanosa. Actas XIX Congreso Latinoamericano de la Ciencia del Suelo, Mar del Plata, Argentina
Schoeneberger PJ, DA Wysocki, EC Benham, Soil Survey Staff. (2012). Field book for describing and sampling soils, Version 3.0. Natural Resources Conservation Service, National Soil Survey Center, Lincoln, NE, 300 p. https://www.nrcs.usda.gov/Internet/FSE_DOCUMENTS/nrcs142p2_052523.pdf
Soil Survey Staff (1975) Soil taxonomy: a basic system of soil classification for making and interpreting soil surveys. Soil Conservation Service, U.S. Department of Agriculture Handbook N0436, Washington DC, 754 p
Soil Survey Staff (1999) Soil taxonomy. A basic system of soil classification for making and interpreting soil surveys. United States Department of Agriculture. Agriculture Handbook No 436, 869 p
Soil Survey Staff (2006) Claves para la Taxonomía de suelos. USDA, Dpto de Agricultura USA, 10 Ed. Traducción, Ortiz Solorio CA, Gutiérrez MC, 339/331 p. http://www.suelos.org.ar/adjuntos/Keys_soil_taxonomy_10_edition_2006_espanol.pdf
Soil Survey Staff (2010) Claves para la Taxonomía de suelos. USDA, Dpto de Agricultura USA, 11 Ed. Traducción de: Ortiz-Solorio CA, Gutiérrez-Castorena MC, Gutiérrez-Castorena EV, 365 p. https://www.academia.edu/13369610/Keys_to_Soil_Taxonomy_11th_edition_Spanish_translation
Soil Survey Staff (2014) Claves para la Taxonomía de suelos. USDA, Dpto de Agricultura USA, 12 Ed. Traducción, Ortiz Solorio CA, Gutiérrez Castorena MC, Gutiérrez Castorena EV, 410 p. https://www.nrcs.usda.gov/Internet/FSE_DOCUMENTS/nrcs142p2_051546.pdf

Sosa D (2012) El agua, excesos y déficits, en la producción agrícola de secano y pecuaria dentro de la cuenca inferior del río salado. Unpublished doctoral thesis, Universidad de la Coruña, España, 236 p. https://core.ac.uk/download/pdf/61906252.pdf

Stiff HA (1951) The interpretation of chemical water analysis by means of pattern. J Petrol Technol 3(10):15–17

Taboada M, Damiano F, Lavado RS (2017) Suelos afectados por sales e inundaciones en la Pampa Deprimida y el Oeste Bonaerense. In: Taleisnik E y Lavado RS (eds) Ambientes Salinos y Alcalinos de la Argentina, Buenos Aires, Editorial Orientación Gráfica, pp 56–87

Taleisnik E, López Lauenstein D (2011) Leñosas perennes para ambientes afectados por salinidad. Una sinopsis de la contribución argentina a este tema. Ecología Austral 21:3–14. https://pdfs.semanticscholar.org/72c0/f150d45698322212b6438fc054e97d37ebce.pdf

Tomar OS (2007) Bio-drainage: performance of various tree species as effective Bio-drain material: 213–221. In: Ambast SK, Gupta SK and Singh G (eds) Agricultural land drainage: reclamation of waterlogged saline lands. Central Soil Salinity Research Institute, Karnal, India, 231 p

Tóth J (1962) A theory of groundwater motion in small drainage basins in central Alberta, Canadá. J Geophys Res 67(11):4375–4387

Tóth J (1995) Hydraulic continuity in large sedimentary basins. Hydrogeol J 3(4):4–16

Tóth J (2000) Las aguas subterráneas como agente geológico: causas procesos y manifestaciones. Boletín Geológico y Minero, Instituto Tecnológico GeoMinero España 8:49–26

Tóth J (2016) The evolutionary concepts and practical utilization of the Tóthian theory of regional groundwater flow. Int J Earth Environ Sci V1:111, 11 p. https://doi.org/10.15344/2456-351X/2016/111

Uranga M (2017) Manejo diferencial por ambientes en suelos clase I y II del sudeste de Córdoba: Una visión práctica. Resúmenes Simposio Fertilidad 2017, Rosario, Buenos Aires, pp 51–54. https://www.fertilizar.org.ar/subida/evento/Simposio2017/ActasSimposio2017.pdf

WRB, World reference base for soil resources (2014) International soil classification system for naming soils and creating legends for soil maps. FAO, 203 p. http://www.fao.org/3/i3794en/I3794en.pdf

Zapata-Hernández R (2004) Química de la acidez del suelo. Universidad Nacional de Colombia, Medellín, Colombia, 197 p. http://www.bdigital.unal.edu.co/1735/1/9583367125.1.pdf

Zárate MA (2003) Loess of southern South America Quaternary. Sci Rev 22:1987–2006

Zárate M, Rabassa J (2005) Geomorfología de la provincia de Buenos Aires. XVI Congreso Geológico Argentino, La Plata, Capítulo VIII:119–138

Chapter 9
Regional Groundwater Flow Systems: Their Role in Conserving the Marismas Nacionales Biosphere Reserve in Nayarit, Mexico

Alessia Kachadourian, Debora Lithgow, Edgar Mendoza, and Rodolfo Silva

Abstract The Marismas Nacionales de México, located on the Pacific coast of the state of Nayarit, is a protected area, internationally recognized as a RAMSAR site. Twelve rivers pass through the reserve, producing complex fluvial and geomorphological networks, which contain an abundance of ecosystems, rich in biodiversity. While there are studies on the general hydrology of these marshes, mainly from hydrographic and ecosystem perspectives, a fuller systemic understanding of the hydrological functioning of the area is needed, given the environmental importance of the area. Based on the Tothian theory of regional groundwater flow systems, this paper presents an analysis of the association between the hydrological features and ecosystems of the Marismas Nacionales. By integrating observations and measurements from various disciplines, the ecosystem interconnectivity of the area is described and proposed as a basis for future studies and sustainable interventions in the Marismas Nacionales and other, similar environments.

Keyword Biodiversity · Ecosystem services · Environmental hydrogeology · Groundwater recharge-discharge · Tóth's theory

9.1 Introduction

Despite recent advances in research, it is still common today for surface water and groundwater to be referred to as separate entities, parts of a water cycle that is divided into phases, with no acknowledgement of the uniqueness of water itself. When the planetary hydrogeological cycle is comprehended, the perception of water changes and appears as a single entity that simply changes place, physical state and/or physical-chemical composition in each of the stages of its cycle. In its atmospheric

A. Kachadourian (✉) · E. Mendoza · R. Silva
Instituto de Ingeniería, Universidad Nacional Autónoma de México, Edificio 17, Ciudad Universitaria, 04510 Mexico City, Mexico
e-mail: alessiakm@gmail.com

D. Lithgow
Red de Ambiente y Sustentabilidad, Instituto de Ecología, AC 91073 Xalapa, Mexico

M. M. Alconada-Magliano (ed.), *Intensified Land and Water Use*, Springer Earth System Sciences, https://doi.org/10.1007/978-3-030-65443-6_9

and surface phases, water is more evident for human beings. Its functions, main mechanisms, effects and manifestations, such as rain, snow, dew, evapotranspiration, runoff is well known. However, water in its underground phase constitutes 97% of the liquid fresh water on the planet, it is the main type of water in the environment and the source of the perennial water bodies that are observed on the Earth's surface. The perennial volume, or base flow, of a river, lake or lagoon comes from the emergent groundwater that provides water on a permanent basis, regardless of the local rainfall regime, and is linked to the regional water flow system, which may begin hundreds of kilometres away.

Tóth's theory of regional groundwater flow systems (Tóth 1970, 1971, 1999, 2009, 2016) allows to verify the unicity of water and how the groundwater phase binds so many of the elements of the environment. Groundwater is ever present, and the only natural mechanism that transports and interconnects matter and energy to and from the Earth's surface, defining local and regional environments, linking the geoforms, soil, vegetation and even the climate of any environment.

The Tothian theory (Tóth 2000) explains how groundwater circulates in flows that are determined by the regional geological context and identifies natural features that are produced by groundwater, though they may be modified later by environmental conditions. Through the analysis of permanent surface water, soil type, vegetation, geomorphology and climate, these surface indicators of groundwater flow systems can be identified (Tóth 1999).

The underground part of the water cycle begins with infiltration in recharge zones, and flows towards discharge zones, where the water emerges to the surface (Tóth 2009). In Fig. 9.1, which is based on Tóth´s theory (Tóth 1962, 1963, 1999), the zones of the groundwater flow systems that can exist in a certain environment (*recharge*, *throughflow* and *discharge*) and the hierarchy of the flows (*local*, *intermediate* and *regional*) are shown. There is no scale in this diagram, as it depends on the geological context; the depth, length and magnitude of the velocity of groundwater flows,

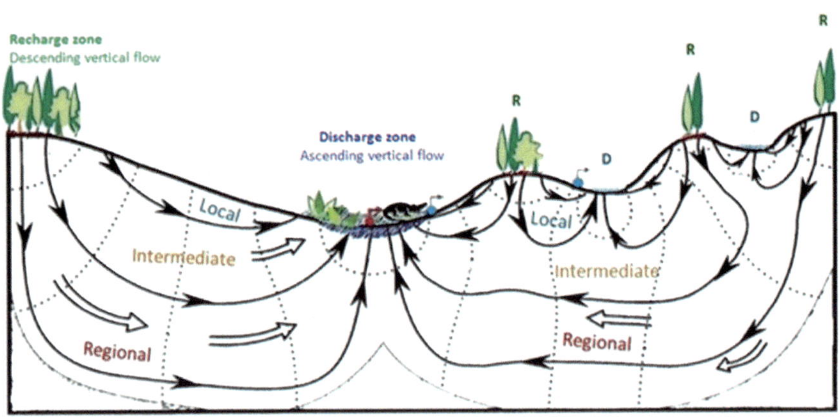

Fig. 9.1 General diagram of groundwater flow systems (modified from Tóth 1999)

determined by the geology of the area, as well as the degree of interaction between water and rocks, and the local-regional environmental dynamics induced by human intervention.

Tóth´s theory of regional groundwater flows explains how the water that circulates, forms and supports the biosphere reserve of the Marismas Nacionales is just one entity, with different phases, all interrelated. However, at present, there is no understanding of the complete hydrological dynamics of this important area. Located on the Mexican Pacific coast (Fig. 9.2), the Marismas Nacionales are an area of outstanding natural wealth, which has unique landforms and valuable ecosystems. Twelve rivers pass through the area, the dynamics of which produce a complex coastal hydrological system containing a large extension of mangroves and other estuarine wetlands, which develop along a plain of frontal dunes that extend inland for more than 15 km.

Within the biosphere, the reserve of the Marismas Nacionales is the largest extension (133,850 ha) of mangroves on the Mexican Pacific coast, as well as a considerable area of rainforest and coastal dry forest, (CONANP 2013). The region was formed by complex hydro-sedimentary dynamics, tidal, riverine and underground. The preservation of the area depends on maintaining these fluxes. Land-use changes have accelerated since the middle of the twentieth century, pressurising the ecosystems via

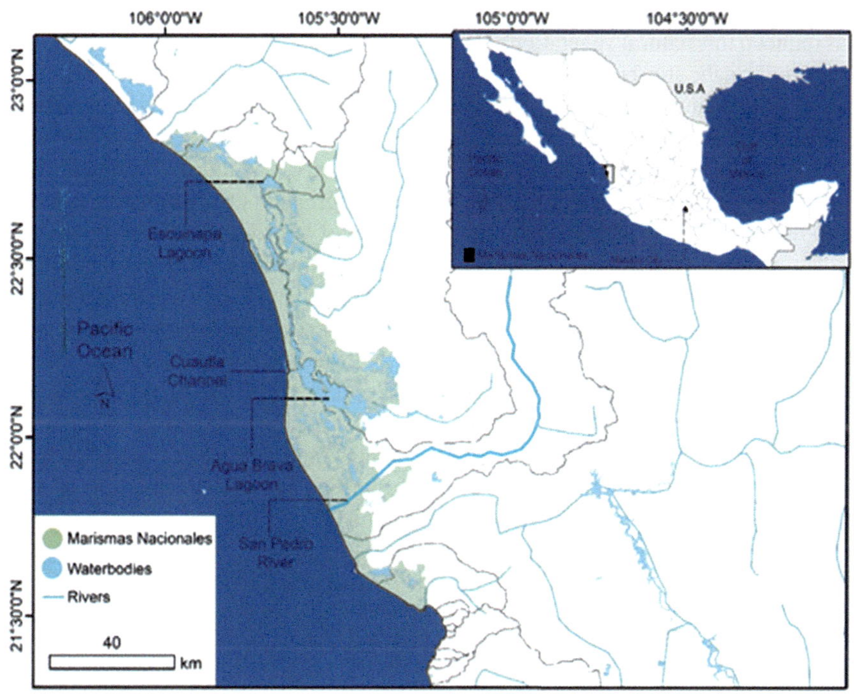

Fig. 9.2 Location of the Marismas Nacionales

the overexploitation of natural resources, such as timber and fisheries, the substitution of natural ecosystems for crops and aquaculture ponds and the construction of infrastructure that interrupts the flows of water and sediment.

The first step in understanding the role that groundwater plays in the conservation of the Marismas Nacionales was the characterization of the zones of the groundwater flow systems in the region. The importance of the marsh and its degree of vulnerability are presented. Then, hydrogeological analyses were carried out on the hydrographic and piezometric features, the soil cover, the vegetation and the topoforms of the reserve.

9.2 Degree of Conservation of the Marismas Nacionales

The biosphere reserve of the Marismas Nacionales is composed of several floodable ecosystems that developed due to the accumulation of sediment in the confluence of the alluvial plains of twelve rivers, that is, the Navarrete, Sauta, El Palillo, Grande de Santiago, San Pedro-Mezquital, Bejuco, Rosamorada, San Francisco, Acaponeta, Las Cañas, Escuinapa and Baluarte streams.

In the past, the alluvial plains were covered by rainforest, but advancing agriculture has reduced this natural vegetation. The coastal wetlands are one of the most valuable formations in the area because of the ecosystem services they provide. These wetlands can be divided into herbaceous and arboreal (mangroves). The herbaceous wetlands cover over 175,000 ha and are in a relatively good state of conservation. However, at the edges of the marshes advancing agriculture and aquaculture are replacing and fragmenting this ecosystem. The extension of mangrove in the reserve has fallen by 50% due to land-use changes and the interruption of hydro-sedimentary fluxes.

Throughout the Marismas Nacionales, there is a generalized loss of naturalness, which is greater inland, where the alluvial plains are seriously altered, and decreasing towards the coast, while the deltaic fronts are less degraded, although there are some patches on the coastal plain which have been altered as well. This loss of naturalness is sometimes referred to as degradation, the response of the environment to all the complex mechanisms occurring in the landscape (Simensen et al. 2018). Anthropic interventions and consequences that generate degradation in the Marismas Nacionales include the building of artificial barriers that obstruct the normal flow of matter and energy in the ecosystems. Bare soil, excavations, deforested areas (including secondary vegetation), burned areas, crops, live fences, are indicators of human interventions that are producing long term changes in the land use of specific parts of the reserve.

In most of the degraded areas, the alterations have been caused by agriculture. The greatest alterations have been induced by aquaculture, where shrimp ponds (artificial water bodies) modify the natural flow of matter and energy between the delta and the lagoon area. In the deltaic and ridge areas, where there is less agricultural activity, there is less alteration, especially in the flood and hypersaline areas. In the lagoons and channels between the dune ridges, known as dune slacks, the main economic activity of the local population is fishing. Traditional fishing is less likely to transform the landscape than aquaculture or agriculture. Since the 1940‘s intensive aquaculture has been established in some flood-prone areas, transforming the landscape at an alarming rate (Lithgow et al. 2019).

The severe deterioration of the mangrove areas in the Marismas Nacionales was attributed only to the Cuautla Channel (Flores-Verdugo et al. 2014). However, the deterioration beyond the area influenced by the channel indicates diverse hydro-sedimentary imbalances. Undoubtedly, the Cuautla Channel produces one of the main hydro-sedimentary imbalances in the region, but other projects, such as the Aguamilpa Hydroelectric Dam and concurrent socio-economic activities, have also contributed to the degradation of the mangroves (Blanco et al. 2014). Figure 9.4 shows the original course of the channel and the 1976 coastline position.

The alteration of the ecosystems in the Marismas Nacionales affects their ability to maintain ecological processes and a diverse community of organisms, and thereby has a direct impact on the ecosystem services they provide.

Perturbation of the wetlands could induce losses of high ecological values by reducing the rich biodiversity of the area. The socio-economic impact of these perturbations will affect the acute dependency of the local population on the mangrove-associated fisheries. It could also affect the health and resilience of interconnected ecosystems.

An approximation of the economic value of the ecosystem services in the Marismas Nacionales by Mendoza et al. (2012) considered the average value per hectare per ecosystem in US dollars, updated in 2018. It was found that the coastal lagoons and the mangrove area are the most valuable in terms of ecosystem services (US$ 26,000 ha^{-1}. $year^{-1}$). The coastal beaches and foredunes were estimated to have an approximate value of US$ 18,000 ha^{-1}. $year^{-1}$, when taking into account protection services against extreme hydrometeorological events (e.g. hurricanes), and recreation services.

9.3 Hydrographic Structure

The hydrological structure of the Marismas Nacionales has diverse geometries and dynamics, but is a single system connected to the ocean. To date, only the hydrological structure of the surface has been characterized. To understand how the hydrological system is interconnected with the ocean, it is necessary to complete the picture by characterizing and incorporating the underground phase.

The surface hydrological network is made up of 2,480 km, which include the main rivers as well as the artificial channels within the reserve (Fig. 9.4). This exceptionally high number is because there are over 200 natural channels between the relict dune ridges, and many large artificial channels. Most of the artificial interventions in the natural drainage network are related to agricultural irrigation, and many are connected to the natural streams, often modifying the natural patterns of runoff in the area. The mild slopes of the coastal plain and the delta cause stagnation of the water flows. The considerable dissection of the terrain by the hydrological network and the low topography of the coastal plain make the area especially vulnerable to rising sea levels (Fig. 9.4). Water bodies occupy a total area of 170 km^2; the vast majority being permanent (168 km^2).

In a large part of the study area there are ridges of about 15–100 m in length, lying roughly parallel to the coast on the 14 km wide coastal plain. The average height of the ridges is 1 m, and the highest is 2 m. The dune ridges are perpendicular to the San Pedro river, inducing the natural formation of lagoons between the ridges.

For example, naturally, the connection between the Agua Brava and Toluca lagoons with the sea tends to erode the ridges, however this process has been intensified by projects designed to facilitate navigation from the lagoons to the ocean, such as the Cuautla Channel (Figs. 9.3 and 9.4).

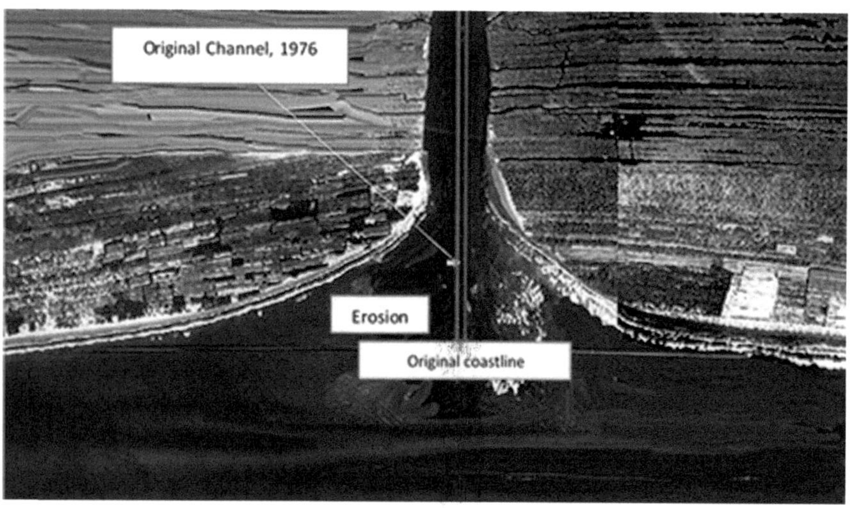

Fig. 9.3 The Cuautla channel, showing the width of the original channel and the coastline recession since 1976 (adapted from Flores-Verdugo et al. 2014)

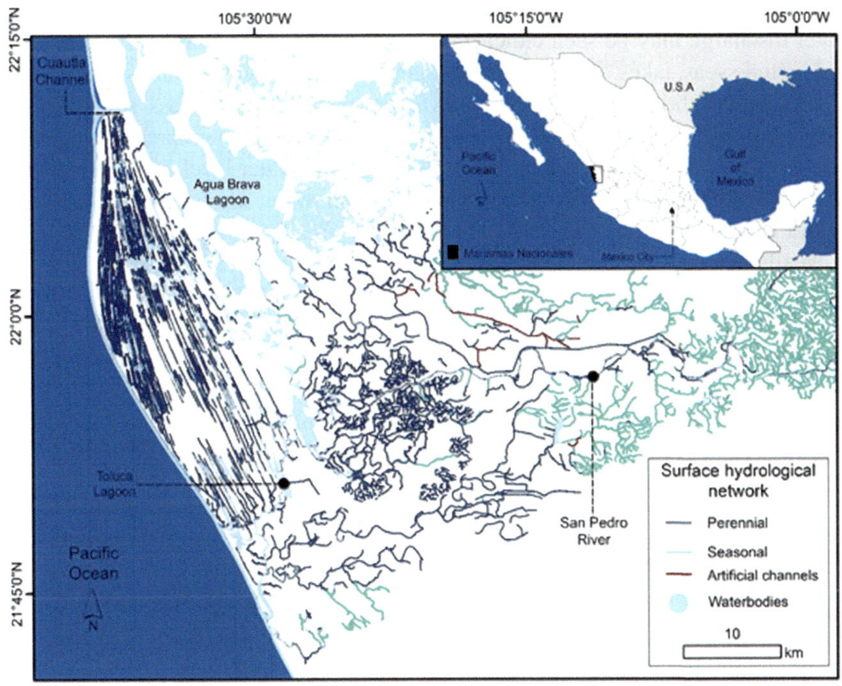

Fig. 9.4 The surface hydrological network of the Marismas Nacionales

9.4 Systemic Hydrological Interconnections

Currently, no complete explanation of the hydrodynamics of the Marismas Nacionales has been carried out; only data on surface hydrology is available.

The permanent, simultaneous in situ interactions between groundwater and the environment determine the distribution of the products of that interaction. It is now recognized that groundwater flow systems are the most important general geological agent in a great number and variety of natural processes and phenomena (Tóth 2016). Identifying surface indicators is the first step in understanding the complete dynamics of the groundwater in the Marismas Nacionales, as these bear witness to the interactions of the underground flows.

To understand the underground path of water, it is easier to start at the end of a groundwater flow, where the underground water emerges onto the surface, the point or zone of discharge. There are four main features of a ***discharge zone*** (Tóth 1971): (i) positive potential slope, (ii) relatively low position, (iii) allochthonous physical-chemical composition of the water and (iv) allochthonous temperature of the water.

The aspect and intensity of these features depends on the hydrogeological context: air temperature, slope of the relief, physico-chemical composition of the rocks, permeability of the rocks, vegetation, land use, etc. Natural features of groundwater discharge may be seen either directly on the surface, or in a diffuse manner, and include seas, lakes, lagoons, springs, geysers, floodplains, swamps, quicksands, salt flats, landslides and mudslides, swamps, depressions, excess soil moisture, hydrophilic vegetation and halophyte vegetation. The permanent presence of water, waterlogging, mud and flooding and a shallow static water level depth are the first and main surface indicators that a site is in a groundwater discharge zone.

In the ***recharge zone*** the water table can vary greatly throughout the year and it is usually at a considerable depth. According to the size of the study area, there may be no signs of permanent surface water in relation to local or regional conditions. Throughflow zones are those between the discharge and recharge zones of groundwater flow systems. The main feature of a throughflow zone is that horizontal flows dominate due the water´s mechanical energy mostly does not vary, while in discharge zones vertical upward flows dominate, and in the recharge zones, the main flows are downward and vertical (Tóth and Hayashi 2010).

9.4.1 Surface Indicators of Groundwater

Based on Tóth´s theory, a systematic, non-linear analysis was carried out of features on the surface related to groundwater flow systems. The following order of features shows their respective importance:

(i) Natural perennial water on the surface, including springs
(ii) Groundwater depth
(iii) Soils
(iv) Vegetation and land use
(v) Topoforms

For each indicator the oldest datasets available were used at scales of 1:250 000 (continental relief, hydrography, depth of piezometric level, edaphic coverage, hydrology, potential wetlands, land use and vegetation) and 1:1 000 000 (climate and topoforms). This official geographic information was used (CONABIO 2013; CONAGUA 2016, 2019; INEGI 1980, 1991, 1992, 1993, 1995, 2000, 2001, 2005, 2006, 2008, 2009, 2010, 2012). For the analysis of the attributes of the original maps INEGI, their respective cartographic interpretation guides and data dictionaries were used (INEGI 2001a, 2009a, 2010a, 2014, 2014a, 2016). The criteria of the World Soil Resources Reference Base (IUSS-FAO 2007) and the Keys to Soil Taxonomy (Soil Survey Staff 2014) were used to analyse soil edaphic properties.

The analyses were made for the polygon known as the Priority Area of the Marismas Nacionales ("Región Terrestre Prioritaria Marismas Nacionales") (Arriaga et al. 2000), which includes the RAMSAR wetland site 732. Priority Areas are those with rich ecosystem diversity and a viable potential for conservation measures, as designated by the CONABIO (Comisión Nacional para el Conocimiento y Uso de la Biodiversidad), WWF (World Wildlife Fund), USAID (United States Agency for

International Development), TNC (The Nature Conservancy), INECOL (Instituto Nacional de Ecología A.C.) and FMCN (Fondo Mexicano para la Conservación de la Naturaleza).

Each surface indicator was reclassified in terms of the characteristics of the zones of the groundwater flow systems. This hydrogeological reclassification is shown in five new maps.

(i) Naturally occurring perennial water on the surface. The information for this surface indicator was taken from maps with spatial information regarding surface water. Features that are directly associated with discharge zones are lakes, lagoons, marsh, swamp, perennial rivers, springs, muddy terrains, wetlands and floodplains. From all the available, official, spatial information regarding surface perennial water a new map was created showing that 88% of the surface area of the Priority Area has perennial water. Applying a 1 km margin around the points and polygons of perennial water found in a topoform associated with discharge the new map identifies locations of potential discharge processes.

This map is presented in Fig. 9.5. The discharge zones of groundwater flow systems are shown by blue shading. Springs are the clearest, most direct indication of the processes of groundwater discharge. A spring is where water emerges directly from the ground, and occurs where there is enough hydraulic conductivity in the rocks and the groundwater table is shallow. Where they occur in a geoform related to discharge, this confirms that the area is one of discharge. Where they occur in topoforms not related to discharge, they are assumed to be points of discharge.

The presence of thermal springs in, or near, confirms that the discharge zones can have regional flow system characteristics.

In the Priority Area the main perennial rivers are the San Pedro, Grande de Santiago, Bejuco and Rosamorada streams, whose respective base flows are the proportion of stream discharge sustained by groundwater discharge. All perennial rivers travel through sections of discharge, throughflow and recharge. Where rivers are in discharge topoforms, this shows that the systemic process of discharge is dominant.

(ii) Groundwater depth. The geo-referenced register of the piezometric depth, which indicates the groundwater depth at a given point, is important in analysing the surface indicators as a whole. The following classification is suggested:

a. Thermal water with static water level ≤ 3 m in discharge topoform
b. Thermal water with static water level between > 3 m and ≤ 11 m in discharge topoform
c. Thermal water with static water level between > 3 m and ≤ 11 m not in discharge topoform
d. No thermal water with static water level ≤ 3 m in discharge topoform
e. No thermal water with static water level between > 3 m and ≤ 11 m in discharge topoform
f. No thermal water with static water level between > 3 m and ≤ 11 m not in discharge topoform
g. Static water level > 11 m

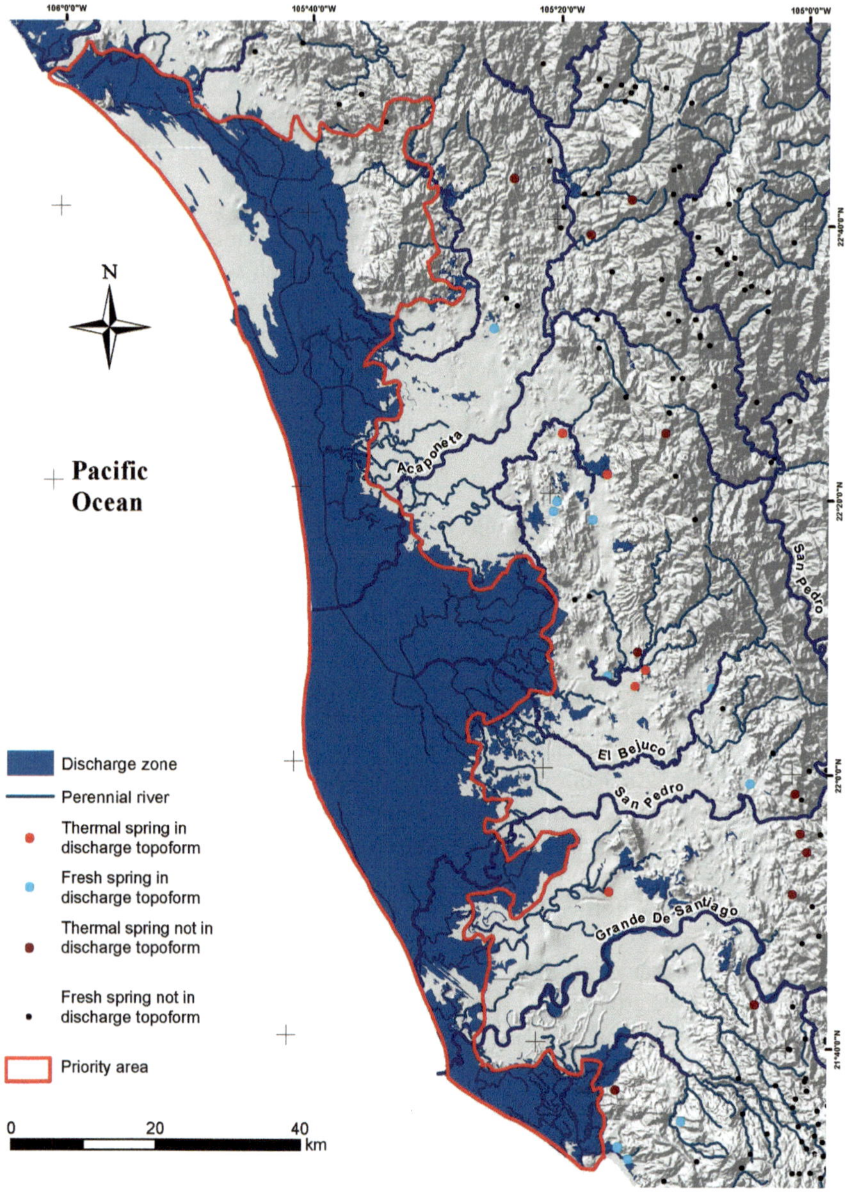

Fig. 9.5 Permanent surface water as an indicator of hydrogeological dynamic

A groundwater depth of less than, or equal to 3 m is a direct indicator of a point or zone of discharge where water emerges, as it is the average limit for evapotranspiration to occur in the subsurface. The range 2–11 m is the interval in which natural temporal groundwater fluctuations have been observed.

There are more points with a shallow depth in discharge zones than in recharge or throughflow zones. A static depth of > 11 m in a regional topoform associated with recharge is a surface indicator of a potential recharge zone of groundwater flow systems. Figure 9.6 shows the publicly registered groundwater depths classified.

The discharge points found close to the coastline at ≤ 3 m show that the groundwater dynamics do not stop at the coast, there are maritime discharge zones (Moore 2010). Submarine groundwater discharge (SGD) provides at least 50% of the fresh groundwater received by the sea (Povinec et al. 2007; Post et al. 2013).

(iii) Soils. The third surface indicator of groundwater in the Marismas Nacionales is the edaphic coverage, which is present in 86% of the area studied (Arriaga et al., 2000). The soil types described in the maps available were analysed using the Reference Soil Groups and principal and supplementary qualifiers (IUSS-FAO 2007; Soil Survey Staff 2014). The Marismas Nacionales have 11 types of soils (INEGI 1991, 2014) which were analysed and grouped into 3 categories:

a. Soils associated with discharge zones: Fluvisol, Histosol, Lixisol, Solonchak and Vertisol
b. Soils associated with recharge zones: Leptosol, Luvisol and Regosol
c. Soils found in indistinct zones: Arenosol, Cambisol and Phaeozem

In the official Mexican spatial information of soil coverage, the groups are described with main qualifiers of the soil. Those associated with discharge processes are endogleyic, epigleyc, ferric, fluvic, gleyc, hypersodic, hyposalic, hyposodic, rheic, salic and vertic (INEGI 1991, 2014). The rest of the edaphic characteristics may occur as a result of both recharge and discharge processes, so they were assigned as indistinct. Figure 9.7 shows the soil classifications resulting from the analysis of soil attributes that show the processes of groundwater flow in each polygon of the soil cover of the priority area and its surroundings: 70% of the edaphic cover is related to discharge processes (*discharge zone*), 14% to *discharge- throughflow* zone, 7% to *throughflow* zones and 5% to *throughflow– recharge* zone. The 4% of the edaphic cover is defined as being of *indistinct* relation and 1% of the soils are typical of *recharge zones.* Outside the main area, two other classifications are observed *major discharge with minor recharge surface*, indicates that 60% of the polygon is discharge soil, and 20-40% is a soil with conditions associated with groundwater recharge; and *major recharge with minor discharge surface* category is an edaphic polygon with ≥ 60% covered by a soil type associated to recharge zones and that 20-40% of the polygon has a typical groundwater discharge soil.

(iv) Vegetation. The maps of Vegetation and Land Use (INEGI 1980, 1993, 2009a, 2014a, 2016) were used to produce Fig. 9.9. This shows the edaphic coverage of the Marismas Nacionales: 14% of the soil cover is classified as water bodies and 68% vegetated by 9 types of vegetation (INEGI, 1980, 1993) associated with:

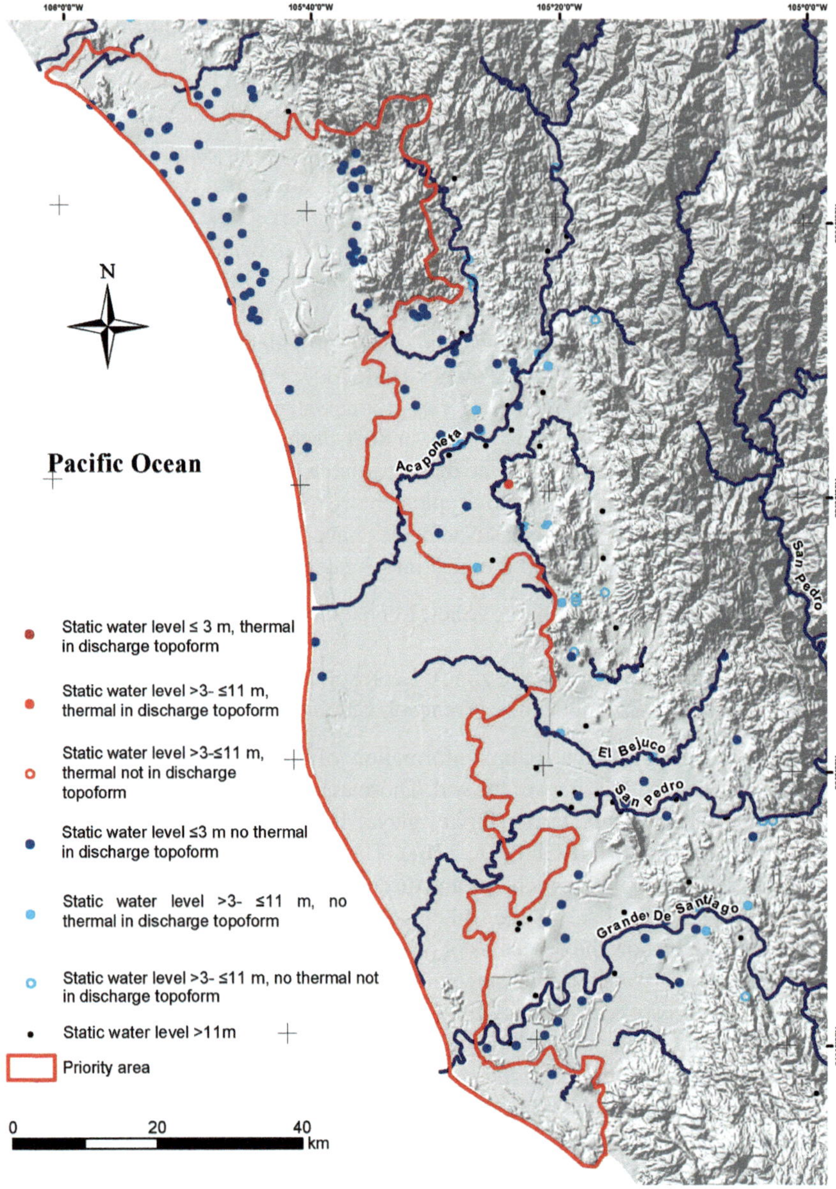

Fig. 9.6 Groundwater depth

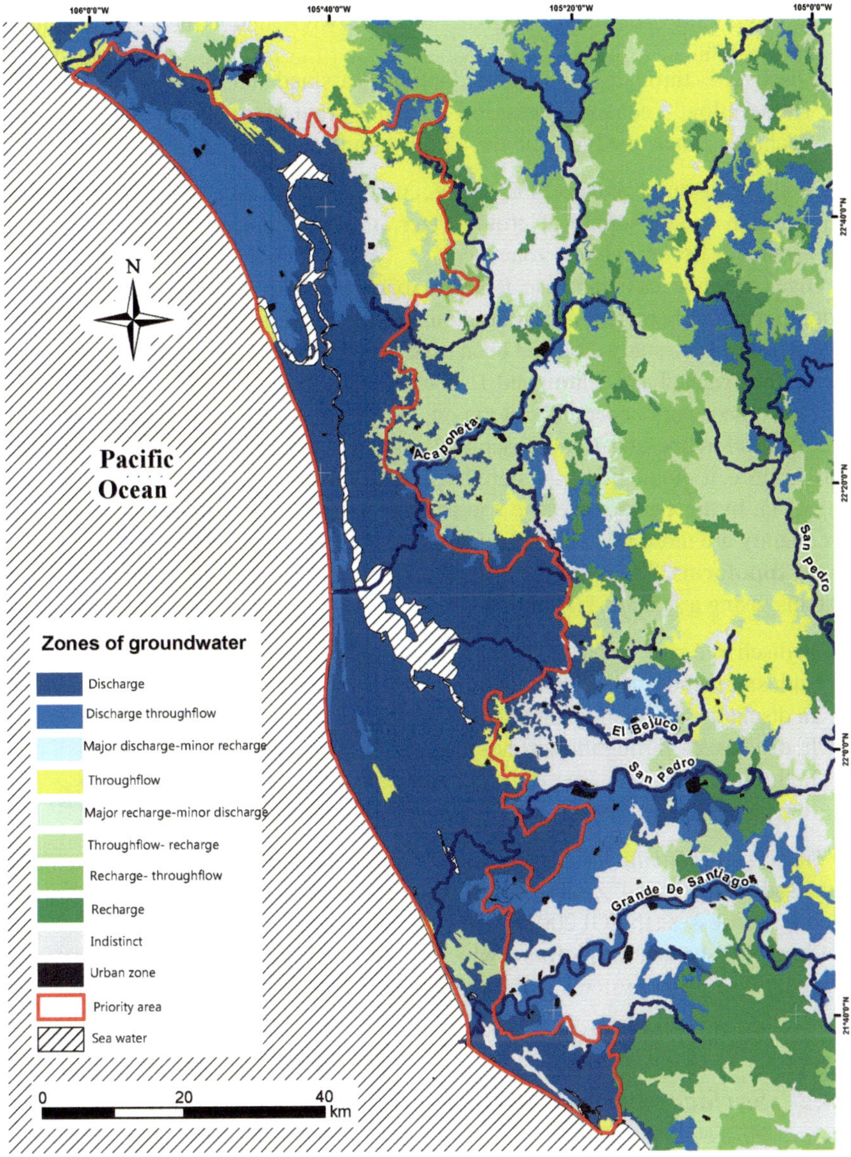

Fig. 9.7 Soils as a hydrogeological surface indicator

a. Discharge zones: halophilic vegetation, mangroves, coastal dune vegetation and palm grove
b. Discharge - throughflow zones: halophilic grassland and spiny lowland forest
c. Throughflow - recharge zones: low deciduous rain forest
d. Recharge zones: medium sub deciduous rain forest

Of all the "natural" cover, 74% is associated with *discharge zones*, types of vegetation which need a supply of groundwater at a shallow depth, 14% is associated with *recharge* or *throughflow* zones.

Given the changes in land use, it is wise to use the oldest land use data available in order to make a characterization based on natural or original vegetation cover. In the 1980s, 18% of the study area had been transformed, 15.7% for seasonal agriculture, 0.85% for irrigated agriculture and 1.45% for cultivated pastures.

Tóth's theory of conceptual criteria on soils and vegetation should be considered as a first reference and a more detailed systemic analysis should be carried out. Furthermore, it should be remembered that groundwater flows are multi-dimensional (Fig. 9.1). In this perspective, the soil and vegetation types with the highest degree of certainty are those related to the discharge and recharge environments (Fig.9.8).

(v) Topoforms. The relief features characterized regarding the groundwater flow systems, using a scale of 1:1 000 000 were:

a. In discharge zones, coastal plains with swamps, coastal salt plains, coastal plains with salt lagoons and deltaic plains; 71.6%
b. In discharge swamping zones, floodable plains with barriers, 5.3%
c. In discharge - throughflow zones, coastal plains with highlands, 2.6%
d. In throughflow - recharge zones, highlands with plains, 3.1%
e. In recharge zones, mountains, highlands and hills with valleys, 14.7%

The term *swamping discharge* zone was used as in the dictionary from the metadata used, the "flooded" characteristic is indistinct for both tidal and/or continental flooding processes (INEGI 2001).

Figure 9.9 shows the topoforms which are surface indicators of the distinct zones of groundwater flow systems in the study area.

9.4.2 *Surface Variables in the Definition of Hydrological Zones*

Each of the individual surface indicators provides preliminary information about the zones of the groundwater flow systems. As previously mentioned, groundwater is the natural agent with the greatest capacity to integrate the greatest amount of natural and environmental components, from the water depths to the atmosphere. For this natural reason of the systemic dynamics of water is that the surface indicators are analysed in conjunction.

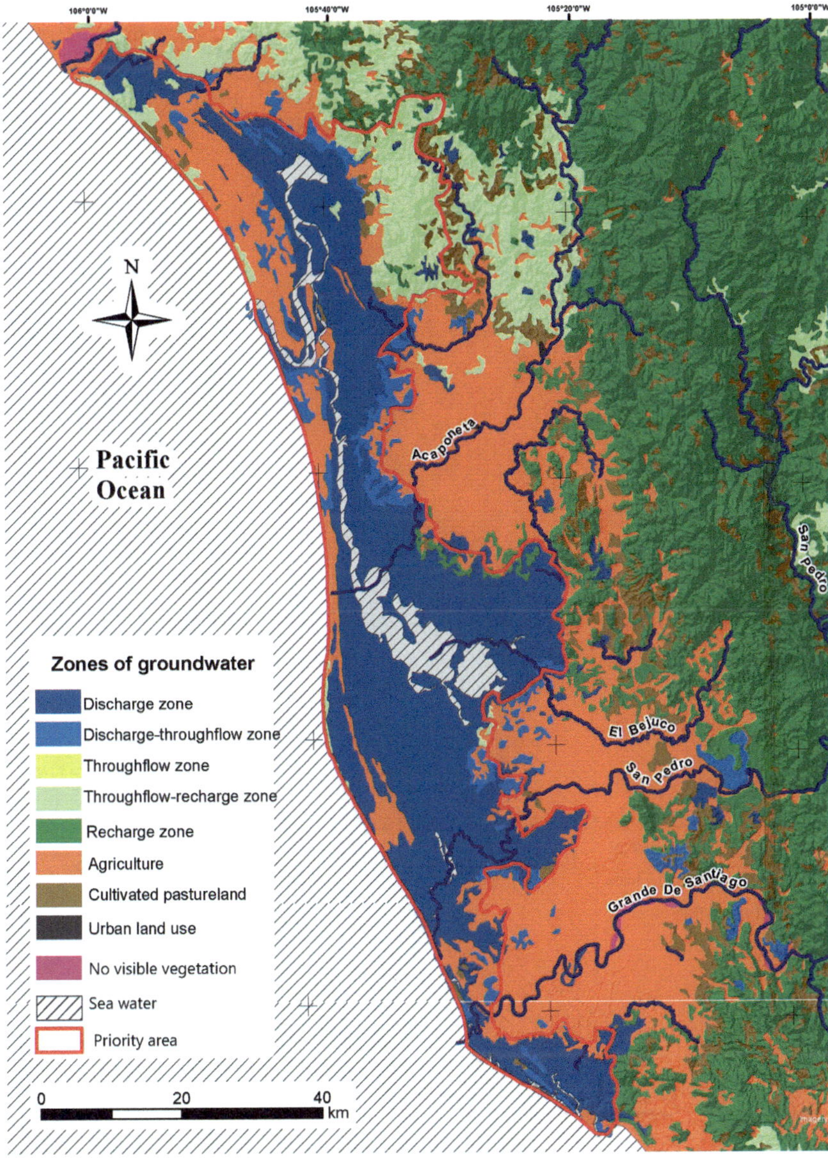

Fig. 9.8 Vegetation as a hydrogeological indicator at the water surface

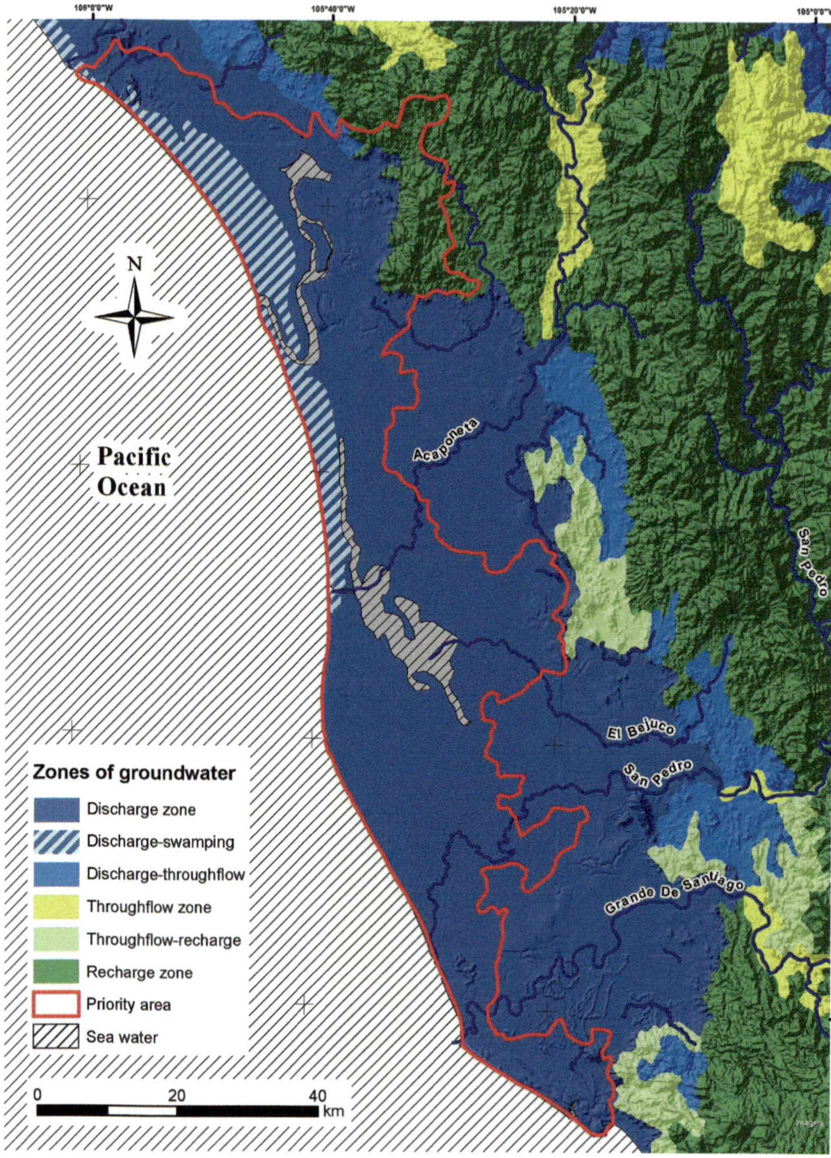

Fig. 9.9 Topoforms as a hydrogeological indicator at the water surface

The attributes of each map were combined into a single feature class inheriting all the attributes of the intersecting features. A comprehensive analysis of all conditions defined for each of the surface indicators was conducted for a final association of groundwater flow assigned to each feature. The areas that had previously been associated with discharge in the "Permanent surface water as an indicator of hydrogeological dynamic" (Fig. 9.5) and "Soils as an indicator of hydrogeological dynamic" (Fig. 9.7) maps were directly associated with discharge; the remaining combinations of unique conditions were systematically analysed separately, defining the zones of the groundwater flow systems, from which the final "Superficial evidence of the presence of groundwater flow systems" map was obtained (Fig. 9.10). This process has limitations as it combines different thematic datasets with different spatial resolution and accuracy and should be interpreted with caution and only in a general context approach.

Approximately ninety per cent (90.33%) of the Priority Area contains surface indicators associated with processes of *discharge* from groundwater flow systems. The features reported in the remaining ~ 10% of this area are associated with *discharge – throughflow*, 2.79%; *throughflow*, 1.51%; *throughflow - recharge*, 4.79% and *recharge* zone with, 0.58%.

In the discharge areas shown in Fig. 9.10, water is permanently present, either at the surface or at a shallow depth. However, the only way to confirm the presence of shallow groundwater at a point in these discharge areas is by using local and up to date records. This is so for two reasons: (i) the origin and temporality of the data for each of the surface indicators, as indicated in the methodology, and (ii) the constant dynamics of the groundwater flow systems, mainly due to human activity, direct or indirect.

At present, groundwater discharge and recharge are mainly affected by water extraction through drilling and wells, soil evapotranspiration and agricultural activity, loss/alteration of natural vegetation, alteration of the natural channels of perennial runoff, soil waterproofing and soil compaction-collapse. Climate change will also affect the volumes and patterns of precipitation, and therefore affect the groundwater flow systems.

Around 90% of the Priority Area studied has been classified as a zone *of discharge*, and of that 88% has perennial water on the surface. In other words, 90% of the spatial polygons of the spatial attributes of all the surface indicators evaluated were directly associated with the discharge *zone*. The complex hydrographical network has produced the ecosystems and topoforms that are found here; it is important to establish the associations which exist between these and the surface indicators of groundwater discharge. Table 9.1 shows the surface of the Priority Area associated with the discharge zone and the degree of correspondence of each indicator in the surface individually associated with the discharge process of groundwater flow systems.

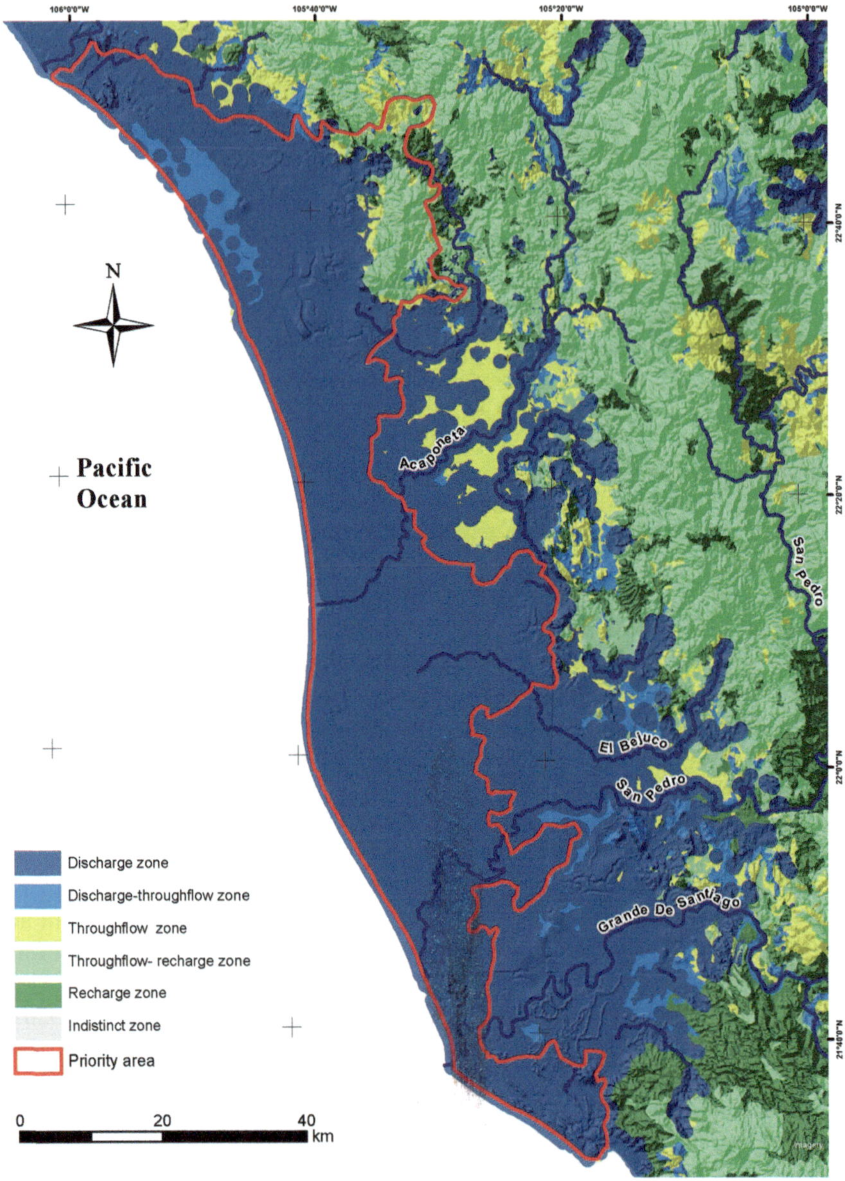

Fig. 9.10 Superficial evidence of the presence of groundwater flow systems

Table 9.1 Surface indicators and association degree in the discharge zones

Final association zone	Surface indicator					Marismas Nacionales (%)
	Surface water	Soil	Vegetation	Groundwater depth	Topoforms	
Discharge	Discharge	Discharge				**60.00**
	Discharge	Discharge-throughflow				7.80
	Discharge	Discharge/Discharge-throughflow				**68.00**
	Discharge		Discharge			**57.00**
	Discharge	Discharge	Discharge			**47.50**
	Discharge	Discharge	Discharge		Discharge	**46.50**
				Discharge		4.56
				Discharge	Discharge	2.00
				Discharge	Discharge/Discharge-throughflow	4.43
	Discharge	Discharge	Discharge	Discharge	Discharge/Discharge-throughflow	0.17

It is worth noting that in 47% of the Priority Area, the edaphic coverage and the vegetation are associated on the maps with the discharge of groundwater flow systems in parallel with the presence of perennial water on the surface or the subsurface. This means, the cartographic features for each surface indicator are characteristic of processes of discharge from groundwater flow systems. While 46.5% of the area associated with the discharge zone registers that all the surface indicators, except "Groudwater depth", are individually classified as discharge zone. This is caused by the scarce record of piezometric depth that exists in the study area, as in the rest of Mexico. It is important to consider the dates when the maps were elaborated. That is, the years and dates of field registration and/or satellite images for each of the maps used are not from the same date. This may mean that in a zone that has been mapped a Solonchak type soil (typical soil of the discharge processes) in the dry season, it can be registered as flooding zone in other maps.

In relation to the recharge zones, Fig. 9.10 shows that 5.37% of the priority area presents manifestations associated with recharge processes (4.79% throughflow - recharge, and recharge zone 0.58%). The importance of the zones does not lie in their territorial extension, but in their location with respect to the environment of interest. In the case of the priority area of the Marismas Nacionales, the zones associated with recharge processes indicate that the recharge that occurs in them supplies the priority area as well as environments outside the polygon. And that evidently, the recharge processes that mostly feed the study area happen outside the established polygon.

It is worth remembering that in smaller recharge/throughflow zones, discharge processes also occur, and in discharge zones processes of recharge/throughflow may also occur locally.

9.5 Environment—Groundwater Interactions

After developing the spatial analysis of the surface indicators, it must be integrated to the existing knowledge about the ecosystem and environmental functioning of the Marismas Nacionales (Blanco et al. 2014; Flores-Verdugo et al. 2014; Lithgow et al. 2019; Silva et al. 2019), allowing to solve questions about the functioning and the ecosystem interconnection through the water dynamics. In addition, however, there will surely be clearer, and perhaps more challenging, questions as the content of this study is assimilated.

To understand natural and environmental systems it is important to start from the origins, development and behaviour of each of the components analysed. By characterizing the zones of groundwater flow systems in the study area, questions about the causes and mechanisms of environmental impacts related to groundwater begin to surface. Many of these have direct effects on the ecosystems in the area, and, therefore, on the ecosystem services they provide.

- In almost all the study areas there is evidence of the processes and features typical of groundwater flow system discharge zones. The recharge zones of this water are beyond the polygon studied; most of the water which supports the ecosystems of the Priority Area of the Marismas Nacionales originates outside the conservation area. Is it possible to conserve the Marismas Nacionales without conserving the zones of recharge and throughflow that feed it?

- The evidence of discharge zones off the coast indicates that the marine ecosystems are dependent on the discharge of groundwater from inland.
- The ecosystems found in the Priority Area depend directly and heavily on groundwater.
- The ecosystems of the Marismas Nacionales are under great pressure from agriculture and aquaculture activities. The effects of these activities are related to inappropriate farm management practices, such as periodic flooding and draining, lack of water treatment, lack of control and monitoring groundwater well operation, as well as water extraction from springs and rivers.
- Even with the naked eye, it is easy to see the degree of change and pollution generated by agriculture and aquaculture. However, the main cause of changes related to all uses of water, is the extraction of underground water via pumping wells, both within and beyond the polygon.
- Pollution of and via groundwater affects all the interdependent ecosystems, both on land and off the coast. For example, phosphates, and other pollutants infiltrate into streams and groundwater inducing mangrove mortality in the surrounding areas and downstream. Despite the lack of samples of polluted groundwater, a significant increase in superphosphates and other pond fertilizers has been detected downstream during harvest season.
- Given the high number of channels between the low ridge dune systems, this coastline is very vulnerable to an increase in sea level. Therefore, in order to maintain the cohesive forces between the sediment particles on the coast, it is vital to maintain groundwater flow systems, since the present equilibrium is due to the presence of water from direct or diffused discharge.
- In the best case scenario, where it is possible to reduce or eliminate the processes causing water pollution within the conservation area, the question arises: Can we guarantee that the groundwater flow systems which discharge in the polygon are not agents of pollution originating beyond the polygon, in zones of recharge?
- Apart from global factors, the Marismas Nacionales sustainability is highly vulnerable to local changes in sedimentary dynamics because the system relies on sediment inputs from inland and sediment transport by waves. The permanent contribution of the sediments from inland occurs through the rivers and their respective baseflow, i.e. through the discharge of groundwater.

Consequently, land or coastal infrastructure and human activities alters and may eventually eliminate hydrogeological interconnections in the area. Therefore, the effect of hydrological and biological processes on the energy and matter fluxes need to be understood. An accurate diagnosis of these relationships is critical in designing conservation and restoration strategies of the system.

9.6 Final Considerations

From the environmental information available, new maps were developed linking environmental features with groundwater flow systems, using Toth's theory. The present work achieves the first contribution or utility of the spatial analysis of surface indicators of groundwater flow systems: making groundwater visible. Phase or condition of water that is so difficult for people to see and understand, apparently.

This work offers new information concerning the natural dynamics of the Marismas Nacionales, based on the application of modern hydrogeological theories to existing databases. The reinterpretation and spatial integration of the main biophysical components on the surface shows the interconnections between groundwater flow systems and the ecosystems within the Marismas Nacionales. This has a direct bearing on the ecosystem services they can provide.

This work should continue, in order to identify the recharge and throughflow zones and their links to the discharge zones. In this way it is possible to know exactly where the continental water recharge that supports the Marismas Nacionales occurs by the groundwater flow systems with discharge zones located inside the conservation polygon.

While from the outset it was obvious that the Marismas Nacionales can be considered a groundwater discharge zone, it is necessary to define the hierarchy of groundwater flow systems, in order to understand the natural dynamics of an ecosystem or specific environment. By defining the recharge and throughflow zones it becomes clearer that a given feature worthy of conservation depends not only on its immediate surroundings. For example, the mangroves of the Marismas Nacionales are recognized as a RAMSAR site and depend on the groundwater supplied to them from beyond the polygon in which they are found.

On the other hand, from the evidence presented by this work the Marismas Nacionales is an area that has potential to produce clean electric energy by salinity gradient technology, since it registers a relevant coastal freshwater discharge zone with a gentle topographic slope.

By understanding the relationship between natural and anthropogenic factors in the dynamics of change in the region, the consequences of actions such as natural or artificial groundwater extraction can be evaluated in specific areas.

Acknowledgements The authors thank their respective institutions for support. We are also grateful to the CEMIE-Océano (grant number 249795) for the financial support. We also are thankful to Jill Taylor for providing language help.

References

Arriaga L, Espinoza JM, Aguilar C, Martínez E, Gómez L, Loa E (2000) Regiones terrestres prioritarias de México. Comisión Nacional para el Conocimiento y uso de la Biodiversidad, México
Blanco M, Flores-Verdugo F, Ortiz-Pérez MO, de la Lanza-Espino G, López-Portillo J (2014) Diagnóstico funcional de marismas nacionales. Comisión Nacional Forestal 190 p
CONANP-Comisión Nacional de Áreas Naturales Protegidas (2013) Programa De Manejo Reserva De La Biosfera Marismas Nacionales Nayarit. Mexico City, Mexico, p 199
CONAGUA-Comisión Nacional del Agua (2016) Wetlands of the Mexican Republic - National Inventory of Wetlands. Mexico
CONAGUA-Comisión Nacional del Agua (2019) Registro Público de Derechos de Agua, México
CONABIO- Comisión Nacional para el Conocimiento y Uso de la Biodiversidad (2013) Distribution of mangroves in Mexico in 1970–1981, México
Flores-Verdugo F, Amezcua F, Kovacs JM, Serrano D, Blanco-Correa M (2014) Changes in the hydrological regime of coastal lagoons affect mangroves and small-scale fisheries: the case of the mangrove-estuarine complex of Marismas Nacionales (Pacific coast of Mexico). Fisheries management of Mexican and Central American estuaries. Springer, Dordrecht, pp 81–91
INEGI-Instituto Nacional de Estadística y Geografía (1980) Conjunto de Datos Vectoriales de Uso de Suelo y Vegetación Escala 1:250 000 Continuo Nacional Serie I, México
INEGI-Instituto Nacional de Estadística y Geografía (1991) Conjunto de datos vectoriales edafológicos. Escala 1:250000. Continuo Nacional, Serie II, México
INEGI-Instituto Nacional de Estadística y Geografía (1992) Continuo Nacional de Topografía. Escala 1:250000, Serie II, México
INEGI-Instituto Nacional de Estadística y Geografía (1993) Conjunto de Datos Vectoriales de Vegetación Primaria Escala 1: 1 000 000. Nivel I and II (1993–09), México
INEGI-Instituto Nacional de Estadística y Geografía (1995) Conjunto de Datos Vectoriales Topográficos. Escala 1:250 000 Serie II, México
INEGI-Instituto Nacional de Estadística y Geografía (2000) Conjunto de Datos Vectoriales Topográficos. Escala 1:1 000 000, Corrientes y Cuerpos de agua, México
INEGI-Instituto Nacional de Estadística y Geografía (2001) Conjunto de datos vectoriales Fisiográficos. Continuo Nacional, Serie I, Sistema topoformas, México
INEGI-Instituto Nacional de Estadística y Geografía (2001 a) Diccionario de Datos Fisiográficos (Vectorial) Escala 1:1 000 000, Sistema Nacional de Información Geográfica, México
INEGI-Instituto Nacional de Estadística y Geografía (2005) Conjunto de Datos Vectoriales Climatológicos Escala 1:1 000 000 Humedad de Suelo, México
INEGI-Instituto Nacional de Estadística y Geografía (2006) Red Hidrográfica Digital de México. Serie I Escala 1:250 000, México
INEGI-Instituto Nacional de Estadística y Geografía (2008) Conjunto de Datos Vectoriales Unidades climáticas Escala 1:1 000 000, México
INEGI-Instituto Nacional de Estadística y Geografía (2009) Cuerpos de Agua Escala 1:50 000, México
INEGI-Instituto Nacional de Estadística y Geografía (2009 a) Guía para la interpretación de cartografía uso del suelo y vegetación Escala 1:250 000 Serie III, México
INEGI-Instituto Nacional de Estadística y Geografía (2010) Continuo Nacional de Aguas subterráneas Escala 1:250 000 Serie II, México
INEGI-Instituto Nacional de Estadística y Geografía (2010 a) Documento técnico descriptivo de la red hidrográfica escala 1:50 000. Edición 2.0 México
INEGI-Instituto Nacional de Estadística y Geografía (2012) Continuo Nacional Humedales Potenciales Escala 1:250 000 Serie I, México
INEGI-Instituto Nacional de Estadística y Geografía (2014) Diccionario de Datos Edafológicos escala 1:250 000 Versión III, México

INEGI-Instituto Nacional de Estadística y Geografía (2014a) Diccionario de datos de Uso del Suelo y Vegetación, Escala 1:250 000, Versión III, México
INEGI-Instituto Nacional de Estadística y Geografía (2016) Anexo A Catálogo de Tipos de Vegetación Natural e Inducida de México, México
IUSS Working Group WRB (2007) World reference base for soil resources 2006, first update 2007. World soil resources reports No. 103 FAO, Rome
Lithgow D, de la Lanza G, Silva (2019) Ecosystem-based management strategies to improve aquaculture in developing countries: case study of Marismas Nacionales. Ecological Engineering 130: 296–305
Mendoza-González G, Martínez ML, Lithgow D, Pérez-Maqueo O, Simonin P (2012) Land use change and its effects on the value of ecosystem services along the coast of the Gulf of Mexico. Ecol Econ 82:23–32
Moore WS (2010) The effect of submarine groundwater discharge on the ocean. Annual Review of Marine Science 2(1):59–88
Post VG, Kooi J, Person H, Shemin MG, Edmunds W (2013) Offshore fresh groundwater as a global phenomenon. Nature 504(7):1–8
Povinec PP, Aggarwal PK, Kulkarni KM, Groening M, Han LF (2007) Isotopic characterisation of groundwater-seawater IAEA–CN–151/131 International Atomic Energy Agency, Marine Environment Laboratories. Monaco, International Atomic Energy Agency, Wein, Austria
Silva R, Chávez V, Bouma TJ, van Tussenbroek BI, Arkema KK, Martínez ML, Oumeraci H, Heymans JJ, Osorio AF, Mendoza E, Mancuso M, Asmus M, Pereira P (2019) The incorporation of biophysical and social components in coastal management. Estuar. Coast 42(7):1695–1708
Simensen T, Halvorsen R, Erikstad L (2018) Methods for landscape characterisation and mapping: A systematic review. Land use policy 75:557–569
Soil Survey Staff (2014) Claves para la Taxonomía de suelos. USDA, Dpto de Agricultura USA, 12 Ed. Traducción, Ortiz Solorio CA, Gutiérrez Castorena MC, Gutiérrez Castorena EV, 410 p https://www.nrcs.usda.gov/Internet/FSE_DOCUMENTS/nrcs142p2_051546.pdf
Tóth J (1962) A theory of groundwater motion in small basins in central Alberta. Canada. J. Geophys. Res. 67:4375–4387
Tóth J (1963) A theoretical analysis of groundwater flow in small drainage basins. J Geophys Res 68:4795–4812
Tóth J (1970) A conceptual model of the groundwater regime and the hydrogeologic environment. J Hydrol 10(2):164–176
Tóth J (1971) Groundwater Discharge: A common generator of diverse geologic and morphologic phenomena. International Association of Scientific Hydrology Bulletin 16(1):7–24
Tóth J (1999) Groundwater as a geologic agent: an overview of the causes, processes, and manifestations. Hydrogeoly J 7(1):1–14
Tóth J (2000) Las aguas subterráneas como agente geológico: causas procesos y manifestaciones. Boletín Geológico y Minero Instituto Tecnológico GeoMinero, Spain 8:49–26
Tóth J (2009) Gravitational systems of groundwater flow. Theory, evaluation, utilization. University of Alberta, Canada; Eotvos Loránd University, Hungary; Cambridge University Press, the Edinburgh Building, Cambridge UK
Tóth J, Hayashi M (2010) The theory of basinal gravity flow of groundwater and its impacts on hydrology in Japan. Journal of Groundwater Hydrology 52(4):335–354
Tóth J (2016) The evolutionary concepts and practical utilization of the Tóthian Theory of Regional Groundwater Flow. Int J Earth Environ Sci. 1: 111 doi:http://dx.doi.org/10.15344/ijees/2016/111

Chapter 10
A Holistic Approach to the Estimation of Economic Losses Due to Water Stress in Agriculture in Buenos Aires Province, Argentina

Raúl Jorge Rosa

Abstract This chapter estimates the direct impact of water stress on the 4 most important crops in the province of Buenos Aires, Argentina. The aim is to highlight the importance of analyzing such events to measure the impact in monetary terms that allows decision makers to visualize the extent of the damage in order to promote actions necessary to minimize their consequences. It is noted that the estimated direct impact of the monetary value of the 140 mm spread between the wet and dry periods in real water use of these crops at prices in 2013/2018 amounts to US$3,578 million at the average yields of the dominant soil types. No other direct impact of agricultural activities is valued. Of the water problems, drought is the least studied and underestimated its effects over time. It is a difficult phenomenon to manage, and appropriate measures for mitigating its impacts are not easy to implement. For the region analyzed it is considered necessary to quantify temporally and geographically the impacts of the phenomenon based on a homogeneous and systematic methodology, with a holistic approach that includes the whole environment, also deepening the interdisciplinary studies of these phenomena to improve their management, combining input technologies and promoting the development of process technologies.

Keywords Monetary valuation · Water stress · Agriculture · Cropwat · Green Water Footprint

10.1 Introduction

The cyclical processes of droughts and floods have caused significant economic losses in the province of Buenos Aires, Argentina, over the years and form a scenario of extreme fragility in the face of such extreme hydrological events. A historical review

R. J. Rosa (✉)
Centro Interdisciplinario de Investigaciones Aplicadas al Agua y al Ambiente. UNLP, Administración Agraria, Facultad de Ciencias Agrarias y Forestales, Universidad Nacional de La Plata, Calle 60 y 119 s/N (1900), La Plata, Argentina
e-mail: rjr@agro.unlp.edu.ar

M. M. Alconada-Magliano (ed.), *Intensified Land and Water Use*, Springer Earth System Sciences, https://doi.org/10.1007/978-3-030-65443-6_10

can be found in numerous researches conducted by specialists such as Ameghino (1884), Durán (1987), Ras (1994), and Moncaut (2003), among others.

Of the total water on the planet, about 0.4% is available freshwater, and from all sectors of the economy the agricultural system is the most sensitive to water scarcity, since it is clearly a necessary input for food production. For this, as the population increases, this essential input conditions the growth and food security of the population, and therefore its quantification and monitoring of its use for sustainable resource management becomes relevant. "*Water scarcity is increasing, as is salinizatio*n, groundwater pollution and degradation of water bodies and related ecosystems..." (FAO 2011).

Of the water problems of the province of Buenos Aires, drought is the least studied and its effects over time have been underestimated. It is a difficult phenomenon to manage, and the right measures to mitigate its impacts are not easy to implement. For the design, management, and evaluation of better quality public policies, decision makers require a clear understanding of the impact of the problem, clear proposals for resolution, benefit/cost ratios, sustainable social and environmental impact to be part of the public investment decision agenda.

To dimension the impact of water stress on agricultural activity has the objective, as is mentioned in chapter 40 of Agenda 21, of contributing to generate information for decision making. This may collaborate to awareness in the generation of institutional capacities that allow to manage events in a proactive way, rather than the traditional approach focused on the implementation of actions during the emergency phase to deal with the impacts detected.

It should be clarified that, of the total of the water counted as input into the production function of the rainfed crops analyzed in this chapter, is rainwater. For this reason, they are subject to climate variability regarding their availability in space and time. It is also relevant to analyze the whole of climate and environmental factors, water (surface and groundwater), soil, crops, and technologies in a comprehensive manner.

However, the water available for crops depends not only on local climate factors, but also on factors such as soil type, local and regional geomorphology, and the in-implemented cultivation and management. This is particularly important in areas where groundwater is recognized at shallow depths, as is the case in large sectors of the province of Buenos Aires. In the regions called "Flooded Pampa" and "Sandy Pampa" in the province of Buenos Aires, there has been recognized the existence of groundwater flows that depend on local rainfall (local flows with low salinity), and mainly that depend on rainfall occurred over long distances (intermediate flows with moderate to high salinity), according to the theory of operation of groundwater (Alconada-Magliano et al. 2011). Consequently, there are changes in the ease with which crops can access groundwater, both because of the depth at which it is located and its quality. While these aspects are not included into the evaluation of this chapter, it highlights the importance of considering in the future the variations associated with the operation of groundwater.

It is necessary to understand the importance of analyzing the biophysical reality of economic processes to understand natural processes and the need to quantify them. The physical limits to the current model of economic development have been studied for many years. Humans are subject to certain restrictions imposed by nature and the environmental environment with which it interacts. Climate change, future uncertainty in water supply, population growth, and increased demand for water will continue to increase the need for effective and cost-effective resource conservation measures. Despite the importance of promoting water conservation in agriculture, little has been done to integrate the hydrological, economic, institutional, and conservation policy dimensions.

It is becoming increasingly clear that sustainable watershed management requires higher levels of integration between groups of natural and social scientists, land and water users and managers, planners and policy makers of all spatial scales. Multiple policy drivers, encompassing urban and rural communities and their relationship to land and water use, have led to the need for an integrated decision-making framework that operates from nationally strategic to local basin level. The lack of integration between policies has resulted in the uncertain outcome of their actions. The combined use of space technologies, scenarios, indicators, and multi-criteria analysis is increasingly used to enable better integration of sustainable watershed management (Macleod et al. 2007).

As the European experience points out, most attempts to manage drought and water scarcity and their related impacts focus on reactive crisis management approaches that are ineffective, inopportune, and not sustainable in the long term. There is now a tendency to move toward proactive management approaches to increase the resilience and sustainability of affected regions. This transition from crisis to risk management is challenging, as there need to be institutional capacities of governments, institutions, and individuals, which in many countries do not exist (European Environment Agency 2012).

Wilhite et al. (2001) developed ten steps planning process for drought and suggested *Organizational Structure for Drought Plan,* and this process evolved based on the incorporation of their experiences and lessons learned. According to Urquijo-Reguera (2015) who studied 6 cases in Europe (Portugal, The Netherlands, Switzerland, Italy, Spain, Greece) (Urquijo Reguera et al. 2017), a risk management plan should include elements such as: (i) *definition,* delimitation, and study of the problem, i.e., the different types of drought and their perceptions; (ii) k*nowledge* of their impacts and vulnerability factors for each affected context and sector; (iii) *a set of measures* covering different problems and operating at different levels of management, where monitoring systems and drought plans are particularly relevant measures; (iv) *efficient participation processes and transparent resource* allocation mechanisms; and (v) *institutional ability* to act, implement measures, and learn from experience through evaluation.

Ostrom (2008) noted that institutional theorists must not recognize what environmentalists have long recognized; *the complexity of what they study and the need to recognize the dynamic, self-organizing, and non-linear aspects, as well as the multiple objectives, the temporal and spatial scales related.*

This chapter aims to highlight the magnitude of the monetary value due to water stress on the 4 most important agricultural crops in the province of Buenos Aires,

Argentina. It addresses the problem based on its direct effects, estimation of declining yields and its valuation in monetary terms, in order to present quantitative evidence that allows decision makers to dimension this problem.

10.2 Technology and Productivity Evolution

According to Andrade et al. (2017), at the global scale, rainfed crops yield on average about 70% less than irrigated crops (FAO 2016), so irrigation, which is accompanied by increased use of other inputs, can have a strong impact on production, depending on the chances of irrigated surface expansion. Irrigation in the province of Buenos Aires is not significant and its chances of widespread development for extensive crops are very scarce. The crops of wheat, corn, soybeans, and sunflower analyzed are dependent on precipitation and the management of technologies for the best use of water (water retained in the soil and/or the uptake that they can carry out from groundwater).

Gleick (2003) defined the *efficient use of water* as the minimum amount necessary to meet a specific purpose; and *water productivity* as the amount of measurable product per unit of water that is used to produce it. This author points out a number of problems that hinder analyses in relation to water use, because systematic collection of uses is not frequent; some are difficult or not possible to quantify them; there are global and regional disparities in collection; and many data collected are inaccurate. Water use is not yet understood, its measurement is difficult and reported inappropriately. As a result, quantitative estimation methods of water consumption are used to assess the allocation at the global or regional level.

Agricultural production in the province of Buenos Aires grew for the period 1969/70 to 2018/19 266% with a 91% increase in the area harvested. Much of this productive increase was due to the generation and adoption of new technologies, as well as improvements in efficiency due to economies of scale and scope (Lema 2015). The evolution of agricultural yields of the four crops analyzed, from recorded data from 1969/70 to 2018/19, has grown at an annual rate of 2.10% in wheat, 2.59% in corn, 1.67% in soybeans, and 1.41% in sunflower (Fig. 10.1 and 10.2, respectively).

According to a report by Bragachini (2018), the decade of 1997 to 2008 (Fig. 10.3) was important on the issue of technological adoption, years that also coincide with the economic crisis of 2001 in Argentina. In addition, the years subsequent were important for the sector in economics, and it had an impact on technological adoption due to the good prices of commodities, as shown in Fig. 10.4 (constant CPI Base dollars 1982–84, updated to 2018).

However, Lema (2015) analysed the evolution of the growth rates of production and productivity of Argentine agriculture, in the context of trade and fiscal policies applied to the Argentine agricultural sector, concluding that an important association can be established between policies and sectorial behavior. From the data it is observed that both production and productivity grew at higher rates in the period

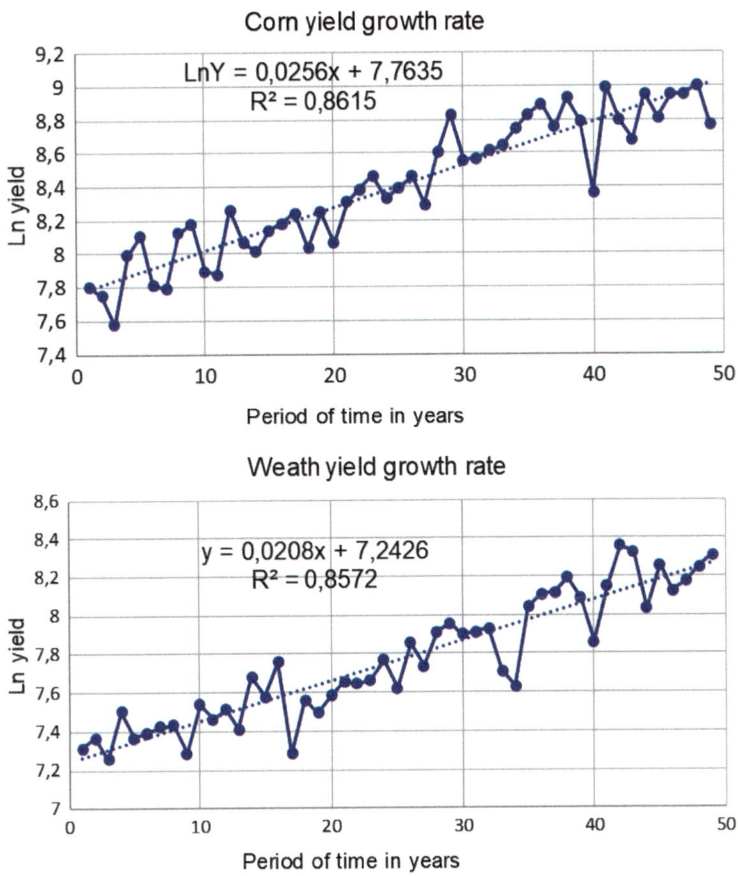

Fig. 10.1 Corn and wheat yield growth rate

1989–2001, compared to the period 1962–89 and decreased in 2002–07 and 2008–13. When compared to other countries in the region, Lema (2015) concluded that agriculture was producing at lower rates than the region and that productivity growth was at similar rates to the average, albeit with a declining trend. As noted Cristini et al. (2009) that bias toward agricultural production with cost-saving technologies (e.g. direct soybean planting) are optimal economic responses to distortive taxation mechanisms. That is, farmers try to reduce the production of goods who are more exposed to taxes or quantitative export restrictions, skewing production towards cost saving technologies.

Moreover, Viglizzo (2014) pointed out that the great vector of innovation in crop farming has been input technologies, which are tangible technologies that materialize in the form of improved seeds, fertilizers, pesticides, machinery, computers, geographic positioning systems, remote sensors, etc. Within them, there are many years of accumulated knowledge and are easy to access and apply, following fairly

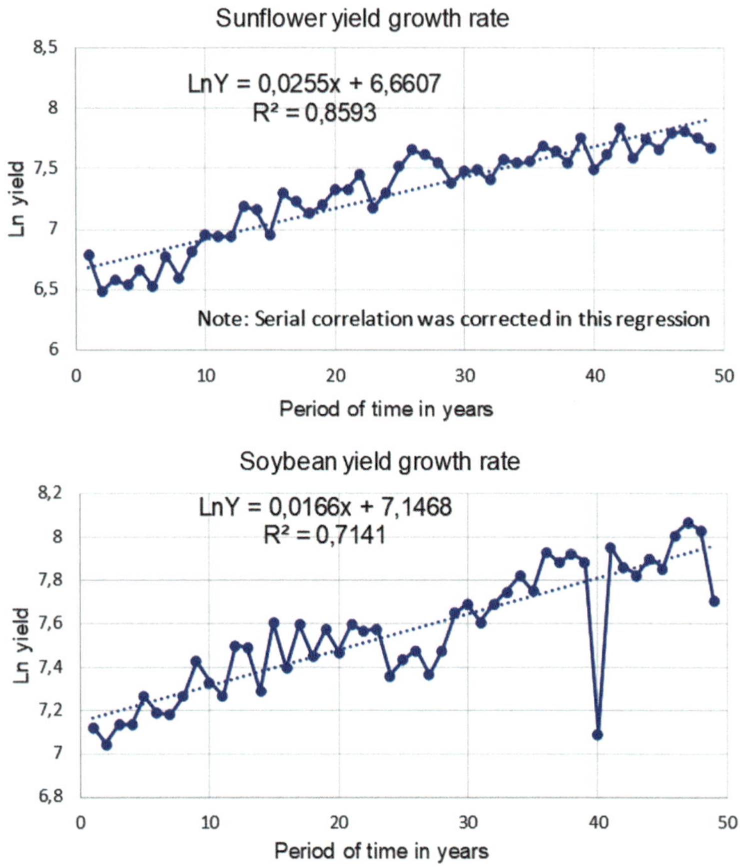

Fig. 10.2 Sunflower and soybean yield growth rate

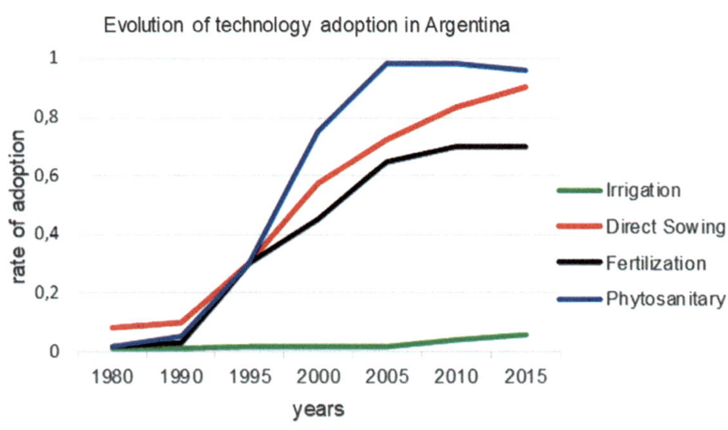

Fig. 10.3 Adoption of technology in agriculture (Adapted from Bragachini 2018)

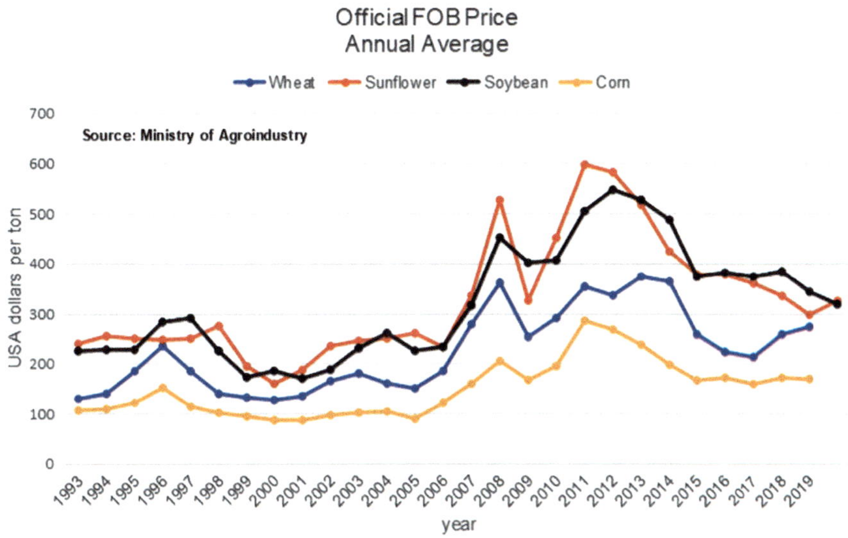

Fig. 10.4 FOB price evolution

simple *standardized protocols*, unlike what happens with livestock, which appears more associated with *process technologies*, that is, those based on the use of intellectual supplies. Viglizzo (2014) noted that because these technologies are intangible (since they are based on knowledge and experience), they cannot be purchased on the market as if they were material supplies, although experts can be hired to do that work, but they require an intellectual cost and a dedication of personal time. These types of process technologies require both training and the availability of timely and accessible information for communication so that they can be widely implemented.

Evidence of some association is therefore seen in *producer responses* to the adoption of *technologies* for increased production in trade and *fiscal policies, international prices and the development of supplies technologies* among others, and *not so much in process technologies.*

Irrigation technology does not seem to be massively adopted in extensive rainfed crops for the province of Buenos Aires in the short to medium term. However, *green water* productivity can be increased through improvements in rainfall capture efficiency (Andrade and Caviglia 2015) and focused more on the approach to *process technologies*. The functioning of groundwater flows could be included in the latter concept and should be studied in order to consider in agricultural management decisions.

10.3 The Concept of Drought

The clear definition of a concept from a conceptual and operational point of view is important for addressing a problem. The concept of drought is used by different disciplines and therefore defined in various ways.

A revision of the definition emphasizes that the four classic definitions (Mishra and Singh 2010) i.e., *hydrological, meteorological, socioeconomic, and agricultural drought*; considered appropriate to include the concept of *groundwater drought*, which poses a major challenge for resource planners.

For example, these prior definitions are not clearly established, as the effect on crops of variations in the water availability associated with the operation of groundwater flow systems (Töth 2000) is considered. That is, what are the water flow systems that manifest themselves in a particular environment (i.e., coexist in a certain place), and how these flows are modified in their quality and depth by their local and regional relations. These relationships are possible due to the existence of regional hydraulic continuity (Engelen and Jones 1986; Töth 1995; Carrillo-Rivera 2000), particularly evident in sedimentary environments such as in the province of Buenos Aires SAGyP-INTA (1989). Whereas this is not always easy to define, due to the intensity of studies required and high costs, it is clear the need to consider groundwater as a variable that specifically affects the availability of water for crops, as well as the selection of sustainable production schemes in many environments of the province of Buenos Aires (Alconada-Magliano et al. 2009).

On the other hand, the definitions are mainly based on hydrometeorological variables, so that they do not allow adequate quantification of economic losses. Therefore, estimating or measuring the impact of drought is a difficult task, since it does not depend exclusively on the dispersion of rain to normal values, but in addition to precipitation involves other variables, such as atmospheric demand for humidity, temperatures, winds, soil types, crop phenology, water availability in the soil, etc. Wilhite and Glantz (1985) defined the *4 types of drought* as follows:

Meteorological drought: lack of precipitation over a region for a certain period of time.

Hydrological drought: the hydrological drought is related to a period with inadequate surface and subsurface water resources for established water uses of a given water resource management system.

Agricultural drought: it refers to a period with declining soil moisture and consequent crop failure without any reference to surface water resources.

Socioeconomic drought: drought is associated with failure of the water resource system to meet water demands and thus associating droughts with supply and demands for an economic element such as water.

Each concept of drought has different impacts and on different dimensions. Most drought research emphasizes physical aspects over socioeconomic features.

10.4 Drought Indicators and Indices

Indicators are variables or parameters used to describe drought conditions. For example, *precipitation, temperature, river flows, groundwater and reservoir levels, soil moisture, and snow mantle.* Indices are usually computerized numerical representations of the severity of droughts, determined by climate or hydrometeorological data, including the indicators listed. They aim to analyze the qualitative state of droughts in the environment over a given period of time. The World Meteorological Organization's handbook of drought Indicators and indices describes a number of these indicators and indices, classified by ease of use.

In relation to the study area, several authors have carried out a geographical historical analysis in relation to periods where rainfall was lower than normal values, as Scarpati and Capriolo (2013) have stated "…in terms of *droughts,* monthly and annual rainfall was well below the average precipitation values in the years 1879, 1883, 1910, 1917, 1918, 1921,1925, 1929, 1930, 1936, 1937, 1938, 1939, 1950,1952, 1954, 1955, 1956, 1959, 1960, 1962, 1963,1967, 1968, 1970, 1971, 1972, 1974, 1975, 1976,1977, 1979, 1981, 1983, 1988, 1989, 1995, 1996, 1997, 1998, 1999, 2004, 2005, 2006, 2007, and 2008". These authors also indicate that the "*most serious periods were in 1974, 1989, 1995, 1996, 1999 and 2008. In 1996, the area affected by the drought was 67% of the total agricultural area of the province and in the last quarter of 1996 and mid-1997, many shallow lakes were completely dried up, including the very important 'Laguna Las Barrancas', which had not happened in 60 years. On 11 January 1997, at the Castelli county the course of the Río Salado was cut short, remaining so for a month. The drought of 2008 affected the whole country, nearly one million head of cattle died and in Buenos Aires the national government declared the 'Agricultural Disaster' condition, highlighting among the areas most affected, the counties of Patagones, Villarino, Bahía Blanca and Puán (all in the southern portion of the province). Rainfall decreased from 40 to 60%, soybean harvest reduced by 30%, wheat crop by 20% and losses above US$ 700 million were estimated.*"

There are institutions that monitor drought events from certain indices and indicators. The Project for the Design and Implementation of a Drought Information System for South America (SISA), which will be carried out by the South American Regional Climate Center (CRC-SAS), is being launched and will be coordinated by the South American National Meteorological of Argentina. The project is funded by Euroclima+, an EU program to support the implementation of the Paris agreement commitments and falls under the "*Disaster Risk Reduction: Droughts and Floods*" component of the Paris agreement.

Losses for different areas, time periods and varied methodological approaches have also been estimated. For example, Thomasz et al. (2018), using a linear regression model for soybeans in the province of Buenos Aires for the year 2009 and 2012, estimated a loss of 2638 million U$S, where 87 of the 95 counties of the province of Buenos Aires were affected in that year. Likewise, by 2012, with 8 of the 95 counties affected, the loss was estimated at U$S292 million for the entire crop.

It should be clarified that there is no methodological uniformity for the estimation of damages that allow geographical and temporal comparisons of the different events. It is considered relevant to highlight the importance of determining the economic impact and its relevance to the provincial and regional economy.

10.5 The Water Footprint

Article 40 of Agenda 21 (ONU 1992) raised the need to generate indicators and information for decision-making, with the aim of "*creating or strengthening local, provincial, national and international mechanisms to ensure that sustainable development planning in all sectors is based on reliable, timely and usable information.*" For this purpose, indicators of sustainable development, prevention of global use of indicators of sustainable development, efficiency of data assessment and analysis methods, and the establishment of a comprehensive information framework, the use of the ability to disseminate traditional information should be developed.

Hoekstra and Chapagain (2008) have shown that identifying the type of water used in the processing of products can help a better understanding of the global character of freshwater. It also contributed to the quantification of the effects of consumption and trade in goods and services on the use of water resources. In this way, it is proposed that better knowledge can be the basis for an improvement in the management of our planet freshwater resources. The concept of *Water Footprint* (Hoekstra 2003) is an indicator of the use of freshwater not only in the direct use of water by a consumer or producer, but also in the indirect use of the resource for the production of a product. From the above, the Water Footprint is defined as the volume of water used directly or indirectly (consumption and contamination) to produce a particular product, quantifying that amount along the whole supply chain.

Three types of Water Footprint (green, blue, and grey) were defined, depending on the origin of the water, and considering the pollution that is generated when producing said product, each of its components being geographically and temporarily specified.

When it comes to agricultural products, the *Blue Water Footprint* refers to the evapotranspiration of irrigation water from crops (surface and groundwater, in lakes, rivers and aquifers). For its part, the *Green Water Footprint* refers to water from rain that is retained in the soil (effective precipitation), which does not drain or recharge aquifers, and which is evapotranspirated during the growth of a plant species (consumptive use). Finally, the *Grey Water Footprint* is defined by the volume of water required to dilute contaminants to natural water concentration levels, or to reach the levels set out in water quality standards (Hoekstra et al. 2011).

Therefore, the evaluation and quantification of the Water Footprint of a product or service is an analysis tool that helps to understand how activities and products relate to water scarcity and pollution, associated impacts, and possible measures conducive to ensuring that such activities and products contribute to sustainable use of freshwater.

10.6 Estimating Water Consumption in Rainfed Farming

Given the interest of governments in using Water Footprint accounting as a basis for the formulation of sustainable strategies in relation to the use of water resources and the design of public policies for efficient management, the Buenos Aires province, through the OPDS (Office for Sustainable Development) and the IFC (Federal Investment Council), convened the National University of La Plata (UNLP) to estimate the green water footprint of the top 4 agricultural products exported by the province of Buenos Aires, following the methodology based proposed in the "Manual of Evaluation of the Water Footprint" (Hoekstra et al. 2011).

10.6.1 Characteristics Considered to Perform the Calculation of Water Footprint

The Province of Buenos Aires is part of the Pampean Region of Argentina, constituting the area with best conditions for the development of agricultural activities, in particular agricultural production. Most of it is located in the so-called "*corn core zone*," as well as the main *wheat regions* of the country. However, the productive characteristics are not the same throughout the provincial territory, since there are notorious climatic, building and physiographic differences, which have determined a wide variety of uses and productive systems. For this reason the different *agro-climatic sub-regions* of the province were characterized, where it was represented in *three situations for each particular crop and agroecological zone*: (i) a year with average precipitation values, (ii) a year with a dispersion toward a drier period of 30% from the average, and (iii) a year with 30% dispersion of precipitation toward a wetter period. The precipitation in a dry, normal, and wet year is defined as precipitation with 20, 50, and 80% chances of excess, respectively. The objective of this approach is to evaluate the differences in the representative values for these conditions in the crops studied.

10.6.2 Climate

In relation to climate, the data necessary for the calculation of the main parameters required were taken based on the availability of information. The 30 year series of the National Weather Service (Servicio Meteorológico Nacional, SMN) was used for the calculation of EVTo (reference evapotranspiration) (Allen et al. 1998), corresponding to 46 locations in the province of Buenos Aires, using the CropWat program (Smith 1992; CROPWAT 8.0), taking into account temperature, relative humidity, wind, and daylight hours data, among other parameters. Subsequently, an extrapolation was made to the rest of the neighboring localities that lack data, evaluating and validating the relevance of the parameter.

With the entry of rainfall data from the last 5 years, taken from the database of the Ministry of Agriculture, Livestock and Fisheries of the Nation, the effective rain was calculated from the EVTo calculated above. In addition, in order to calculate drier, normal, and wetter periods, available rainfall data from the aforementioned national government distribution for the period 1996/2011 was used, covering virtually all the counties of the province of Buenos Aires and, in addition, representing the closest scenario for the production of the main crop, which is soybean.

10.6.3 Soil

The Pampean region is essentially a broad plain consisting of modern unconsolidated wind sediments, covering the Chaco-Pampean Plains, characterized by a humid temperate climate and natural prairie vegetation, a combination of factors that has given rise to one of the most conducive areas in the world for the production of grains and meat (Imbellone et al. 2010). The soil orders described in the Province of Buenos Aires are Molisoles, Alfisoles, Entisoles, Aridisoles, and Vertisoles (Soil Survey Staff 1975; SAGyP-INTA 1989; Moscatelli and Pazos 2008). These soils result from the interaction of 5 forming factors (original material, climate, living organisms, time, and relief). How they combine, in direction and intensity, determines the type of soil (Porta et al. 1994). In relation to the climate, the two elements that best correlate with the properties of the soils, are rain and temperature. However, for some building properties, these climatic elements affect more for their extreme conditions than for the averages of a region. Regarding the other factors that form the soil, the relief stands out in the province of Buenos Aires, which affects the development of the soil and vegetation, due to the changes in the functioning of the groundwater. As previously mentioned, groundwater is close to the surface in large sectors of the province, mainly in the "Sandy Pampa" and "Flooded Pampa" regions (SAGyP-INTA 1989).

10.6.4 Regionalization

The regionalization criterion that is adopted to define the Water Footprint of the main exportable crops of the province of Buenos Aires can be made according to the characteristics of the landscape, considered as a socio-economic-environmental system, and/or from the particularities of crops, *wheat, soybeans, corn, and sunflower.* To estimate *Green Water Footprints* in areas with homogeneous characteristics, regionalization by the Agricultural Risk Office of the Ministry of Agriculture was displayed there, where the 8 homogeneous Agroeconomic Zones (regions) are presented there based on soil characteristics, climate and use, which are recognized in the province of Buenos Aires (Fig. 10.1). The *Agroeconomic Zones* (AEZ) are as follows: AEZ

(1) Irrigation and arid livestock area; AEZ (2) Mixed zone of the South West of Buenos Aires province; AEZ (3) Mixed zone of the south central area of Buenos Aires province; AEZ (4) Livestock zone of the Rio Salado Basin; AEZ (5) North-eastern Zone of Buenos Aires province; AEZ (6) Mixed central area of Buenos Aires province; AEZ (7) Mixed area of Northwestern of Buenos Aires province and Southern Córdoba province; and AEZ (8) Agricultural core area of Northern Buenos Aires province, Southern Santa Fe province, and Southeastern Córdoba province (SAGPyA 1998).

10.6.5 Procedure for Calculating the Water Footprint

The methodology proposed in the handbook "The Water Footprint Assessment Manual: Setting the Global Standard" (Hoekstra et al. 2011) was used. In relation to the software used for calculations, it was decided to work with two methodological approaches.

(i) For the calculation throughout the provincial territory, the CropWat software was based, focusing on extensive use for the entire province and based on secondary data and expert estimates regarding yields for each zone per main soil Series (taxonomic category in Soil Survey Staff 1975).

(ii) On the other hand, we worked with a second methodological approach based on the AquaCrop software (FAO 2017) focusing on case studies on primary information provided by the Argentine Association of Regional Agricultural Experimentation Consortia (AACREA), based on own field trials, which allowed for further modeling and a check against the first methodological approach.

It should be noted that the Water Footprint Assessment Manual states that "*Green Water refers to precipitation on land that does not cause runoff or is added to groundwater but is kept on the ground or its surface or vegetation. Finally, this is the part of the precipitation that will evaporate or that the plants perspire.*"

Consistent with the above definition, the *Green Water Footprint* of the growing process of a crop (HHVERDE, $m^3\ tn^{-1}$) is calculated as the Green Water requirement by the crop (RACVERDE, $m^3\ ha^{-1}$) divided by its yield (Y, $tn\ ha^{-1}$). That is:

$$HH_{VERDE} = \frac{RAC_{VERDE}}{Y}$$

This formula represents the ratio between a volume of water used by the crop, and the mass produced. To estimate the water use (demand) of the crops analyzed, as previously mentioned, the software developed by the Land and Water Development Division (AGL) of the Food and Agriculture Organization of the United Nations (FAO), called CropWat 8.0, was used.

There are also other methodological approaches to the calculation of the Water Footprint with the life cycle approach based on the international ISO standard in 2014:

"ISO 14046:2014 – Environmental Management – Water Footprint – Principles, Requirements and Guidelines."

10.6.6 Green Water Footprint Estimation

The Green Water Footprint was calculated for first-occupancy *corn, wheat, sunflower, and soybean* crops using CropWat software at 380 polygons in the province of Buenos Aires, distributed in AEZ 2 to 8 (Fig. 10.5), covering an area of agricultural soils of 11,660,044 ha, and considering precipitation of drier, normal, and wetter years. In total, 4560 water footprint values were estimated, corresponding 1140 to each crop (for more reference see Rosa et al. 2013). Weighted average results are shown in Table 10.1.

It was found that the calculated water footprint for each of the crops has great variability in terms of its values, even within the same AEZ. As it is well known, biomass production in any crop or plant community is strongly determined by the amount of water available in the soil. This depends upon the revenue (rain, groundwater

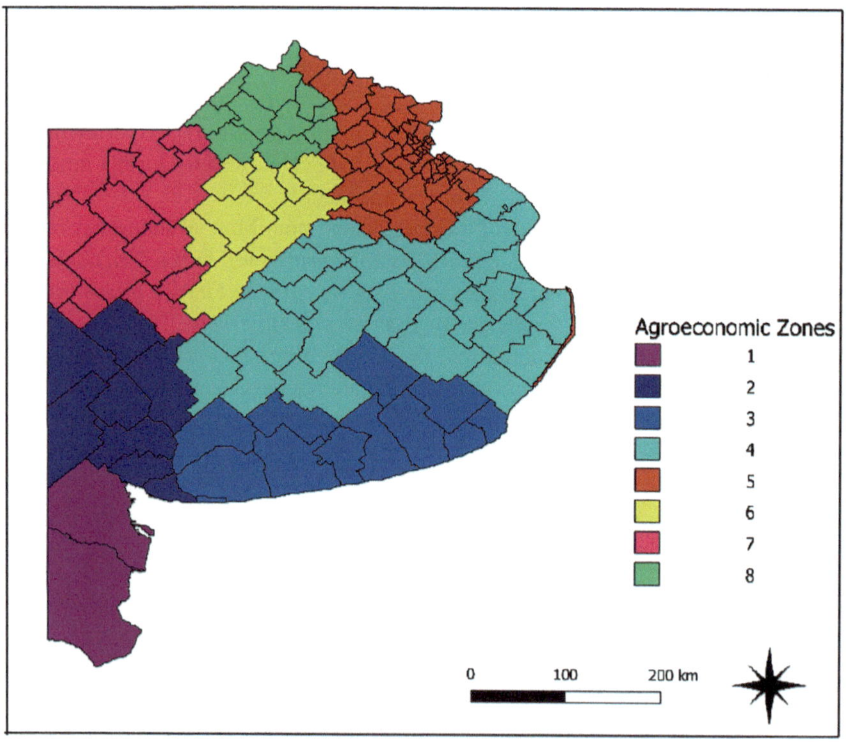

Fig. 10.5 Agro-Economics Zones of Buenos Aires (Adapted from SAGPyA-ORA 2005)

Table 10.1 Water footprint in $m^3.tn^{-1}$ for drier, normal, and wetter periods **a** crop of corn and wheat, **b** crop of sunflower and soybean

(a)	Corn			Wheat		
AEZ	Dry	Normal	Wet	Dry	Normal	Wet
2	1394	1265	1185	1510	1516	1517
3	1175	980	910	672	672	672
4	1110	1040	994	955	938	938
5	794	717	690	1021	1021	1021
6	817	778	740	1099	1100	1100
7	827	769	739	1132	1133	1132
8	620	575	558	828	828	828
(b)	Sunflower			Soybean		
AEZ	Dry	Normal	Wet	Dry	Normal	Wet
2	3531	3570	3630	1852	1953	2012
3	2622	2677	2703	1492	1576	1624
4	2863	2916	2989	1741	1808	1851
5	2431	2479	2469	1504	1584	1619
6	2355	2390	2409	1545	1568	1638
7	2780	2809	2828	1644	1695	1722
8	1850	1883	1984	1193	1254	1278

flows, irrigation), the soil own storage capacity (proportion of thick elements and soil porosity), and the density and depth of the plant root system, which determines the volume of soil used, relative to the total.

On the other hand, the needs or demand for water by crops depend heavily on climatic conditions, which are synthesized in potential evaporation, or atmospheric demand for water, a parameter that integrates the effects of the amount of radiation, humidity, ambient temperature, and wind speed. The total foliar area and the architecture of the plant and crop also determine the actual expenditure against a certain atmospheric demand. Consequently, the plant develops its foliar architecture based on its genetic characteristics and the availability of resources during growth. The technology used also plays an important role.

As it is known, it can be concluded from the above that environmental conditions strongly condition the water expenditure of plants and strongly influence biomass production levels. Also, another conclusion that a priori emerges as expected, is the *negative correlation between Water Footprint and soil quality* (higher soil quality lower water footprint). In relation to the comparison between water footprints and climatic periods evaluated for the same crop, no significant differences were found.

Unsurprisingly, the result of the estimates showed that the Water Footprint does not provide relevant information for different water stress situations in the same soil type. This is because water footprint is represented in a ratio where water productivity affects yields and it is expressed by an amount of water used by cubic meter in the numerator per ton of yield in denominator (m^3 tn^{-1}). Thus, for a crop in the same soil maintaining the same production practices, the ratio only reflects the variation expressed by the productivity factor of the water (Ky factor), given its availability from the formula presented in the following item (Eq. 10.1).

The Water Footprint is an interesting indicator when it comes to sensitizing the importance of water to produce goods and services. In any case, there is no agreement upon the method for incorporating the "concept of scarcity" into the determination of the Water Footprint (Perry 2014). In fact, the terms drought and scarcity are sometimes used interchangeably as if they were synonymous (Schmidt et al. 2012). Water scarcity basically refers to permanent deficit situations, where there is an imbalance between supply and demand, i.e., contributions are not sufficient to meet society demands. This situation is greatly aggravated during periods of drought and therefore drought and scarcity manifest the same in practice (Van Loon 2015). It is especially important to distinguish between these two concepts since being their different causes, they require different measures.

10.6.7 Monetary Estimation of Water Stress

Cropwat software calculates the yield decrease factor based on FAO Irrigation and Drainage Paper nº 33 (Doorenbos and Kassam 1979) based on the following equation:

$$\left(1 - \frac{Y_a}{Y_m}\right) = K_y\left(1 - \frac{ET_{c-aj}}{ET_c}\right) \tag{10.1}$$

Equation 10.1 CropWat Performance Decrease Calculation

Where Y_a and Y_m refer to current yield and maximum yield, and ETc_{aj} and ETc crop evapotranspiration under non-standard conditions and crop evapotranspiration under standard conditions (mm day^{-1}). The K_y is the yield response factor (dimensionless).

This equation is the basis for the calculation in the different phenological phases of the crop to determine the impact of water (variable factor) on additional production. The modelling assumed the default values estimated by FAO for the different phenological phases of the crops. These factors are key, as they have a decisive influence on the results.

Water consumption was calculated for each crop for each climatic period (wetter, normal, and drier). The differential yield modeled with the Cropwat for each period was calculated from the maximum estimated yields for the dominant soil Series from the expert knowledge (expected yield for wet period and expected yield for dry period related to this maximum). The differential yield between periods was then multiplied by the FOB (Free On Board) price (U$S tn^{-1}) and the value for different conditions of water stress was obtained for each crop evaluated.

Assuming the production differential that is obtained by contrasting each climatic period (drier and wetter period expressed by 140 mm real use of water and decreased

Table 10.2 Total amount in US dollars of the differential in production

Diferential in production valued in U$S (wet condition vs dry condition)					
	Differential in kg per hectare (wet-dry)	Hectares	Production (Tn)	Price U$S.tn $^{-1}$ (2007-2011)	Total dollar amount U$S
Corn	2523	900,536	2,271,806	203	461,562,743
Soybean	768	4,943,945	3,796,060	411	1,562,002,753
Wheat	1419	2,459,206	3,489,363	263	918,714,393
Sunflower	550	1,042,134	573,006	449	257,084,981
Total					3,199,364,870

yield due to this water stress), and multiplying that value by the FOB price (U$S tn^{-1}) and the average area harvested from the evaluated campaigns (2007/2011), the total amount in gross value of the production not obtained due to the production losses (yield differential due to water scarcity) was calculated. The results are presented in Table 10.2.

It can be observed that, between drier and wetter conditions where the average difference is approximately 140 mm of water use for all crops throughout the period, the total amount in lost production due of reduction of yield estimated (Table 10.2) for that difference for all crops in the province of Buenos Aires, was valued at US$3199 million (Rosa et al. 2013). Updating crop distribution and prices for the period 2013/2018 this value amounts to U$S3578 million dollars. This simulation assumes the spatial coverage of this water deficit difference for the four crops for the entire province of Buenos Aires, with the consequent relative decrease in yields between both ends (wetter and drier). This would represent on average production of the last five years for 33% for corn, 27% for soybeans, 41% for wheat, and 33% for sunflower. It is likely that using the average values of difference in yields between wetter and drier periods will overestimate the impact due to the simplification of the number of *soil Series* used for analysis (the predominant Series of each cartographic unit was used). For the assignment of the data required by the CropWat program, the Series that occupies more than 50% of each cartographic unit was selected (detailed procedure in Rosa et al. 2013). On the other hand, the area occupied by each Series in the total of the province was determined and the most representative ones were described. Thus, *61 soil Series* were selected (detailed procedure in Rosa et al. 2013), each occupying more than 75,000 hectares (ha) throughout the province. This reduced the dispersion of yields on soils not included in the analysis and therefore their impact on monetary values. In any case, this estimate aims to give an idea of the magnitude of the impact with a methodological approach and with the information available.

In a work carried out by CICPES/INTA (Table 10.3, Lema et al. 2018), it was estimated the decrease in yields in soybeans and corn for the Pampean region based on the intensity of drought defined from the Standardized Precipitation Index, the following decreases in yields. Yield reduction dispersion was estimated for each

Table 10.3 Yield reduction of different drought events

	Yield	Soybean (tn.ha^{-1})	Corn (tn.ha^{-1})
	Min yield	2.5	8.3
	Avg yield	3.8	9.5
	Maxi yield	4.2	10.3
		Yield reduction	
Moderate draught	Min probable yield	−8%	−21%
	Avg probable yield	−29%	−22%
	Max probable yield	−10%	−19%
Severe draught	Min probable yield	−60%	−52%
	Avg probable yield	−55%	−56%
	Max probable yield	−40%	−50%
Extreme draught	Min probable yield	−100%	−79%
	Avg probable yield	−74%	−70%
	Max probable yield	−52%	−62%

Source Lema et al. (2018)

drought scenario based on the Standardized Precipitation Index by determining the minimum, average, and maximum yields for each scenario.

Following with the study (Rosa et al. 2013) the yield decrease due to the water stress of 140 mm of actual water use was calculated relative to the theoretical maximum for each agro-economic region and the results are presented in Table 10.4.

Table 10.4 shows the decrease in yields expressed as a percentage of the maximum yield (the maximum yield is the average yield of each type of soil in each AEZ) for each climatic situation (wetter, normal, and drier) and subsequently, the decrease in yield expressed as a percentage taking as a reference the wetter situation versus the drier situation.

The simulated values in this work showed the variation of yields in different conditions but it is observed that percentages obtained in yield reduction among drier and wetter conditions are similar to the average values in moderate drought scenarios in reference to the study taken as a reference to compare (see Table 10.3). Therefore, losses in production in severe and extreme drought events could be greater than those estimated by circumscribing it to a defined geographical area, duration, and intensity.

Table 10.4 Reduction of yield due to water stress

	Yield reduction	
	Soybean (tn.ha^{-1})	Corn (tn.ha^{-1})
Max yield	3.72	9.03
Avg yield in wet scenario	3.25	7.08
Avg yield in normal scenario	2.93	5.99
Avg yield in dry scenario	2.47	4.57
Reduction from max	%	%
Wet scenario	13	22
Normal scenario	21	34
Dry scenario	34	49
Reduction Dry-Wet (140 mm)	24	35

10.7 Final Considerations

Addressing and managing the drought phenomenon requires a clear definition of concepts, monitoring of indicators and indices to quantify it and the adoption of a proactive strategy for its management. It is a phenomenon that is difficult to manage, as well as implementing appropriate measures for mitigation. For the design, management, and evaluation of better quality public policies, decision makers require a clear understanding of the impact of the problem, clear proposals for resolution, benefit/cost ratios, sustainable social and environmental impact to be part of the public investment decision agenda.

The quantification of water stress in the estimated agricultural sector highlights the size of the economic losses of this phenomenon. Despite the efforts made in the province of Buenos Aires, it is necessary to deepen the development of comprehensive studies with interdisciplinary methodological approaches that allow detecting the needs for research, monitoring of variables and institutional capacity building to address this problem.

Adopting a risk management approach is geared toward impact prevention and mitigation through planning and with a long-term vision. The development of comprehensive and interdisciplinary studies, taking advantage of the increasing availability of new technologies at low costs, permitted us to monitor key variables at the regional level. Linking the advancement of Information and Communication Technology with investments in technological infrastructure in the territory; with research and development; capacity building; and with the participation of users in the adoption of technologies, it will contribute to deepen knowledge of processes and develop technologies to minimize the impacts of climate events.

In summary, the importance of measuring and generating indicators to deepen scientific knowledge of phenomena and processes; cooperate inter-institutionally and with all actors involved; developing and incorporating new technologies into rural areas and investing in non-structural measures could contribute to reducing the impact of climate variability with very satisfactory cost–benefit ratios for a sector with a high capacity to generate foreign exchange.

References

Alconada-Magliano MM, Bussoni A, Rosa R, Carrillo Rivera JJ (2009) El bio-drenaje para el control del exceso hídrico en Pampa Arenosa, Buenos Aires, Argentina. Investigaciones Geográficas, Boletín del Instituto de Geografía, UNAM 68:50–72. http://www.redalyc.org/articulo.oa?id=56912236005

Alconada-Magliano MM, Fagundo Castillo JR, Carrillo Rivera JJ, Hernández PG (2011) Origin of flooding water through hydrogeochemical identification, the Buenos Aires plain, Argentina. Environ Earth Sci 64(1):57–71

Allen RG, Pereira LS, Raes D, Smith M (1998) Crop evapotranspiration. Guidelines for computing crop water requirements. FAO Irrigation and drainage paper 56. http://www.fao.org/3/x0490e/x0490e00.htm and http://www.fao.org/3/a-x0490s.pdf

Ameghino F (1884) Las secas y las inundaciones en la provincia de Buenos Aires. 5° Edición, Ministerio de Asuntos Agrarios de pcia Bs As, 62 p. http://naturalis.fcnym.unlp.edu.ar/repositorio/_documentos/sipcyt/bfa000991.pdf

Andrade F, Caviglia O (2015) Incrementando la productividad de los recursos. Productividad del agua'. Acta IV Simposio Nacional de Agricultura, Paysandú, Uruguay, pp 123–130

Andrade F (compiler), Taboada M, Lema D, Maceira N, Echeverría H, Posse G, Prieto D, Sánchez E, Ducasse D, Bogliani M, Gamundi JC, Trumper E, Frana J, Perotti E, Fava E, Mastrángelo M (2017) Los desafíos de la agricultura argentina. Satisfacer las futuras demandas y reducir el impacto ambiental. Ediciones INTA, Ciudad Autónoma de Buenos Aires, Argentina, 127 p. https://inta.gob.ar/sites/default/files/lib_desafiosagricultura_2017_online_b.pdf

Bragachini M (2018) Adopción de Tecnología en el Sector Agropecuario Argentino en los últimos 28 años, Red de Agricultura de Precisión, INTA, 12. http://www.agriculturadeprecision.org/articulos/maquinaria-agricola/Adopcion-Tecnologia-Sector-Agrop-Argentino-ultimos-28-anios.asp

Carrillo-Rivera JJ (2000) Application of the groundwater-balance equation to indicate interbasin and vertical flow in two semi-arid drainage basins. Hydrogeol J V8(5):503–520

Cristini M, Chisari O, Bermudez G (2009) Agricultural and macroeconomic policies, technology adoption and agroindustrial development in Argentina: old and new facts. Revista Argentina de Economía Agraria, XI, pp 95–114

CROPWAT 8.0. http://www.fao.org/land-water/databases-and-software/cropwat/es/

Doorenbos J, Kassam, AH (1979) Yield response to water. FAO irrigation and drainage paper no. 33. Rome, FAO. http://www.fao.org/3/i2800e/i2800e02.pdf

Durán D (1987) Sequías e inundaciones. Propuestas, OIKOS, Asociación para la promoción de los estudios territoriales y ambientales, 198 p

Engelen GB, Jones GP (1986) Developments in the analysis of groundwater flow systems. IAHS 163:67–106

European Environment Agency (2012) European environment agency: water resources in Europe in the context of vulnerability, EEA 2012 state of water assessment, EEA Report No 11/20122, Publications Office of the European Union, Luxembourg, 92 p. https://doi.org/10.2800/65298

FAO (2011) El estado de los recursos de tierras y aguas del mundo para la alimentación y la agricultura. La gestión de los sistemas en situación de riesgo. Ed Mundi-Prensa, Madrid, España, 338 p. http://www.fao.org/3/a-i1688s.pdf

FAO (2016) FAOSTAT, Statistical database. Statistics division. Food and Agriculture Organization of the United Nations, Roma. http://www.fao.org/

FAO (2017) AquaCrop software. AquaCrop training handbooks. Book II: Running AquaCrop. http://www.fao.org/aquacrop/software/aquacropstandardwindowsprogramme/es/ and http://www.fao.org/documents/card/en/c/7512bf78-fd88-48a7-bbd9-415240289c88/

Gleick PH (2003) Water used. Annu Rev Environ Resour V 28:275–314. https://doi.org/10.1146/annurev.energy.28.040202.122849

Hoekstra AY (2003) Virtual water trade. Proceedings of the international expert meeting on virtual water trade, value of water research report series N12. Delft: IHE Delft, The Netherlands, 248 p. https://www.waterfootprint.org/media/downloads/Report12.pdf

Hoekstra AY, Chapagain AK (2008) Globalization of water: sharing the planet's freshwater resources. Blackwell Publishing, Oxford, p 224

Hoekstra AY, Chapagain AK, Aldaya MM, Mekonnen MM (2011) The water footprint assessment manual. Setting the global standard. Earthscan, London-Washington, DC, 199 p. https://waterfootprint.org/media/downloads/TheWaterFootprintAssessmentManual_2.pdf

Imbellone P, Giménez JE, Panigatti JL (2010) Suelos de la región pampeana. Procesos de Formación, Editorial INTA, Argentina, Buenos Aires, 288 p. https://inta.gob.ar/sites/default/files/script-tmp-inta_-_suelos_de_la_regin_pampeana.pdf

Lema D (2015) Crecimiento y productividad total de factores en la agricultura. Argentina y países del Cono Sur 1961–2013. Serie de informes técnicos del Banco Mundial en Argentina, Paraguay y Uruguay N° 1, 63 p. https://ucema.edu.ar/6/sites/default/files/2016/novedades/paper_lema_2015.pdf

Lema D, Amaro IB, Pace I (2018) Impacto de la sequía sobre los márgenes brutos esperados de soja y maíz en la región pampeana: ¿En qué situación los aumentos de precios compensarán las pérdidas de rendimientos? Informe CICPES, Instituto de Economìa, INTA, 9 p. https://inta.gob.ar/sites/default/files/inta_cicpes_instdeeconomia_lema_impacto_sequia_margenes_brutos_0.pdf

Macleod CJA, Scholefield D, Haygarth PM (2007) Integration for sustainable catchment management. Sci Total Environ V 373(2–3):591–602

Mishra AK, Singh VP (2010) A review of drought concepts. J Hydrol Elsevier V 391(1–2):202–216. https://doi.org/10.1016/j.jhydrol.2010.07.012

Moncaut CA (2003) Inundaciones y sequías tienen raíces añejas en la pampa bonaerense (1576–2001). Chapter 1: 27–48, In: Maiola OC, Gabellone NA, Hernández MA (eds) Inundaciones en la región pampeana, Honorable Cámara de Diputados de la provincia de Buenos Aires, Editorial UNLP, Argentina

Moscatelli G, Pazos MS (2008) Los suelos de la Argentina y su Geografía. http://www.madrimasd.org/blogs/universo/2008/04/17/89382

ONU (1992) Agenda 21. United nations conference on environment and development. Rio de Janeiro, Brazil, 351 p. https://sustainabledevelopment.un.org/content/documents/Agenda21.pdf

Ostrom E (2008) The economic analysis of the environment. J Compilation Inst Econ Aff 6(1):81–98. https://onlinelibrary.wiley.com/doi/full/10.1111/j.1468-0270.2008.00840.x

Perry C (2014) Water footprints: path to enlightenment, or false trail? Agric Water Manag Elsevier V 134:119–125. https://doi.org/10.1016/j.agwat.2013.12.004

Porta J, López-Acevedo M, Roquero C (1994) Edafología para la Agricultura y el Medio Ambiente. Editorial Mundi Prensa, Madrid, España, p 804

Ras N (1994) Crónica de la frontera sur. Academia Nacional de Agronomía y Veterinaria, Editorial Hemisferio Sur, Buenos Aires, Argentina

Rosa RJ, Alconada M, Andreau R, Darwich N, París H, Presutti M, González A, Lombardo A, Lusarreta P; Manis E, Micheloud, Moyano J, Galbusera S, Tinghitella G (2013) Huella hídrica de los productos agrícolas en las exportaciones de la provincia de Buenos Aires. La Plata, Argentina:

Consejo Federal de Inversiones, 368 p. http://biblioteca.cfi.org.ar/wp-content/uploads/sites/2/2013/01/huella-hidrica.-informe-final.pdf
SAGPyA (1998) Zonificación Agroeconómica y Sistemas Productivos Predominantes, Sub Proyecto Riesgo y Seguros Agropecuarios, Secretaría de Agricultura, Ganadería, Pesca y Alimentos, PROSAP ARG 96/006. http://www.ora.gov.ar/archivos/zonificacion%20y%20sistemas%20productivos.pdf
SAGPyA-ORA (2005) Oficina de Riesgo Agropecuario de la Secretaría de Agricultura. http://www.ora.gov.ar/riesgo_zonas.php
SAGyP-INTA (1989) Mapa de suelos de la provincia de Buenos Aires (Escala 1:500.000). Proyecto PNUD ARG/85/019, Buenos Aires, 544 p
Scarpati OE, Capriolo AD (2013) Sequías e inundaciones en la provincia de Buenos Aires (Argentina) y su distribución espacio-temporal. Investigaciones Geográficas, Elsevier V 82:38–51. https://www.sciencedirect.com/science/article/pii/S0188461113727857
Schmidt G, Benítez JJ, Benítez C (2012) Document: working definitions of water scarcity and drought. Version 4.0. European Commission, Inecsa-Inarsa s.a. and Typsa, Madrid, España
Smith M (1992) CROPWAT, a computer program for irrigation planning and management. FAO Estudio de Riego y Drenaje 46, Organización de las Naciones Unidas para la Agricultura y la Alimentación, Roma
Thomasz EO, Vilker AS, Rondinone G (2018) The economic cost of extreme and severe droughts in soybean production in Argentina. Contaduria y Administración 64(1):1–24
Töth J (1995) Hydraulic continuity in large sedimentary basins. Hydrogeol J 3(4):4–16
Töth J (2000) Las aguas subterráneas como agente geológico: causas procesos y manifestaciones. Boletín Geológico y Minero, Instituto Tecnológico GeoMinero España 8:49–26
Soil Survey Staff (1975) Soil taxonomy: a basic system of soil classification for making and interpreting soil surveys. Soil Conservation Service, U.S. Department of Agriculture Handbook 436, Washington DC, 754 p
Urquijo-Reguera J (2015) La gestión de la sequía: contribuciones para su evaluación. Universidad Politécnica de Madrid. https://www.fundacionaquae.org/wp-content/uploads/2019/01/6_TESIS_Julia-Urquijo_final-2015-1.pdf
Urquijo Reguera J, Pereira S, De Stefano L (2017) A methodology to assess drought management as applied to six European case studies. Water Resour Dev 33(2):246–269
Van Loon AF (2015) Hydrological drought explained. Wiley Interdiscip Rev: Water 2(4):359–392.
Viglizzo EF (2014) Sistemas ganaderos y tecnología: Estado actual y prospectiva, Chapter 2. In: Cauhepé MA (ed) Manejo de la Cría Vacuna en la Región Pampeana. Editorial Hemisferio Sur, Buenos Aires, pp 19–37
Wilhite DA, Glantz MH (1985) Understanding the drought phenomenon: the role of definitions. Water Int V10(3):11–120
Wilhite DA, Hayes MJ, Knutson C, Smith KH (2001) Planning for drought: moving from crisis to risk management. J Am Water Resour Assoc V36(4):697–710

Chapter 11
Water Security and Groundwater: The Absence of Scientific Criteria in Groundwater Management Through Three Case Studies in Mexico

Gonzalo Hatch-Kuri, Julio César Sánchez-Angulo, Juanalberto Meza-Villegas, and Yussef Ricardo Abud-Russell

Abstract The water debate seems to be summarized in two fully opposite positions: (i) the Dublin Conference of 1992 perspective favors movements that defend against the introduction of market economies in the management of water, arguing the production of "water scarcity"; and (ii) the groups calling for water as a "common good," recognizing that there is enough fresh water access but an unequal appropriation. However, both lack to propose the required scientific research to guarantee a hidden basic concept: water insecurity. Countries such as China, Japan, Australia, and Canada, have implemented methodologies as the Tóthian groundwater flow system (Tóth 1962, 1999, 2016), concept that provides with solid interdisciplinary analyses of the related ambient components. In Mexico the situation is different, various political-administrative factors have prevailed over the scientific understanding of the functioning of groundwater. Through three case studies: Querétaro State, and the Northern, and Southern international boundaries of Mexico (transboundary groundwater) that reflect challenges in water management; such as the characterization and evaluation of groundwater flow systems shared. This chapter seeks to contribute to

G. Hatch-Kuri (✉) · J. C. Sánchez-Angulo
Programa de Maestría En Gestión Integrada de Cuencas, Facultad de Ciencias Naturales, Universidad Autónoma de Querétaro. Av. Junípero Serra, Antiguo Aeropuerto, Campus Aeropuerto, S/N, C.P76140 Santiago de Querétaro, Querétaro, México
e-mail: ghatch@comunidad.unam.mx

J. C. Sánchez-Angulo
e-mail: julio.angulo08@gmail.com

J. Meza-Villegas
Programa de Maestría En Ciencias Sociales, Facultad Latinoamericana de Ciencias Sociales. Colonia Héroes de Padierna, Carretera al Ajusco 377, C.P. 14200 Tlalpan, Ciudad de México, México
e-mail: juanalberto.meza@estudiante-flacso.mx

Y. R. Abud-Russell
Colegio de Geografía, Facultad de Filosofía y Letras, Universidad Nacional Autónoma de México, Circuito Interior. Ciudad Universitaria S/N, C.P. 04510 Ciudad de México, México
e-mail: yussef.abud@gmail.com

M. M. Alconada-Magliano (ed.), *Intensified Land and Water Use*, Springer Earth System Sciences, https://doi.org/10.1007/978-3-030-65443-6_11

reveal the political and scientific elements that characterize groundwater management in Mexico, and its relation to Water Security.

Keys words Water security · Groundwater flow systems · Aquifer · Transboundary aquifer · Water rights · Mexico

11.1 Introduction

The debate on water seems to be summarized in two fully opposite positions. Firstly, the perspective that surged from the Dublin Conference of 1992, that favors those movements that defend from the increasingly introduction of trading laws, that argue that a concept like "*water scarcity*" should be addressed through market based management policies. Contradictorily, there's another point of view in which the groups calling for water as a "*common good*," recognize that there is enough access to fresh water fresh sources; nevertheless, as a product of socio-economic and political conditions that result in its inequitable distribution, water access appropriation is unequal. Notwithstanding, both proposals overlook the required scientific research that guarantee the fundamental principle of instrumental condition for the fair livelihood of people inducing a hidden basic concept: *water insecurity*.

One of the most accepted descriptions of **Water Security** is the one defined by UN-Water (2013), which indicates that it is "the capacity of a population to safeguard sustainable access to adequate amounts of water of acceptable quality for sustenance, well-being and development sustainable socio-economic; to ensure protection against water-borne pollution and related disasters, and to preserve ecosystems, in a climate of peace and political stability."

This definition is broad but also ambiguous, and has led to a fruitful debate in academia about its possible interpretations and implications for water management policy in each of the nation-states (Cook and Bakker 2012). Precisely, a positive aspect of this debate is that it recognizes the importance of water in operating two other complex systems, that is, energy and food, thus posing an interdependent triangle between water security, energy security, and food safety.

The Organization for Economic Cooperation and Development (OECD) published its own definition, which explicitly remarks the notion of risk and water vulnerability: "water security consists of keeping four risks associated with water at acceptable levels: (i) risk scarcity, such as lack of sufficient water (in the short and long term) for beneficial uses for all users; (ii) the risk of inadequate quality for a given purpose or use; (iii) the risk of excesses (including floods), understood as exceeding the normal limits of a hydraulic system (natural or built) or the destructive accumulation of water in areas that are not normally submerged; and (iv) the risk of deteriorating the resilience of freshwater systems, due to exceeding the assimilation capacity of surface or underground water sources and their interactions, with the eventual exceeding of acceptable thresholds, causing irreversible damage to hydraulic and biological functions of the system." In this definition, it's possible to appreciate water scarcity as an issue, as well as the importance of punctual maintenance of the infrastructure as a prevention mechanism against the posed risks.

Considering the above, in the context of a research project related to the key role of water security and its impact on the modernization of groundwater management in Mexico at the facilities of the Autonomous University of Querétaro (2018-2020) with financing from the National Science and Technology Council, it sought to answer questions related to the modernization of the official groundwater management schemes in Mexico and the challenges faced to meet water security according to the most accepted concepts. In this sense, it was considered that one of the most prominent elements in water security is, precisely, groundwater.

Little is understood of its natural planetary distribution and circulation, through the hydrological cycle itself groundwater constitutes 97% of physically accessible continental freshwater (Rivera 2008). In Mexico, it is estimated that more than 75% of this supply source is used for sectors such as the public urban water systems, agricultural (considering irrigation) and industrial sectors (CONAGUA 2019). However, its management is strongly related with the political decisions of the authorities. Conversely, hydrogeologists such as Tóth (1999, 2016) and Freeze and Cherry (1979), among others, argue that groundwater management should be sustained on the strict application of scientific evidence. This exposes the connection and the lateral and continual hydraulic movement of water through the varying porosity of the rocks (aquifers) of the subsoil in three dimensions: local, intermediate, and regional.

Government actions regarding education, training, and scientific application of the Tóthian Flow System (TFS) need to be openly recognized in the quest to control and minimize the undesirable environmental effects (drought and flooding, regulate water quality, recognize transboundary flows, and soil subsidence). Through a comprehensively established and concealed scientific framework, the law-making processes and water regulation implementation would have better criteria to address phenomena related to water insecurity.

Countries such as China, Japan, Australia, and more recently Canada, have implemented methodologies as the Tóthian Flow System (TFS) concept to provide water policymaking with solid interdisciplinary analyses of the related ambient components (original soil and vegetation, presence/absence of surface water, geomorphological and geological framework, groundwater chemistry and isotopic composition, and groundwater flow hydraulics) all of which have a shared interrelationship resulting in a clear, and well-founded scientific description and understanding of groundwater functioning (Tóth 1962, 1999, 2016).

Through three case studies, this contribution reveals results related to the aforementioned research project from an approach that, although it comes from the social sciences, has sought a scientific explanation in the recent contributions of hydrogeology in order to have an interdisciplinary look at the problem here addressed.

11.2 Science and Policy: The Invisible Linkage of Groundwater Management in Mexico (1948-2018)

In Mexico, governmental data estimate that groundwater supplies 60 million people (the last census showed almost 130 million Mexicans) and irrigates two million hectares, 75% of the water supplied in Mexico comes from groundwater (CONAGUA 2019). Despite the strategic groundwater value, it has not been possible to reform groundwater management toward a democratic model that guarantees an equitable access. Two of the main obstacles for democratization are the authoritarian nature of the political system and lack of sound scientific research about groundwater.

Carrillo-Rivera et al. (2016) show the methodological and scientific limitations for groundwater evaluation in Mexico, which could be related to the ever growing number of legal, environmental and social conflicts over this water. Schmidt and Hatch-Kuri (2012) suggest that water, especially transboundary groundwater, is a national security issue; Rivera and Candela (2018) and Megdal (2018) indicated that a democratic and equity model in water rights is needed in order to guarantee adequate institutional articulation between the agencies responsible for government decentralization, transparency in decision-making and participation of all those interested (Fig. 11.1).

Based on the above, it is necessary to identify how the components that define groundwater policy in Mexico are articulated. That is to understand the implications of the linkage between legal instruments that contain provisions for its use and distribution, the scientific-technical groundwater evaluation methodologies role, the spatial dimension in the national territory, and the political criteria. To that regard, political control over groundwater management disregards scientific criteria,

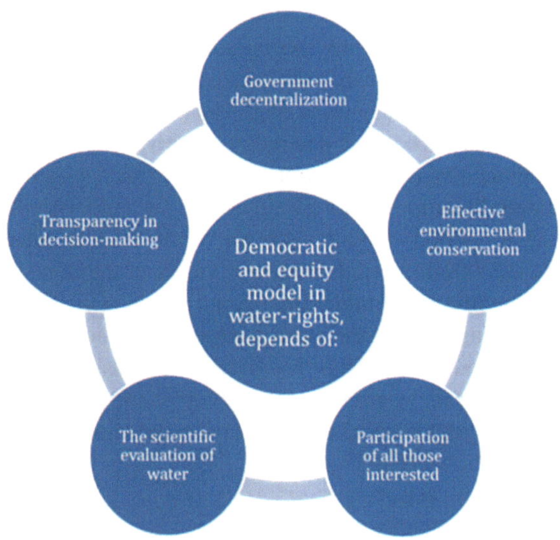

Fig. 11.1 Democratic and equity model for water rights (groundwater) according to Rivera and Candela (2018) and Megdal (2018)

endowing discretionary decisions on authorities and allowing political expediency among the main users of water, with possible underlying unethical consequences.

President Enrique Peña Nieto (2012–2018) made three meaningful decisions related to groundwater management. The first was the Presidential Decree of April 5, 2013 with which practically all groundwater extraction was banned (termed a "*veda*" in Mexico) in the entire national territory, drilling of new wells was prohibited, as well as obtaining or renewing water concessions. Entitlements are defined as "the use, exploitation or utilization of the volume that is given in the concession, which must be respected by concessioners. Water entitlements are unbundled from property title, they can be traded, leased or transferred and last between 5-30 years with the expectation of periodic renewal," it's unclear the renewal process but the COTAS (Comités Técnicos de Agua Subterránea, i.e., Technical Committees Groundwater) play a certain role (OECD 2013). The second decree was on October 31, 2017, and the National Water Commission (CONAGUA) created the Technical Committee for Groundwater Management (COTEMA), a council board responsible with analyzing and implementing alternative solutions to problems on groundwater management (pollution, groundwater depletion, reduction of base flow to rivers, soil subsidence, yield reduction of wells, and seawater intrusion). The third one was the decree, March 23, 2018, reversing the provisions of the April 5, 2013 decree, establishing a temporary suspension until December 31, 2018, in order to regulate, extend, and grant new water concessions and groundwater allocation. Between these two decrees, unlike the first ten decrees enacted by former president Miguel Alemán Valdés, groundwater provisions went unnoticed almost entirely for the public opinion.

The Peña administration failed to structurally reform the water legal framework, due among other things to the deregulatory policy (Hatch-Kuri et al. 2017) geared to maintain and expand water supply to, specially, big private users. The 2018 decree was the legal instrument used to renew and grant new concession and water entitlements without proper scientific technical support. CONAGUA (2018a) supports the notion that groundwater management is strictly a technical issue, but the decrees manifest an "invisible" political component: the absolute power of the President to decide the water rights policy and groundwater conservation policy in Mexico.

All groundwater decrees issued by the Federal Executive Power from 1948 to 2018 were reviewed, to determine management and the main classification (banned, reserve zone, or regulated zone), their justification, and spatial display expressed in two key concepts "*administrative aquifer*" and "*irregular polygon*" which have a potential unethical groundwater management component.

Regarding the interdisciplinary proposal of the *Hydro-Social Cycle* Theory (Linton 2010), the methodological framework consists of a combined approach from Political Geography, Hydrogeology and Political Science. This reveals that the water–power nexus in Mexico is sustained on a crucial aspect of its political system: Presidentialism (Schmidt 1986), which is a determining factor in understanding the actual water rights policy scheme. One key element is the superficial and biased use of scientific-technical elements for decision-making, for example the "*determination of the average annual water availability*" to grant water rights and the symbolic legislation to create rules in paper which are not translated into real life. The main results showed that:

(i) Twelve presidents of the Mexican Republic from 1948–2018 issued 105 provisions for legal groundwater ordering (Table 11.1). Luis Echeverría Álvarez (1970–1976) issued the highest number of decrees (20) but, considering the spatial scope of the legal provisions the 19 decrees issued by Adolfo Ruíz Cortines (1952–1958) had effects on 89 "irregular polygons." Enrique Peña Nieto (2012–2018) decreed a ban on 332 administrative "aquifers" in Mexico (98% of the national territory).
(ii) Groundwater management, seen through presidential decrees, reveals the political decision to use two concepts to define the territorial scope of these provisions, and both of which lack scientific criteria. The first is the "**irregular polygon**," which refers to a geographical area in which the imposition of a banned reserve zone or a regulated zone has been imposed; it is determined by unclear criteria by CONAGUA (Alley 1993:77). At the middle of the twentieth century, the concept "**Aquifer**" was formalized, institutionalized, and characterized by administrative and political purposes, since its definition and spatial dimension considers only the surface of the geometry of the hydrogeological aquifer, without considering its depth and thickness. This way water administration fails to consider the determination of the systemic functioning of groundwater flows and their interdependent relationship with other environmental components.
(iii) The groundwater presidential decrees reveal a rhetoric that changed over time. From 1948 to 1972, the presidents justified the issue of banning decrees or reserve zones, to expropriate (public interest or public domain declaration) for

Table 11.1 Number of drilling ban decrees imposed in Mexico 1948-2018

Presidents of the Republic	**Government period**	**Decrees**	**Irregular poligons**
Miguel Alemán Valdés	1946-1952	10	11
Adolfo Ruiz Cortinés	1952-1958	19	89
Adolfo López Mateos	1958-1964	17	42
Gustavo Díaz Ordaz	1964-1970	12	52
Luis Echeverría Álvarez	1970-1976	20	63
José López Portillo y Pacheco	1976-1982	17	43
Miguel de la Madrid Hurtado	1982-1988	8	19
Enrique Peña Nieto	2012-2018	2	332

works for artificial lighting (pumping wells), which included the infrastructure and previously acquired groundwater rights. In 1972, the first presidential decree that justifies the imposition of a ban to "conserve and protect the balance of water in aquifers" was recorded, although the concept aquifer was not yet defined by law, at present, it refers to a groundwater definition. This justification has a direct relationship with the agreements of various international environmental conferences, such as the Stockholm International Conference (1972), where it was agreed that the States will develop environmental protection policies, which include water.

(iv) The groundwater legal decrees for the decades 2000–2020, even though supposedly are based on environmental and groundwater conservation, they actually respond to water management rights policy, which responds to the logic of economic growth without scientific rigor, and it delivers water rights according to discretionary presidential criteria even though, in many cases, it contradicts the fundamentals of preserving the environment. Installation of industrial zones occurs under this water allocation scheme, in regions where aquifers were declared "overexploited" thus, governing water management policy might lack ethical considerations in favor of political criteria, because it doesn't consider the rights of the original people, or the well-being of society at large.

Mexican groundwater management based upon the absence of solid scientific technical aspects, has a close relationship with the application of regulatory criteria reflecting negligible updating of the scientific methodology for its correct assessment according to international standards. These rules have facilitated that the serving president can have political control and discretionary management in providing water to users, as some are more influential than others, reinforcing inequality. The lack of thorough conceptual and scientific criteria for concepts such as "irregular polygon" and "aquifer" (administrative), also marginalize the objectives shared by Mexico's adherence to international policies, such as actions against climate change; which initially could be unethical.

Contrary to the usual argument that groundwater management is exclusively a technical issue, groundwater management in Mexico is based on political decisions geared to creating policies of patronage and political loyalty among water users to support the discretional acts of authority. Water policy is part of the symbolic legislation that shows existing laws and regulations which are hardly applied, either because of political commitments or due to corruption. This contradicts the international commitments made by Mexico and affects water preservation and environmental conservation. In sum, this discretionary policymaking in water allocation is used by the government to justify the legal system and discretionary decisions; and this in turn reinforces the unethical use of concepts that lack consistency and move away from a proper scientific application of modern hydrogeology, political geography, and democracy.

11.3 Groundwater Rights Concentration in the Amazcala Valley Aquifer, Querétaro

In Mexico, groundwater management is determined in first instance by the definition of the concept "aquifer," present in the third article of the National Water Law (LAN, *Ley de Aguas Nacionales*):

> *Any geological formation or set of formations hydraulically connected to each other, through which water circulates or is stored water can be extracted for usage and exploitation and whose lateral and vertical limits are conventionally defined for purposes of evaluation, management and administration of the national waters in the subsoil* (LAN 2020).

This concept divides Mexican territory into 653 administrative polygons which in turn have facilitated the implementation of the groundwater policy. In principle they resemble surface dimensions of a possible unverified geological formation, represented in turn, by a watershed of a conventional type (political-administrative) (Carrillo-Rivera 2014). In this regard, the Official Mexican Standard 011-CONAGUA-2015 (NOM, *Norma Oficial Mexicana*) establishes that the annual groundwater availability average is determined by the arbitrary polygonal delimitation of each of the previously defined administrative aquifers. This official water balance technique is applied to administrative aquifers, incorporating the measurement of "water volumes" while overlooking chemical quality and the groundwater flow system approach (Tóth 1999), and neglecting the environmental impacts. One of the most significant problems implicated by applying the NOM 011-CONAGUA-2015 is that the water balance is considered through mathematical equality, therefore, the operation of the hydrological cycle resembles the behavior of financial resources, that is to say, how much water enters the system (administrative aquifer) and how much water can be extracted from it. This situation, which ultimately has benefited unequal distribution of water among users, encourages those who have access to greater technological resources to pump and extract greater volumes of water, a situation that would also indicate the concentration of groundwater rights to the detriment of other users (Wester and Hoogesteger 2011). Concepts such as overexploitation, resulting from the extract-recharge ratio in which the extraction of groundwater exceeds the annual average recharge volume (CONAGUA 2015), are applied to the water sector, but above all to groundwater. In managing administrative aquifers, this situation has fostered an extreme reductionism that disregards understanding the relationship among natural components such as the edaphic layer, topography and vegetation with the functioning of the hydrological cycle and its relationship with groundwater flow systems.

This research considered as a starting point the case study of the Amazcala Valley administrative aquifer in Querétaro State, considered by CONAGUA (2018b) as one of the four "overexploited" aquifers located in the proximity of the Mexico State-Querétaro-San Luis Potosí economic and industrial corridor, which in turn constitutes one of the main sources of supply for the metropolitan region of Querétaro city. Specialists such as Sandoval (2016) consider that this corridor is a geographic space of intense activity and economic growth, which has driven the demographic phenomenon to a sustained growth 4.5 faster for the 1970–2015 period than in previous years, and the urban expansion of Querétaro City to a 35.2% in the same period (INEGI 2010; SDUOP 2015). The resulting accelerated land use change, deforestation and the expansion of residential areas, agro-business and hydroelectric industries have consolidated, among other factors, due to the current policy that governs groundwater rights. In this sense, a conflict that prevents the effective construction of water security in the Amazcala Valley, Querétaro, is the unequal access to water among users. In this way, the current groundwater policy has caused some users to be relegated, while others resist through low-profitability hydraulic technologies installed by the main companies that negotiate with the drilling of wells (Domínguez and Carrillo-Rivera 2007); and on the other hand, users with greater economic power are benefited by concentrating water rights.

According to the official update (2018) on the average annual availability of water in the Amazcala Valley Administrative Aquifer, number 2201, the result of the water balance showed that the average annual recharge was of the order of 34.0 $Mm^3.year^{-1}$, and It was estimated that users extracted a concessioned volume of the order of 54.32 $Mm^3.year^{-1}$; which translated into a deficit of 23.12 $Mm^3.year^{-1}$ (DOF 2018). Considering the above, the initial research question posed how the continuous extraction of water over time from said administrative aquifer would be explained, since there is a significant deficit that would prevent the continuation of the water concession and the metropolitan growth of Querétaro?

From the delimitation of the administrative polygonal of the "Valle de Amazcala" aquifer, the surface water bodies that make up the Chichimequillas river basin were mapped (Fig. 11.2), to perform a rigorous analysis in the Public Registry of Water Rights (REPDA, *Registro Público de Derechos de Agua*). The result facilitated the identification of the users who concentrate the largest number of groundwater concessions and, later, characterize them in order to relate it to the discretional groundwater federal distribution policy. The information was corroborated in the field through semi-structured interviews with various water users.

One of the findings derived from the cartographic analysis indicates that the polygon of the Amazcala Valley Administrative Aquifer, which has an area of 60.57 km^2, coincides by a smaller margin with the area of the Chichimequillas river hydrological basin that reaches a surface of the order of 61.75 km^2, registering a minimum difference of 1.18 km^2 between both. The vertices of the polygon of the aquifer and of the hydrological basin are similar, and the watershed of the administrative aquifer is represented by the hydrological basin of the Chichimequillas River. However, it is highly desirable that this cartographic representation be updated and consider the determination of groundwater flow systems in order to recognize the particularities of water circulation within the hydrological cycle.

Regarding the georeferencing of underground water uses reported in the REPDA, 99% are located in the territory of the Chichimequillas river hydrological basin, not within the administrative aquifer polygon, which indicates that the georeferenced information relative to groundwater users is not clear enough to place it within the groundwater management schemes themselves (Fig. 11.3).

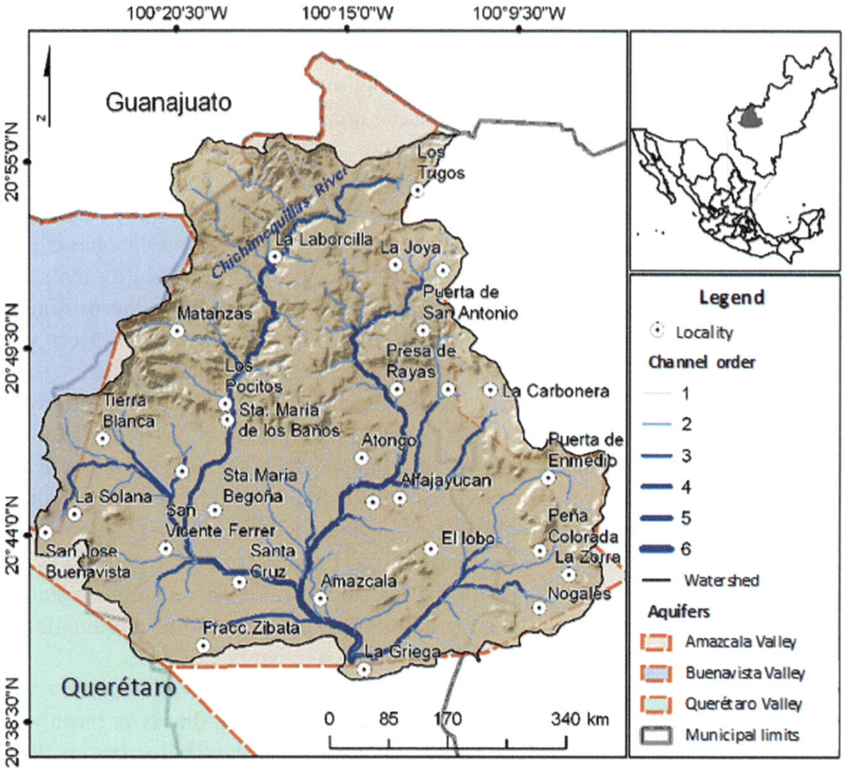

Fig. 11.2 Amazcala valley administrative aquifer in the delimited hydrological basin (adapted from Sánchez et al. 2019)

Another relevant finding that the comparative analysis between the REPDA information and its cartographic representation yielded is that, at least 30 private sector users associated publicly with the agribusiness industry in the region, account for 89% of the total estimate of the recharge average annual water level (34 $Mm^3.year^{-1}$) through the allocation of 106 water concession titles for groundwater use, unlike peasant organizations who have only 43 concession titles for agricultural use for the benefit of a total of 516 families, with an extraction volume of the order of 14.8 $Mm^3.year^{-1}$, which represents 43.7% of the average annual water recharge. The above contrasts with the Local Water Committees, which are organized rural users that manage the wells that provide water, this form of administration represents only 1.1% of the annual water recharge volume used for domestic purposes. Around 12,000 inhabitants of rural areas benefit from this, indicating the existence of a marked inequality in the current distribution of water among users who are registered in the REPDA and who are associated with the Amazcala Valley Administrative Aquifer (Table 11.2).

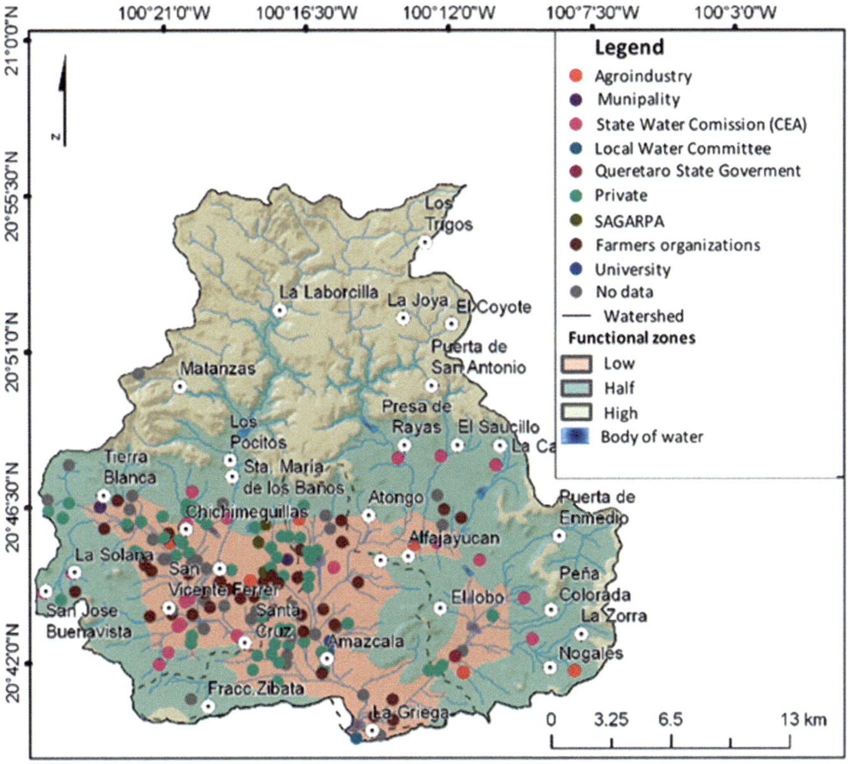

Fig. 11.3 Extraction Well distribution in Chichimequillas Basin, Querétaro (adapted from Sánchez et al. 2019)

Finally, analysis of historic data and field interviews establish a precedent that explains the actual concentration of water rights in the Amazcala Valley. With the implementation of the *San Juan River Irrigation Plan*, in 1947, together with the legal water reforms enacted between 1948 and 1972, some sectors close to the Mexican President during the mentioned periods, were specifically benefited from the water management results. It was verified that since 1950 the surname of the Roiz-Amieva family concentrates 13 water concession titles, which together are equivalent to 12.94% of the total annual water recharge of the Valle de Amazcala Administrative Aquifer. This family partnered with another family under the surname Ruiz-Rubio, dedicated to the local soft drink industry, in order to create real estate businesses responsible for the largest housing development in Querétaro city in the 1980s (Sánchez, 2019).

In addition, an aspect that cannot be ignored is the plenipotentiary powers of the President of the Republic in turn to provide water to users, through the enactment of decrees and the issue of water concession titles, which provoke competition among the users, illegal commodification through the transfer of concession titles,

Table 11.2 Groundwater rights allocation in Amazcala Valley administrative aquifer (adapted from Sánchez et al. 2019)

Owner of water rights	N° Water rights	Groundwater granted ($m^3.year^{-1}$)	No registration of $m^3.year^{-1}$	% Annual water recharge accumulation
Private sector	106	30,443,803.00		89.5
Farmer organization	43	14,849,146.00		43.7
Local Water Committee	3	387,583.00		1.1
University	1	600,000.00		1.8
SAGARPA	4	438,840.00		1.3
State Water Comission (CEA)	19	905,000.00	16	2.7
Municipality	3	800,000.00	2	2.4
Querétaro State Government	1	350,000.00		1.0
No data	32	–		0.0
Total	212	48,774,372.00		
Annual Recharge		34,000,000.00		100.0
Water deficit		14,774,372.00		43.5

in exchange for increasing private investment through the commercialization of land use, which has exponentially gone from "ejidos" to industrial and residential, putting water security at risk now protected in the water markets (Aboites 1998; Sosa et al. 2019).

11.4 Groundwater Management Strategies Comparative Study in the Binational Region "Tijuana-San Diego"

The Tijuana-San Diego binational region is an important zone of economic activity on the Mexico-United States border, which concentrates around 6.5 million people (the highest in the entire border) and registers an estimated annual Gross Domestic Product of 230 billion U.S. dollars (Leyva 2020). Regarding the issue of water, both cities are highly dependent on the supply provided by the Colorado River, a shared water source that is subject to different concessionary uses along its channel from the Rocky Mountains to the Gulf of California in Mexico. During this course more than 80% of the annual volume of the river is used in irrigation districts that have an area of around 16,500 km^2, while the remaining flow (20%) is used to provide water and electricity to the population that inhabits this transboundary basin (Hinojosa-Huerta and Carrillo-Guerrero 2010).

Water dependence on the Colorado River, which derives from an apparent scarcity of local water, constitutes a challenge for binational water security, which is why various alternative strategies for water supply have been proposed, which reveal political, economic, and economic asymmetries, institutional, and even scientific knowledge on water, which exist in both countries. One of the possible alternatives is the use of groundwater from the Transboundary Aquifer System (SAT) called "Tijuana-San Diego" (Fig. 11.4) (UNESCO 2007; Sánchez and Eckstein 2017) that extends under the territory of both cities. However, in both countries there are different definitions, legal instruments, strategies and uses of groundwater, which is why this system is managed differently. With these considerations, this case study sought to answer, through a comparative analysis, questions such as: *What is the role of groundwater as an alternative source of binational supply*? and *What implications do these differences have for local groundwater management*? At this point an analysis is carried out.

To respond to the above, the *Hydro-Social Cycle Theory* was used, based on the intersection of political geography and political ecology (Linton and Budds 2014). The hydrosocial cycle represents the process by which alteration or manipulation of water flows and quality affect social relations and structure, which in turn affect further alteration or manipulation of water (Linton and Budds 2014:175). This theoretical approach studies water from the interpretation that considers that it is not only

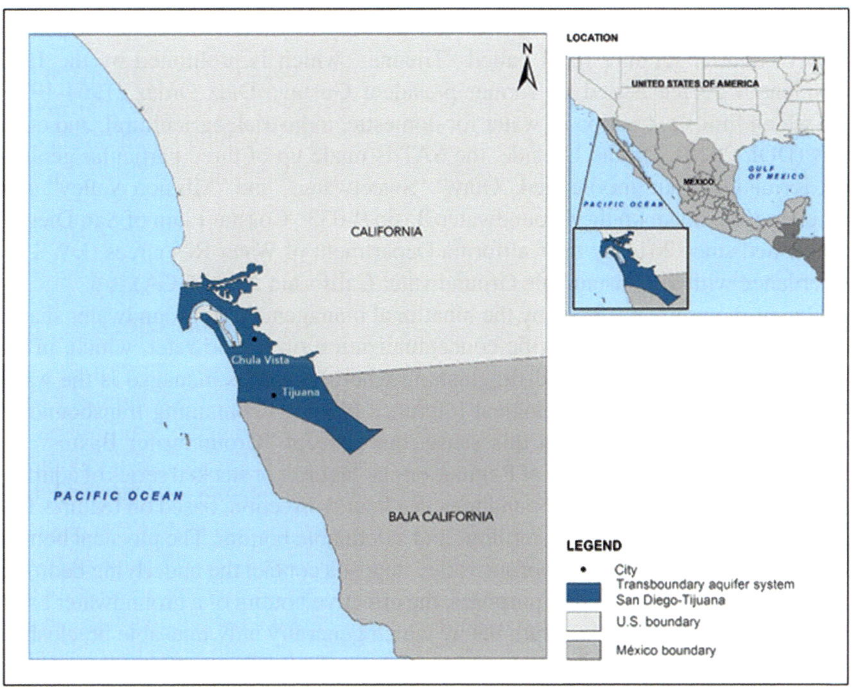

Fig. 11.4 SAT "San Diego-Tijuana" (adapted from IGRAC-UNESCO-IHP 2015)

a chemical compound (H_2O), but that it is also constituted from the social relations that affect hybridization processes (water as a socio-natural "hybrid"), which allows, in the first instance, to question the meanings and scientific discourses that are elaborated on it (Linton 2010:18). In this sense, it is recognized that scientific knowledge related to the characterization and evaluation of the conditions of the hydrological cycle is not neutral, on the contrary, it is subject to interests and power structures that are capable of being characterized and evaluated. The *Hydro-Social Cycle* based this comparative analysis on scientific conceptions regarding transboundary groundwater, and the implications that emerge for the development of strategies and management practices for this source of water supply.

According to the physical characterization reported by UNESCO (2015), the SAT "San Diego-Tijuana" is made up of alluvial materials in the valleys and by conglomerates in the terraces and adjacent hills, and its surface extension reaches the order of the 300 km^2 in Mexican territory and 250 km^2 in the United States. It is estimated that its thickness in Mexican territory reaches 300 meters, while in the San Diego coastal plain it tends to decrease. The salinity of the groundwater flow system increases with depth in Mexican territory and the occurrence of seawater intrusion is recorded in the coastal plain of San Diego (UNESCO 2015). This information was systematized in the subsequent study by Sánchez and Eckstein (2017), who established a classification for the Mexico-United States transboundary aquifers, in order to assess the amount of scientific information related to these, the SAT in question was considered to have of a "reasonable level of confidence."

According to CONAGUA, in Mexican territory the SAT is made up of administrative aquifer number 0201 called "Tijuana," which is prohibited by the 1965 presidential decree issued by former president Gustavo Díaz Ordaz (1964-1970) and which limits extraction of water for domestic, industrial, agricultural, and other uses (DOF 2018). On the US side, the SAT is made up of three particular geological formations (aquifers) called "Otay," "Sweetwater," and "Mission Valley" and, which in turn, make up the Groundwater Basin 9-033 "Coastal Plain of San Diego," established since 2014 by the California Department of Water Resources (DWR) in accordance with the Sustainable Groundwater California Act (SGCA).

One of the problems faced by the binational management of groundwater shared by both countries is the scientific conceptualization of groundwater, which, in the first instance, does not allow distinguishing whether what is managed is the water "per se" or the particular geological reference (aquifer) containing transboundary groundwater flow systems. In this sense, the concept "Groundwater Basins" are defined by the California Code of Regulations as "aquifer or stacked series of aquifers with reasonably well-defined boundaries in a lateral direction, based on features that significantly impede groundwater flow, and a definable bottom. The physical bottom of the basin occurs where the porous valley deposits contact the underlying bedrock. For groundwater management purposes, the effective bottom of a groundwater basin is sometimes defined as the depth below which generally only unusable brackish or saline groundwater can be found. In addition, an aquifer "refers to a three-dimensional body of porous and permeable sediment or sedimentary rock that contains sufficient saturated material to yield significant quantities of groundwater to wells and springs."

In contrast, and as previously noted, in Mexican territory the CONAGUA delimits the aquifers according to the conventional limits defined by the authority on the subject.

The conceptual differences raised have influenced the development of institutional adaptation strategies by the California authorities to solve the problem of dependence on surface water and thus try to overcome the problem posed by local water scarcity. The Sustainable Groundwater Management Act (2014) in California, was approved as an emergency measure because of the damage done by drought over five years; it mandated the creation of county groundwater sustainability agencies. In California, groundwater is considered essential for economic development, which is why the search for sustainable management of these sources has been identified since 2009 through the California Statewide Groundwater Elevation Monitoring Program run by the DWR. Groundwater has been managed since 2014 at a local level through the SGMA, which promoted the creation of local Groundwater Sustainability Agencies (GSAs) for the monitoring and administration of the Groundwater Basins that must report annually to DWR, as well as the preparation of a specific sustainability plan according to the priorities. The aim of this strategy is to address problems such as significant groundwater level declines, groundwater-storage reductions, seawater intrusion, water-quality degradation, land subsidence, and surface water depletions (California Sustainable Groundwater Management 2014).

Additionally, one of San Diego County's priorities is the development of new local supply and storage sources, where it is believed that groundwater can contribute to the regional effort to reduce dependence on water transfers from the Colorado River to through the Metropolitan Water District (MWD), which are estimated to be about 95%, according to the San Diego County Water Authority. This strategy, dating from 1997, was based on the Strategic Plan for Water Supply whose main objective was the planning and development of the local water supply system, which has been complemented with the desalination of groundwater through reverse osmosis in Reynolds Groundwater Desalination Facility in Chula Vista and which produces 37,854,117 L per day^{-1} of potable water.

Tijuana also recognizes the problem of dependence on the Colorado River that began in 1972, which is why the "Río Colorado-Tijuana" Aqueduct was built to bring water to that city and that constitutes more than 90% of the supply. Currently, the city's supply diversification strategy also prioritizes the use of water desalination. Unlike San Diego, the desalination of water from the Pacific Ocean has been proposed through a desalination plant in the neighboring municipality of Playas de Rosarito. This project was subject to criticism due to the nature of its negotiation as well as its economic and environmental infeasibility, which is why it could not be implemented (Cervantes 2020).

The comparative analysis that was carried out in this investigation (Table 11.3) yielded three notable findings regarding the management of transboundary groundwater Mexico-United States:

- – The ***lack of information*** related to the characterization and scientific evaluation of the SAT "Tijuana-San Diego" is manifested even in the reports published by UNESCO (2007, 2015), in which despite the years that have passed, there's no

Table 11.3 Comparative analysis of groundwater management at SAT "Tijuana-San Diego"

Features	**San Diego**	**Tijuana**
Colorado River water reliance	95%	90-95%
Groundwater management unit	Groundwater Basin	Aquifer
Scientific term	Coastal Plain of San Diego	Acuífero 0201 "Tijuana" Aquifer
	(Geological boundaries)	(Administrative boundaries)
Availability	Not overexploited	Drilling ban (*veda*)
Management	Local Water Agencies	Organismo de Cuenca regional (CONAGUA)
Act or Law	Sustainable Groundwater California Act	No groundwater law
Use	Groundwater desalination (Sweetwater) for domestic use	Water public system and industrial use

record of an evaluation update for the particular geological landmark in its American portion. This reveals the lack of interest and scientific cooperation between both countries in the matter, however, more recent studies classify this SAT as having "reasonable" information.

- – One of the ***scientific and conceptual asymmetries*** between both countries is the way of conceiving and managing groundwater. While in Mexico this is synonymous with a polygon that would resemble a particular geological reference without scientific verification, in the United States (California State), the concept "Groundwater Basin" refers to the recognition of an aquifer or group of geological formations defined by its lateral boundaries, thickness and contact with other geological strata. It is clear that both definitions are not yet conceived as strategic in the evaluation of groundwater flow systems. So this has not translated into a true mapping of the Mexico–U.S. transboundary groundwater.
- – In sum, none of the authorities in both cities contemplate the ***shared management*** of the "Tijuana-San Diego" SAT, fundamentally, due to the lack of a transboundary groundwater management vision, from a systemic approach, which hinders the possibility to create a binational strategy for water security that could benefit the region in a context of climate change and economic integration.

Finally, if the interpretation of the *Hydro-Social Cycle Theory* is considered, this suggests that the scientific discourse related to the conceptualization of groundwater is not homologated, on the contrary, there are marked discrepancies. In California, the regulatory management instruments and public policy of groundwater suggest the consideration of scientific evidence on aquifers, while in Mexico, the administrative and political management of groundwater stands out because authority acts do not have sufficient scientific criteria making them discretionary. In this regard, the *Hydro-Social Cycle* indicates that all scientific conceptualization is an act of power relations, which in this case, the lack of interest and cooperation at the local level between specialists and water authorities is observed.

11.5 Groundwater Management Challenges in the "Peninsula de Yucatan-Candelaria-Hondo" Transboundary Aquifer: Calakmul, Campeche

In 1948 an important advance was raised in the groundwater legislation in Mexico, when a Regulatory Law of the Fifth Paragraph of the twenty-seventh Constitutional article was promulgated for Subsoil Water. This policy recognized that these waters could be extracted by wells and that drilling bans ("*vedas*"), would be the governmental mechanism to regulate its use: an imposed polygon on the limits of one or more administrative aquifers. In this conceptual difference between groundwater and the aquifer (geological reference whose properties determine the movement and volume of water that is present), it becomes relevant when the irregular polygon is superimposed on a polygon of administrative characteristics displacing the systemic point of view of groundwater properties.

In December 1975, a year after finding large fields for oil exploitation off the bay of Campeche, former President Luis Echeverría Álvarez (1970-1976) established a "*veda*" on the entire territory of the state of Campeche to regulate water extraction through concession titles. The decree established preference to provide available water for the provision of public water and sanitation services; development of restoration, conservation, or preservation of natural elements and, when the State deems appropriate, exploit water sources under public utility reasons according to the availability indicated by the NOM 011-CONAGUA-2015.

In Campeche there are two administrative aquifers delimited with conventional criteria: "Yucatán Peninsula" and "Xpujil." Both polygons have, according to CONAGUA's own data, an annual availability average of 2,842.7 Mm^3 and 301.7 Mm^3 (2018c), respectively. Thus, this research established as a starting point the case study of Calakmul, Campeche where these two administrative aquifers register high availability of groundwater, the most important source of supply for the environment and users (Gondwe et al. 2010), but whose extraction is limited due to the *veda* presidential decree.

The correlation between the *Hydro-Social Cycle Theory*, used to explain the water–power nexus, and the Groundwater Flow Systems Theory (Tóth 1999) constitutes the methodological basis, in order to analyze the implications of the "*veda*" instrument applied to administrative aquifers in the groundwater management in Calakmul. One of the essential components of this analysis are the implications of the imposition of the "*veda*" which affects a polygon of irregular dimensions with the aim of ordering water use; and the aquifer, as a two-dimensional unit for the purposes of managing groundwater volumes, whose management lacks sufficient evidence to explain the behavior of groundwater flow systems from a systemic perspective (Tóth 1999). In this way, it was sought to answer the questions: *What are the characteristics that define groundwater management in Calakmul, Campeche?* and *What implications does it have with the construction of transboundary water security in southern Mexico?*

One of the most relevant findings is the concentration of water concession titles in activities associated with agricultural production, which amounts to more than 70% of the total volume of groundwater concessioned in Calakmul, with more than 6.6 $Mm^3.year^{-1}$. In contrast, just over 373,000 m^3 are destined to supply the public water and sanitation service, infrastructure covers only 37 localities of the more than 200 distributed in the municipal territory. In this regard, the distribution of the water concessions presents the same asymmetric pattern, for the portion of the polygon of the "Yucatán Peninsula" Aquifer within the municipality of Calakmul, registers 44 water concessions that cover 5.8 $Mm^3.year^{-1}$ none of which are used for environmental protection, as stated for the "veda" in Campeche, while in the portion of the Xpujil aquifer polygon, 5 concession titles cover 1.14 Mm^3.year.

The aforementioned becomes increasingly relevant when it is considered that the portion of the "Yucatán Peninsula" Aquifer polygon is juxtaposed with a protected natural area of more than 7,000 km^2 (Calakmul Biosphere Reserve), while the portion of the "Xpujil" Aquifer polygon is located in an area that concentrates the poorest and more marginalized number of inhabitants in the municipality (Abud 2019). Furthermore, 1.05 Mm^3 of the groundwater concessioned in "Xpujil" is in the hands of four users who use the groundwater for agricultural production.

The foregoing reveals that the interest of the State was to promote the expansion and consolidation of agricultural economic projects in the studied municipality, where the "veda" played a strategic role in water rights allocation. In a broader context, this form of regulation relates to the constitution of large monoculture fields owned by guaranteed water availability in benefit of a vigorously growing oil industry (Abud 2019).

Another relevant finding regarding the characteristics of groundwater management in Calakmul is reflected in the lack of conclusive scientific criteria to determine the operating conditions of groundwater local management of the "Xpujil" aquifer, which ultimately prevents CONAGUA to determine the scope of the underground channels and the hydrogeological dimensions of the aquifer in question. In contrast, UNESCO (2015), recognizes the dimensions of the Yucatán Peninsula-Candelaria-Hondo Transboundary Aquifer based on the determination of groundwater flow systems, recharge and discharge zones, hydrogeological characteristics, topographic limits, or geological and hydraulic properties (Fig. 11.5).

On this matter, Mexico, through the International Boundary and Water Commission (CILA, *Comisión Nacional de Límites y Aguas*), has entered into bilateral agreements with the governments of Guatemala and Belize to manage, from an environmental point of view, surface transboundary channels (Hondo River, Suchiate, Candelaria), but none have considered the evaluation of transboundary aquifers, much less of transboundary groundwater flow systems.

A review of the national water management frameworks in Mexico (National Water Law), Guatemala (it does not have a water law) and Belize (National Integrated Water Resources Act) warns of the lack of scientific concepts that refer to groundwater management units, which hinders the establishment of joint management schemes for transboundary aquifers. The absence of homologated concepts and the precarious scientific evaluation are accompanied by asymmetries in financing and

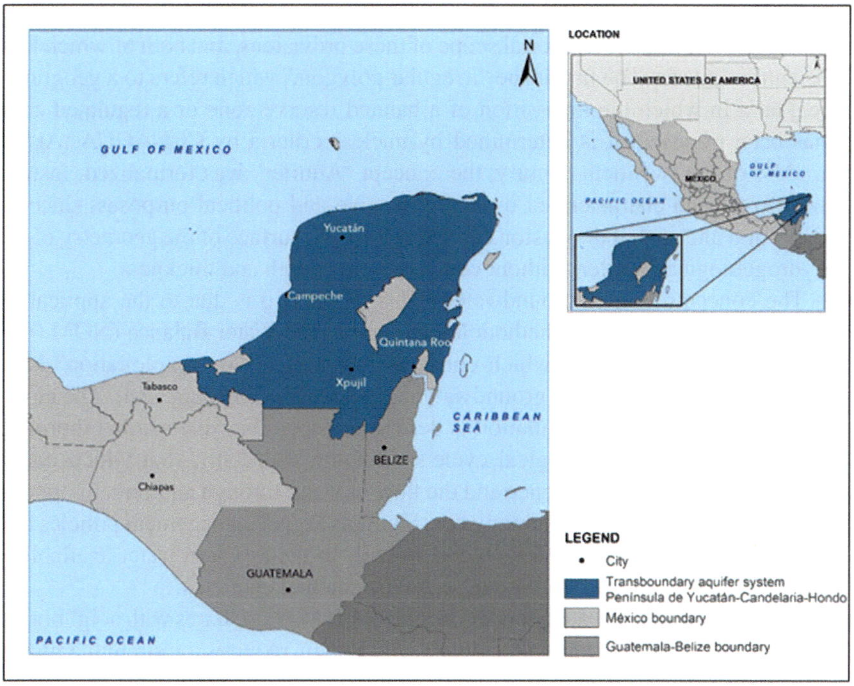

Fig. 11.5 Transboundary aquifer system "Península of Yucatán-Candelaria-Hondo" (Adapted from IGRAC-UNESCO-IHP 2015)

management units. In Mexico, the water policy is sustained with 1,500 million dollars (2017), for its part, Belize does so with 30 million dollars.

In effect, Mexico has not been able to comply with the domestic provision of water to all users in Calakmul, and neither with the shared management of underground channels of transboundary dimensions, due to the lack of recognition of the transboundary dimensions of groundwater and its management unit (aquifer) that it shares with Guatemala and Belize, is a situation that adds another challenge to the process of achieving shared water security.

11.6 Final Considerations

It is recognized that water security is a guideline that has influenced the water policy of Western countries, but its successful application requires prior modernization of the standards applied to water management as a whole. In the case of groundwater, it was found that there is a close relationship with several elements of analysis, such as:

- – Groundwater management in Mexico reveals the political decision to use two concepts to define the territorial scope of these provisions, and both of which lack scientific criteria. The first is the "irregular polygon," which refers to a geographical area in which the imposition of a banned reserve zone or a regulated zone has been imposed; it is determined by unclear criteria by CONAGUA. At the middle of the twentieth century, the concept "Aquifer" was formalized, institutionalized, and characterized by administrative and political purposes, since its definition and spatial dimension considers only the surface of the geometry of the hydrogeological aquifer, without considering its depth and thickness.
- – The concentration of groundwater rights in Mexico is due to the application of questionable scientific methodologies such as the Water Balance (NOM 011-CONAGUA-2015) under which concepts referring to "overexploitation" have substantial implications on groundwater assessment and management. The modification of this form of regulation is desirable since it has disregarded thorough knowledge of the hydrological cycle's behavior, especially, that which corresponds to movement, residence and the flow of water through aquifers. In opposition to what modern hydrogeology contributions argue, management policies and water rights allocation in Mexico corresponds to the assessed water availability more than to its interaction with other environmental elements.
- – The management of transboundary aquifers that Mexico shares with neighboring countries, the United States, Guatemala, and Belize, requires the strengthening of the production of evidence that allows characterizing shared water flow systems from a systemic perspective, through schemes approved international cooperation methodologies that influence the formulation of joint cooperation agreements and treaties for their comprehensive management in order to achieve water security for border populations.

In sum, building effective water security in Mexico requires the incorporation of modern assessment methodologies for groundwater management, as well as the democratization of decision-making in water management, considering the infeasibility of decision concentration Authorities in the Federal Executive Power with effects on a national territory characterized by an almost continental extension (2 million km^2) and different ecosystems where groundwater behaves differently.

References

Aboites L (1998) El agua de la nación: una historia política de México (1888-1946). CIESAS, México

Abud Y (2019) Conflictos por la distribución política del agua subterránea en el Acuífero Transfronterizo "Península de Yucatán-Candelaria-Hondo". El caso de Calakmul, Campeche. Published undergraduate Dissertation, Universidad Nacional Autónoma de México

Alley MW (1993) Regional Ground-Water Quality. Available via https://books.google.com/books?id = ZBmSfpFZNZcC&pg = PA77&lpg = PA77&dq = water + irregular + polygons&source = bl&ots = vnag5ZEEyv&sig = ACfU3U1-sKHK_9rnoJfxvGwtAVU4JDsgMw&hl = en&sa

= X&ved = 2ahUKEwiB1K6mwpDoAhVSnKwKHciTDwAQ6AEwDHoECAwQAQ#v = onepage&q = water%20irregular%20polygons&f = false. Accessed: 10 Oct 2019
Carrillo-Rivera JJ (2014) Por una Gestión Integral del Agua Subterránea (GIAS): Una Propuesta. In: Coloquio Nacional de Agua Subterránea en México, del 7 al 9 de noviembre del 2013, sede Instituto Mexicano de Tecnología del Agua (IMTA), Jiutepec, Morelos. Available via: https://agua.org.mx/wp-content/uploads/2014/04/ColoquioNacional_Agua_Subterranea_en_Mexico.pdf
Carrillo-Rivera JJ, Peñuela L, Huizar-Álvarez AR, Ortega-Guerrero A, Vallejo J, Hatch-Kuri G (2016) Capítulo 10. Conflictos por el agua subterránea. Geografía de México. Una reflexión espacial contemporánea 1:151–166
Cervantes J (2020) El gobernador Bonilla, entre dos aguas. Proceso. Julio 2020:43–45
Cook C, Bakker K (2012) Water security: Debating an emerging paradigm. Glob Environ Change 22:94–102
CONAGUA (2015) Programa Hídrico Regional 2014-2018 de la Región Hidrológico-Administrativa XII Península de Yucatán. Comisión Nacional del Agua, México
CONAGUA (2018a) Comunicado de Prensa: "Instala el Director General de la CONAGUA el Comité Técnico de Manejo de Aguas Subterráneas" [31 de octubre de 2018]. Comunicado de Prensa N. 660–17
CONAGUA (2018b) Registro Público de los Derechos del Agua. Información Estadística. Available in: https://www.gob.mx/cms/uploads/attachment/file/438360/qro_31012019.pdf. Accessed: 28 Jan 2019
CONAGUA (2018c) Determinación de la disponibilidad de la disponibilidad de agua en el acuífero Xpujil, estado de Campeche. CONAGUA, México
CONAGUA (2019) Estadísticas del agua en México. Edición 2019. CONAGUA, SEMARNAT, México
Diario Oficial de la Federación (2018) Available via: http://www.dof.gob.mx/nota_detalle.php?codigo=5406343&fecha=04/09/2015. Accessed 23 Mar 2019
Domínguez J, Carrillo-Rivera J (2007) El agua subterránea como elemento de debate en la historia de México. In: Mayer A. (coord.), México en Tres Momentos 1810–1910- 2010. Instituto de Investigaciones Históricas, Universidad Nacional Autónoma de México, p 177-199 Available via: http://siaps.colmex.mx/documentos/estudios/Agua%20Subterranea.pdf. Accessed 5 Jul 2020
Freeze RA and Cherry JA (1979) Hydrogeology. Prentice-Hall Publications, 604 p
Gondwe BR, Lerer S, Stisen S, Marín L, Rebolledo-Vieyra M, Merediz-Alonso G, Bauer-Gottwein P (2010) Hydrogeology of the south-eastern Yucatan Peninsula: New insights from water level measurements, geochemistry, geophysics and remote sensing. Elsevier, Journal of Hydrology 389:1–17
Hatch-Kuri G, Schmidt S, Carrillo-Rivera JJ (2017) Elementos de análisis de la propuesta de Ley General de Aguas en México a partir del derecho humano al agua y sus repercusiones en el quehacer científico, docente y de la investigación. Revista de El Colegio de San Luis, Año 8(13):30–61
Hendriks J, Boelens R (2016) Acumulación de derechos de agua en el Perú. Anthropologica del Departamento de Ciencias Sociales, Vol 34 (37) Available via http://revistas.pucp.edu.pe/index.php/anthropologica/article/view/15159/1654.Accessed 25 Nov 2019
Hinojosa-Huerta O, Carrillo-Guerrero Y (2010) La Cuenca Binacional del Río Colorado. Cotler H. Las Cuencas Hidrográficas de México, Diagnóstico y Priorización. Instituto Nacional de Ecología, pp 180–189
INEGI. Instituto Nacional de Estadística y Geografía (2010) Censo de Población y Vivienda 2010. Available via: https://www.inegi.org.mx/programas/ccpv/2010/. Accessed 03 Jul 2020
Ley de Aguas Nacionales (2020) Artículo 3. Definiciones, México
Leyva G (2020) Mega Región Tijuana-San Diego, una potencia económica transfronteriza. Available via:https://www.elimparcial.com/tijuana/columnas/Mega-Region-Tijuana-San-Diego-una-potencia-economica-transfronteriza-20190516-0012.html. Accessed: 22 Aug 2020
Linton J (2010) What is water? the history of a modern abstraction. UBC Press, Canada

Linton J, Budds J (2014) The hydrosocial cycle: Defining and mobilizing a relational-dialectical approach to water. Geoforum 57:170–180. https://doi.org/10.1016/j.geoforum.2013.10.008
Megdal SB (2018) Invisible water: the importance of good groundwater governance and management. NPJ Clean Water 1(1):1–5
Norma Oficial Mexicana 011-CONAGUA (2015) Conservación del recurso agua-Que establece las especificaciones para determinar la disponibilidad media anual de las aguas nacionales, DOF: 27/03/2015, SEGOB México
Organización de las Naciones Unidas-AGUA; Water Security and the Global Water Agenda. A UN-Water Analytical Brief, United Nations University (2013) Available via: https://www.unwater.org/publications/water-security-global-water-agenda/. Accessed: 4 Aug 2020
OECD. (2013). Water security for better lives. OECD
Rivera A (2008) Groundwater Sustainable Development in Canada – Emerging Issues Geoscience Canada Volume 35 (2):73–87
Rivera A, Candela L (2018) Fifteen-year experiences of the internationally shared aquifer resources management initiative (ISARM) of UNESCO at the global scale. Journal of Hydrology: Regional Studies 20:5–14
Sánchez J, Hatch G, Luna H (2019) Agua y Poder: el control del agua subterránea en Amazcala, Revista Nthe, Consejo Estatal de Ciencia y Tecnología del Estado de Querétaro. Available via http://nthe.mx/contadores/articulosnthe27-3.php. Accessed: 20 July 2020
Sánchez-Angulo JC (2019) La transición de la organización hídrica: gestión del agua y redes sociales en la microcuenca Amazcala, El Marqués, Querétaro. Unpublished master Dissertation. Universidad Autónoma de Querétaro
Sánchez R, Eckstein G (2017) Aquifers shared between Mexico and the United States: Management perspectives and their transboundary nature. Groundwater 55(4):495–505
Sandoval J, Vázquez M (2016) En la senda del TLCAN: Una visión crítica. Universidad de Sonora. Available via: http://www.rmalc.org/wp-content/uploads/2016/05/Libro-En-la-senda-del-TLCAN.pdf. Accessed: 5 Dic 2019
Schmidt S (1986) El Deterioro del Presidencialismo Mexicano: Los Años de Luis Echeverría. EDAMEX, México
Schmidt S, Hatch-Kuri G (2012) El agua en México. Una cuestión de seguridad nacional. Foreign Affairs Latinoamerica, 12 (4): 89–96
SDCWA (2014) Groundwater. Available in: https://www.sdcwa.org/groundwater. Accessed: 26 Jul 2020
SDUOP (2015) Proyecto del Programa de Ordenación de la Zona Metropolitana de Querétaro. Available in: https://gobqro.gob.mx/sduop/index.php/desarrollo-urbano/Accessed: 16 Jul 2020
Sosa L, Valdivia R, Hernández J, Contreras J, Monroy R (2019) "Análisis de las transmisiones de derechos de agua para el periodo 2006–2016 en Celaya, Guanajuato" in Hernández, V., Valdivia, R., Hernández, J., Cruz, P. and Cruz, G., (comp.). Estudios recientes sobre economía ambiental y agrícola en México primera edición, México, Universidad Autónoma de Baja California Sur. Available in: https://www.researchgate.net/profile/Victor_Hernandez_Trejo/publication/337935379_Estudios_recientes_sobre_economia_EBOOK/links/5df54b5c92851c83647e76be/Estudios-recientes-sobre-economia-EBOOK.pdf#page=348. Accessed: 21 July 2020
Sustainable Groundwater Management Act (2014) State of California. Retrieved from http://groundwater.ca.gov/docs/2014%20Sustainable%20Groundwater%20Management%20Legislation%20with%202015%20amends%201-15-2016.pdf
Sweetwater Authority Groundwater (2020) Available via: https://www.sweetwater.org/265/Groundwater. Accessed 07 July 2020
Tóth J (1962) A theory of groundwater motion in small drainage basins in central Alberta. Canadá. Journal of Geophysical Research 67(11):4375–4387
Tóth J (1999) Groundwater as a geologic agent: an overview of the causes, processes, and manifestations. Hydrogeol J 7(1):1–14

Tóth J (2016) The Evolutionary Concepts and Practical Utilization of the Tóthian Theory of Regional Groundwater Flow. Int J Earth Environ Sci V1: 111, 11 p https://doi.org/10.15344/2456-351X/2016/111

UNESCO. (2007) Sistemas Acuíferos Transfronterizos en las Américas – Evaluación Preliminar, Serie ISARM Américas N°1

UNESCO (2015) Estrategia regional para la evaluación y gestión de los Sistemas Acuíferos Transfronterizos en las Américas. Oficina Regional de Ciencia para América Latina y el Caribe, Montevideo, Uruguay

USGS (2020) Sustainable Groundwater Management. Available via ca.water.usgs.gov/sustainable-groundwater-management. Accessed: 07 Sep 2020

Wester P, Hoogesteger J (2011) Uso intensivo y despojo del agua subterránea: hacia una conceptualización de los conflictos y la concentración del acceso al agua subterránea. In: Boelens RA, Cremers L, Zwarteveen MZ (eds) Justicia hídrica: acumulación, conflicto y acción social. Justicia Hídrica, IEP, PUCP. Lima: 111–133

GPSR Compliance

The European Union's (EU) General Product Safety Regulation (GPSR) is a set of rules that requires consumer products to be safe and our obligations to ensure this.

If you have any concerns about our products, you can contact us on ProductSafety@springernature.com

In case Publisher is established outside the EU, the EU authorized representative is:

Springer Nature Customer Service Center GmbH
Europaplatz 3
69115 Heidelberg, Germany

Batch number: 10732090

Printed by Printforce, the Netherlands